2D MATERIALS FOR PHOTONIC AND OPTOELECTRONIC APPLICATIONS

Woodhead Publishing Series in Electronic and Optical Materials

2D MATERIALS FOR PHOTONIC AND OPTOELECTRONIC APPLICATIONS

Edited by

QIAOLIANG BAO
Monash University, Clayton, VIC, Australia

HUI YING HOH
Shenzhen University, Shenzhen, China

ELSEVIER

WP WOODHEAD PUBLISHING
An imprint of Elsevier

Woodhead Publishing is an imprint of Elsevier
The Officers' Mess Business Centre, Royston Road, Duxford, CB22 4QH, United Kingdom
50 Hampshire Street, 5th Floor, Cambridge, MA 02139, United States
The Boulevard, Langford Lane, Kidlington, OX5 1GB, United Kingdom

Library of Congress Cataloging-in-Publication Data

A catalog record for this book is available from the Library of Congress

British Library Cataloguing-in-Publication Data

A catalogue record for this book is available from the British Library

ISBN: 978-0-08-102637-3 (print)

ISBN: 978-0-12-815435-9 (online)

For information on all Woodhead publications visit our website at https://www.elsevier.com/books-and-journals

Publisher: Matthew Deans
Acquisition Editor: Kayla Dos Santos
Editorial Project Manager: Andrae Akeh
Production Project Manager: Debasish Ghosh
Cover Designer: Mark Rogers

Typeset by SPi Global, India

Contents

Contributors

Kah-Wee Ang
Department of Electrical and Computer Engineering; Centre for Advanced 2D Materials, National University of Singapore, Singapore, Singapore

Qiaoliang Bao
Department of Materials Science and Engineering, ARC Centre of Excellence in Future Low-Energy Electronics Technologies (FLEET), Monash University, Clayton, VIC, Australia

Jianji Dong
Wuhan National Laboratory for Optoelectronics, Huazhong University of Science and Technology, Wuhan, China

Hui Ying Hoh
Key Laboratory of Optoelectronic Devices and Systems of Ministry of Education and Guangdong Province, College of Optoelectronic Engineering; Institute of Microscale Optoelectronics, Shenzhen University, Shenzhen, China

Weida Hu
State Key Laboratory of Infrared Physics, Shanghai Institute of Technical Physics, Chinese Academy of Sciences, Shanghai, China

Xinxin Jin
School of Electronic and Information Engineering, Beihang University, Beijing, China

Qasim Khan
College of Electronic Science and Technology, Shenzhen University, Shenzhen, China

Sayed Ali Khan
College of Electronic Science and Technology, Shenzhen University, Shenzhen, China

Chang-Ming Li
Institute of Advanced Cross-field Science, College of Life Science, Qingdao University, Qingdao; Institute of Material Science and Devices, Suzhou University of Science and Technology, Suzhou, China

Sean E. Lowe
Centre for Clean Environment and Energy, School of Environment and Science, Griffith University, Gold Coast, QLD, Australia

Yuerui Lu
Research School of Engineering, College of Engineering and Computer Science, Australian National University, Canberra, ACT, Australia

John Schaibley
Department of Physics, University of Arizona, Tucson, AZ, United States

Ge Shi
Centre for Clean Environment and Energy, School of Environment and Science, Griffith University, Gold Coast, QLD, Australia

Bannur Nanjunda Shivananju
Institute of Advanced Cross-field Science, College of Life Science, Qingdao University, Qingdao, China; Department of Materials Science and Engineering, ARC Centre of Excellence in Future Low-Energy Electronics Technologies (FLEET), Monash University, Clayton, VIC, Australia

Peng Wang
State Key Laboratory of Infrared Physics, Shanghai Institute of Technical Physics, Chinese Academy of Sciences, Shanghai, China

Lin Wang
Department of Electrical and Computer Engineering; Centre for Advanced 2D Materials, National University of Singapore, Singapore, Singapore

Jiong Yang
Research School of Engineering, College of Engineering and Computer Science, Australian National University, Canberra, ACT, Australia

Xinliang Zhang
Wuhan National Laboratory for Optoelectronics, Huazhong University of Science and Technology, Wuhan, China

Meng Zhang
School of Electronic and Information Engineering, Beihang University, Beijing, China

Han Zhang
Key Laboratory of Optoelectronic Devices and Systems of Ministry of Education and Guangdong Province, College of Optoelectronic Engineering, Shenzhen University, Shenzhen, China

Huijun Zhao
Centre for Clean Environment and Energy, School of Environment and Science, Griffith University, Gold Coast, QLD, Australia

Yu Lin Zhong
Centre for Clean Environment and Energy, School of Environment and Science, Griffith University, Gold Coast, QLD, Australia

CHAPTER 1

Introduction

Hui Ying Hoh[a,b], Qiaoliang Bao[c]
[a]Key Laboratory of Optoelectronic Devices and Systems of Ministry of Education and Guangdong Province, College of Optoelectronic Engineering, Shenzhen University, Shenzhen, China
[b]Institute of Microscale Optoelectronics, Shenzhen University, Shenzhen, China
[c]Department of Materials Science and Engineering, ARC Centre of Excellence in Future Low-Energy Electronics Technologies (FLEET), Monash University, Clayton, VIC, Australia

1.1 Background

Nanotechnology, low-dimensional materials, and quantum engineering are buzz words that have popped up in the 21st century but are more than just trends. The two-dimensional (2D) revolution started more than a decade ago with the discovery and isolation of graphene.[1] However, our fascination in low-dimensional materials is not new, and scientists have been studying these subjects long before modern characterization techniques and computational resources became available. A well-known example is Dr Feynman's 1959 talk, "There's plenty of room at the bottom," in which he shared his vision of atomic-scale manipulation.[2] In another example, graphene, despite being successfully isolated only in 2004, was already being studied back in 1947.[3.]

The major propellant for research activities in low-dimensional materials came in the 1980s, with the development of electron microscopy. It has led to the discovery of low-dimensional materials such as C_{60}, the Buckminsterfullerene,[4] carbon nanotubes (CNTs),[5] and graphene.[1] Each of these materials—from the zero-dimensional C_{60} to the 2D graphene—sparked a revolution. Two-dimensional materials are of particular commercial importance as these hold more promise than their other low-dimensional counterparts, allowing the designed fabrication of layered devices. In addition, many bulk materials contain 2D and 1D subcomponents, and these materials are excellent candidates for the assembly of van der Waals heterostructures.[6] Devices made with these materials can have wide-ranging applications such as photonics, electronics, catalysis, energy, and biological applications.

Today, more than 1000 2D materials are expected to be stable and may be exfoliated without much difficulty.[7] The list of those successfully synthesized grows as novel materials are discovered constantly, and new

2D Materials for Photonic and Optoelectronic Applications
https://doi.org/10.1016/B978-0-08-102637-3.00001-2

knowledge is formed every day. The 2D materials that are currently known, as well as their preparation methods and proposed applications, are summarized in Table 1.1. We will also discuss some of these 2D materials in the following chapters. Note that Table 1.1 is not meant to be exhaustive, for reasons mentioned earlier. For beginners who find it hard to navigate the 2D world, the article by Jacoby provides an easy-to-understand introduction to the various types of 2D materials.[113] The accompanying interactive figure is also an excellent starting point.

Table 1.1 Types of 2D materials, year it was first isolated, methods of synthesis, and proposed applications.

Material in 2D form	First isolated	Methods of production	Proposed applications
Graphene	2004	Mechanical exfoliation,[1, 8] electrochemical exfoliation,[9] chemical methods,[10] liquid-phase exfoliation,[11–13] chemical vapor deposition (CVD),[14–16] epitaxial growth from silicon carbide,[17, 18] unzipping CNTs[19]	Modulators,[20] lasers,[21] photodetector,[22] light-emitting diodes,[23–25] plasmonic device,[26] optical transistor,[27] information storage,[28] bio-related applications[29]
Transition metal dichalcogenides			
Molybdenum sulfide (MoS_2)	2010	Mechanical exfoliation,[30, 31] liquid-phase exfoliation,[32] CVD[33, 34]	Modulators,[35] lasers,[36] photodetectors,[37] solar cells,[38, 39] transistors,[40] valleytronics[41–43]
Molybdenum selenide ($MoSe_2$)	2014	Mechanical exfoliation,[31] liquid-phase exfoliation,[44] CVD[45–47]	Lasers,[44, 48, 49] modulators,[35] photodetectors,[47] solar cells,[39] valleytronics[50]
Molybdenum telluride ($MoTe_2$)	2014	Mechanical exfoliation,[51] solution-phase synthesis,[52] CVD[53]	Photodetectors,[54] transistors,[55] valleytronics,[56] quantum spin Hall devices[53]

Table 1.1 Types of 2D materials, year it was first isolated, methods of synthesis, and proposed applications.—cont'd

Material in 2D form	First isolated	Methods of production	Proposed applications
Tungsten sulfide (WS_2)	2012	Mechanical exfoliation,[57] liquid-phase exfoliation,[35] chemical vapor transport (CVT),[58] CVD[59–61]	Lasers,[59, 60] modulators,[35] solar cells[39]
Tungsten selenide (WSe_2)	2012	Mechanical exfoliation,[31] liquid-phase exfoliation,[62] CVT,[58] CVD[63]	Modulators,[64] solar cells,[62] valleytronics[65]
Tungsten telluride (WTe_2)	2014	Mechanical exfoliation,[66] solution-phase synthesis,[52] CVD[53]	Magnetic sensors[67]
Elements			
Black phosphorus (BP)	2014	Mechanical exfoliation,[68–70] liquid-phase exfoliation,[71, 72] plasma treatment,[73] pulsed laser deposition[74]	Modulators,[71] lasers,[68, 71, 72] photodetectors,[75, 76] solar cells,[72, 77] bio-related applications,[78] field-effect transistors (FETs)[69, 74]
Silicene (Si)	2012	Epitaxial growth[79]	Valleytronics,[80] spintronics,[79] quantum Hall devices[79]
Germanene (Ge)	2014	Epitaxial growth[81, 82]	Valleytronics,[83] spintronics,[83] quantum Hall devices[83]
Stanene (Sn)	2015	Molecular beam epitaxy[84, 85]	Superconductor[84]
Arsenene (As)	2016	Plasma-assisted growth[86]	Solar cells[87]
Antimonene (Sb)	2016	Mechanical exfoliation,[88] liquid-phase exfoliation,[89] molecular beam epitaxy,[90] van der Waals epitaxy[91]	Flexible electronics,[91] supercapacitors[92]

Continued

Table 1.1 Types of 2D materials, year it was first isolated, methods of synthesis, and proposed applications.—cont'd

Material in 2D form	First isolated	Methods of production	Proposed applications
Bismuthene (Bi)	2016	Liquid-phase exfoliation,[93] epitaxial growth[94]	Modulators,[93] photodetectors,[93] quantum Hall devices[94]
MXenes (2D metal carbides and nitrides)			
Ti_3C_2	2011	Chemical exfoliation[95]	Transparent conductor[96]
MXenes	2012	Chemical exfoliation[97]	Lasers,[98, 99] photodetectors,[100] energy storage,[101] water treatment,[101] sensors,[101] flexible electronics[102]
Others			
Hexagonal boron nitride	2004	Mechanical cleavage,[30] chemical synthesis,[103] chemical exfoliation,[104] CVD[105]	Lasers,[106] solid-state lighting,[107] bio-photonics[108]
Molybdenum oxide (MoO_3)	2009	Wet chemical etching[109]	Organic light-emitting diode,[110] gas sensors,[111] field-effect biosensor[112]

1.2 Graphene—The 2D material that led the revolution

This section serves to provide a background since there have been numerous publications on graphene.

1.2.1 Structure and properties

Graphene is a single layer of carbon atoms arranged in a honeycomb lattice (Fig. 1.1). The carbon atoms are bonded via sp^2 bonding, which gives rise to its mechanical strength and stiffness. The p_z orbitals that form the π and π^* bands and the delocalized electrons give rise to its unique electronic

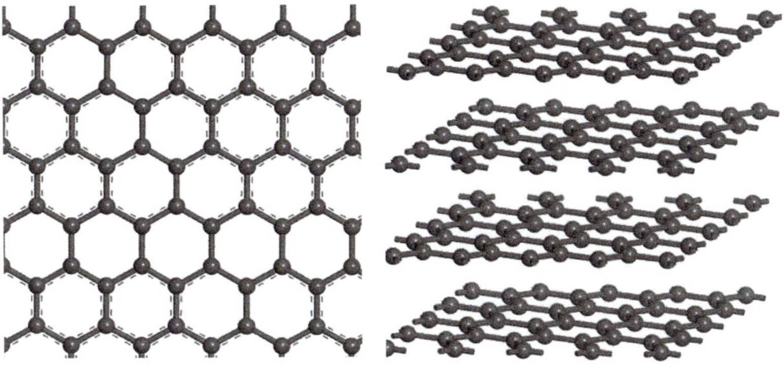

Fig. 1.1 Structure of graphene (*left*) and graphite (*right*).

properties. Graphite, the bulk form, is made up of stacked layers of graphene held together by weak van der Waals forces.

The properties of graphene have been thoroughly discussed in several works.[114–116] In terms of electronic properties, graphene can be briefly described as a zero-gap semiconductor, with the conical-shaped valance and conduction bands demonstrating linear dispersion near the Dirac point (Fig. 1.2).

The allure of graphene to scientists following its discovery is not surprising. Graphene possesses many exceptional properties, such as high carrier

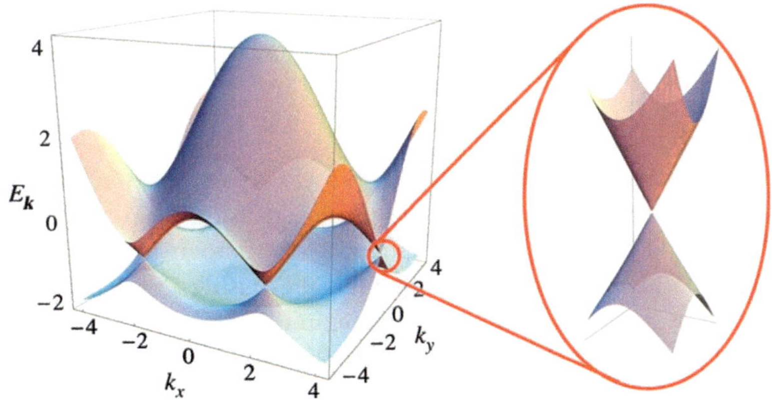

Fig. 1.2 Electronic dispersion of graphene. The expanded section is a zoom-in of the bands close to one of the Dirac points. It shows the conical-shaped valence and conduction bands at the Dirac point.[185] *Reprinted with permission from Castro Neto, A. H.; Guinea, F.; Peres, N. M. R.; Novoselov, K. S.; Geim, A. K., The Electronic Properties of Graphene. Rev. Modern Phys.* ***2009****,* 81 *(1), 109-162. Copyright 2009 by the American Physical Society.*

mobilities (>200,000 cm^2/V s at electron densities of $2 \times 10^{11}/cm^2$),[117, 118] superior thermal conductivity (4800–5300 W/m K),[119] and extreme strength (Young's modulus of 1 TPa).[120, 121] Being only one atom thick, one can imagine that the optical properties of graphene will be just as extraordinary. Indeed, owing to its unique electronic band structure, the optical conductance of pure monolayer graphene is frequency-dependent over a broad range of photon energies.[122] Graphene interacts with light ranging from the far-IR to ultraviolet range,[123] and therefore it offers tremendous promise in broadband photonic applications. The optical transmittance of pure monolayer graphene is also frequency-dependent and can be defined by the fine-structure constant.[124] In the visible light region, the optical absorption of monolayer graphene is around 2.3%, and this adsorption increases proportionally with the number of layers (up to 10).[123, 124] Hence, graphene is also being considered a transparent conductor for optoelectronic applications such as photovoltaic devices and touchscreens. The photonic and optoelectronic applications of graphene have been discussed in several publications, and interested readers are encouraged to refer to Bao et al.[115, 125] and Bonaccorso et al.[126]

1.2.2 Methods of production

Graphene can be produced via a large variety of methods as can be seen in Table 1.1. For the purpose of this book, we will briefly introduce the methods that can be applied to other 2D materials. The review by Choi et al.,[127] as well as that by Avouris and Dimitrakopoulos,[128] is an excellent reference in which the synthesis of graphene and its potential applications are discussed in greater detail.

1.2.2.1 Mechanical exfoliation

The revolutionary work of Geim and Novoselov of Manchester University in 2004 was performed by graphene produced using a simple tool: the Scotch tape.[1] The technique involves repeatedly removing graphene layers from highly oriented pyrolytic graphite (HOPG) using a tape and pressing the tape on a silicon oxide (SiO_2) substrate, until atomic-thin layers of graphene is obtained. This is known as the Scotch tape or sticky tape method, or more formally mechanical cleavage.

Since the sticky tape method produces graphene of high quality but in low quantities, micromechanical techniques with some degree of automation were later proposed, for example, using an ultra-sharp diamond wedge to cut HOPG, which is fed slowly toward the wedge.[129] Another method utilizes a three-roll mill machine with a polymer adhesive.[130] This method acts in the same way as the Scotch™ tape method described before, and single- and few-layer graphene sheets can be obtained through continuous exfoliation. Nevertheless, both methods are not without drawbacks, with the main challenge being the size of flakes; these are still too small for commercial interest. Over the last decade, new methods of mechanical exfoliation have also been proposed, such as the ball-milling and fluid dynamics methods. These methods are discussed in a review by Yi and Shen.[8]

1.2.2.2 Liquid-phase exfoliation/chemical methods

Although the graphene produced via mechanical exfoliation is of excellent quality, this method is too tedious and painstaking. To propel graphene from the laboratory to real-life applications, large-scale production methods are required. Large-scale exfoliation was first explored using organic solvents such as *N*-methyl-pyrrolidone (NMP), whose surface energies match that of graphene, and thus the energy required to exfoliate graphene is balanced by the graphene-solvent interaction.[11] Using this method, graphene dispersions with concentrations up to ~ 0.01 mg mL^{-1}, and monolayer yield of ~ 1 wt%, could be obtained. The use of surfactants to obtain suspended graphene solutions has also been reported.[12] In such cases, aqueous-phase reactions avoid the use of organic solvents, thus reducing the amount of chemical waste in industrial-scale productions.

Electrochemical exfoliation of graphite is another promising method for low-cost mass production of graphene. Parvez et al. obtained high yield of graphene sheets (>85%) via electrochemical exfoliation using aqueous solutions of inorganic salts such as $(NH_4)_2SO_4$, Na_2SO_4, and K_2SO_4.[9] These graphene sheets are less than three layers thick, with a lateral size of up to 44 μm, low degree of oxidation, and high hole mobility (310 cm^2/V s). The authors also demonstrated the simple fabrication of highly conductive graphene films on an A4-sized paper using the exfoliated sheets. Readers interested in the chemical methods for the production of graphene are invited to refer to the review by Park and Ruoff,[10] as well as that by Zhong et al.[131] In addition, the review by Abdelkader et al. highlights significant works on electrochemical exfoliation.[132]

Sonication may be used in conjunction with the aforementioned methods to improve the efficiency of exfoliation. The mechanism and methods are discussed in a review by Ciesielski and Samori.[13] Readers are also encouraged to refer to the review on mechanical exfoliation by Yi and Shen mentioned before for details on exfoliation mechanics and drawbacks of this method.[8] Most notably, sonication can also be employed to exfoliate other layered systems such as BN, MoS_2, WS_2, $NbSe_2$, and TaS_2, thus enabling investigations on other 2D materials

1.2.2.3 *Chemical vapor deposition*

Graphene produced via chemical vapor deposition (CVD), frequently referred to as CVD graphene, was first reported in 2008 and 2009 by two independent groups: Reina et al.[14] and Yu et al.[15] In both works, graphene films were first grown on nickel under ambient pressure, then transferred to other substrates almost intact.

Later, in a breakthrough, Bae et al. reported the roll-to-roll production of large-sized (30-inch diagonal) graphene films.[16] These films are highly conducting and transparent, with a sheet resistance of around 125 Ω per square and optical transmittance of 97.4%. In particular, it was demonstrated that the sheet resistance of p-doped four-layer film can drop to 30 Ω, while maintaining around 90% optical transmittance. These values are comparable to those of commercial transparent electrodes such as indium tin oxide, and thus propel the status of graphene as a transparent conductor. As the attention of graphene research shifts from fundamental investigations to application-based studies, we see more efforts in large-scale production methods. This trend is likely to be similar for other 2D materials.

1.3 Transition metal dichalcogenides

The bulk form of transition metal dichalcogenides (TMDs) has a layered structure like graphite. This is a well-known fact. After the discovery of graphene and its exceptional properties, scientists wondered whether other layered materials can be isolated similarly and whether these possess unique properties dissimilar to their bulk form. The answer is yes.

TMD is a class of materials of the type MX_2, where M is a transition metal and X is a chalcogen (S, Se, or Te). The first TMD to gain attention is MoS_2, although studies involving the rest emerged soon after. TMDs are not to be confused with MXenes, which will be discussed in a later section. Within the TMDs, there are many subsets, for example, the group 6 TMDs, where M is a group 6 metal such as molybdenum (Mo) or tungsten (W). Most current research efforts are concentrated on these group 6 TMDs, but attention on other subsets are gaining traction, such as group 10 TMDs, where M is a noble metal such as palladium (Pd) and platinum (Pt).

1.3.1 Structure and properties

The structure and properties of TMDs have been discussed in several publications. The reviews by Li and Zhu, Wang et al., and Lv et al. are good reads, offering a background as well as discussion on synthesis methods, properties, and potential applications.[133–135] Chhowalla et al. offer a different perspective by considering the chemistry of TMD nanosheets,[136] while Yun et al. explored how the electronic properties of TMDs depend on the thickness of the material and presence of strain.

We will concentrate on group 6 TMDs in this section. As an example, let us consider the structure of MoS_2 since it is the most commonly studied TMD. Each Mo atom is bonded to six S atoms in a trigonal prism, while each S atom is bonded to three Mo atoms (Fig. 1.3). The Mo plane is sandwiched by two S planes, and this S-Mo-S triplane is considered one monolayer of MoS_2. Bulk MoS_2 consists of such stacked layers held together by weak van der Waals forces.

Bulk MoS_2 is an indirect bandgap semiconductor and exhibits negligible photoluminescence. In contrast, a MoS_2 monolayer exhibits strong photoluminescence, indicating that quantum confinement results in a transition from an indirect to a direct bandgap of 1.8 eV.[137, 138] By virtue of this direct bandgap, MoS_2 holds much promise in optoelectronic applications such as energy harvesting, i.e., solar cells. However, initial studies on the carrier mobilities of monolayer MoS_2 are in the range of 0.5–3 $cm^2/V\ s$, too low

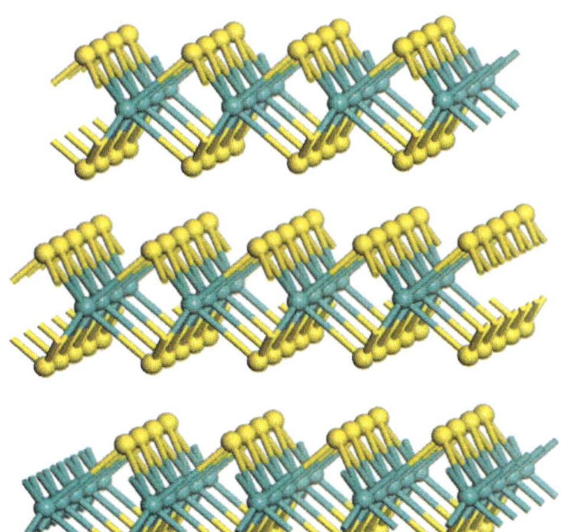

Fig. 1.3 Structure of molybdenum sulfide. Mo atoms are cyan (*dark gray*) and S atoms are *yellow* (*light gray*).

for practical devices.[30] Nevertheless, carrier mobility of $>200\ cm^2/V\ s$ has been reported by Radisavljevic et al.; this is comparable to that of graphene nanoribbons.[40] The authors therefore suggested that MoS_2 may be used with graphene to form heterostructures for the aforementioned applications. Very likely, this idea can also be extended to other group 6 TMDs.

One noteworthy characteristic of monolayer TMDs is that the direct bandgap occurs in two unequaled corners (K and K′) of the Brillouin zone (BZ),[139] unlike other direct bandgap semiconductors in the bulk form such as GaAs. This is unique to hexagonal 2D materials. In monolayer TMDs, due to the lack of inversion symmetry, quasiparticles such as neutral and charged excitons are allowed a new degree of freedom, the valley pseudospin. Moreover, spin splitting at the K and K′ corners has opposite signs due to strong spin-orbit coupling in TMDs, resulting in an effective coupling between spin and valley pseudospin.[139] Strong spin-valley coupling suggests that the spin may be manipulated by modifying the properties of the valley, bringing up prospects in valleytronics. Indeed, Langer et al. recently demonstrated that the valley pseudospin in a monolayer of WSe_2 can be changed by lightwave-driven intraband transport within a few femtoseconds, evidence of ultrafast pseudospin dynamics.[65] As valleytronics is an emergent topic requiring more in-depth discussions, we will revisit this in Chapter 10.

Although TMDs have a similar layered structure as graphite, the presence of heterogeneous atoms and the type of coordination suggest that these may have more than one type of crystal structure (Fig. 1.4). This will have significant implications on their properties, such as electronic properties (semiconducting or metallic), which will in turn be useful in numerous applications. For example, enhanced catalytic activity observed in strained chemically exfoliated WS_2 nanosheets is associated to the metallic phase of WS_2.[140] If this structural phase transitions can be controlled and exploited, in theory an electronic device may be fabricated from a single material, with semiconducting and metallic regions. Duerloo et al. explored this idea by considering mechanical deformations on TMD monolayers in order to investigate phase transitions in the crystal structures of TMDs, and whether the metallic phases are stable or merely metastable.[141] The authors found that these mechanical deformations are indeed possible and that it is highly achievable in $MoTe_2$.

Later, Song et al. provided experimental evidence of the strain-induced semiconductor-metal transition in thin-film $MoTe_2$.[142] Using an atomic force microscope tip, a small tensile strain of 0.2% (contact force of 200 nN) was applied to $MoTe_2$, causing a phase transition from 2H to 1T′, as shown in

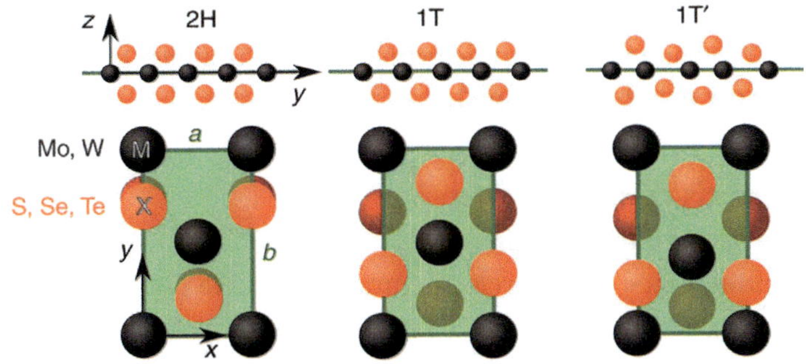

Fig. 1.4 Three different crystalline phases of 2D TMDs. Each phase can be represented in a rectangular unit cell with dimensions $a \times b$. In all three phases the transition metal atoms (black, dark) are sandwiched by chalcogen (*orange*, light) atoms. The 2H (H for hexagonal) phase is often referred to as the trigonal prismatic structure, and the metallic 1T and 1T′ (T for tetrahedral) are called octahedral and distorted octahedral, respectively. The 1T′ phase can be thought of as 1T after a symmetry-reducing distortion. *Reprinted by permission from Springer Nature: Duerloo, K.-A. N.; Li, Y.; Reed, E. J., Structural Phase Transitions in Two-Dimensional Mo- and W-Dichalcogenide Monolayers. Nat. Commun.* ***2014****, 5, 4214, Copyright 2014.*

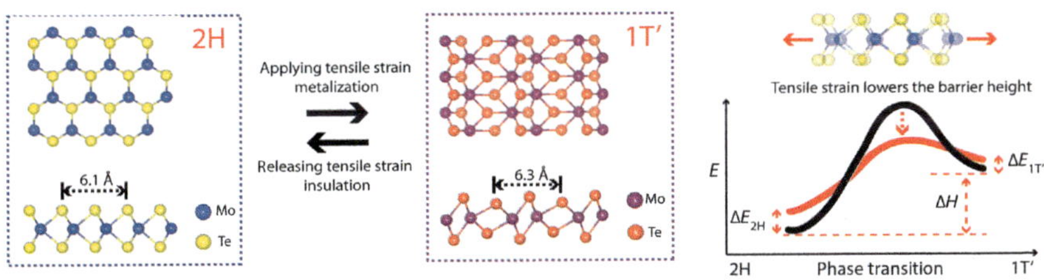

Fig. 1.5 *Left*: Transition of $MoTe_2$ between the 2H to 1T′ phases. Mo atoms are represented by *blue/purple* (dark) circles and Te atoms are represented by *orange/ yellow* (light) circles. Application (release) of tensile strain results in metalization (insulation) of $MoTe_2$. *Right*: Energy diagram depicting reduction of barrier due to the application of tensile strain. *Reprinted with permission from Song, S.; Keum, D. H.; Cho, S.; Perello, D.; Kim, Y.; Lee, Y. H., Room Temperature Semiconductor-Metal Transition of* $MoTe_2$ *Thin Films Engineered by Strain. Nano Lett.* ***2016****, 16 (1), 188-193. Copyright (2016) American Chemical Society.*

Fig. 1.5. The 2H phase is semiconducting and has the commonly known trigonal prismatic structure as shown in Fig. 1.3. The 1T′ phase is metallic with an octahedral-like M-X coordination. Under tensile strain, the activation energy required for the 2H to 1T′ phase transition is reduced, thus allowing it to occur at room temperature under ambient conditions (Fig. 1.5). Another significant observation is that the phase transition is fully reversible after release of the strain, suggesting potential applications in optoelectronic or biological sensing. Since a small amount of stimuli—in this case, strain—results in a significant change in electronic properties due to phase transition, ultra-sensitive sensors may be developed.

1.3.2 Methods of production

1.3.2.1 Mechanical exfoliation

In the beginning stage of research, the unquestionable choice was mechanical exfoliation, as samples produced by this method were of excellent quality with pristine properties. This method is akin to the Scotch™ tape method described in Section 1.2.2.1. This method has been applied to all the group 5 chalcogenides (Table 1.1). As in the case of graphene, to overcome manual mechanical exfoliation, researchers have designed techniques and devices to obtain high-quality products. For instance, MoS_2 flakes were first mechanically exfoliated using tape, then deposited on a viscoelastic gel stamp.[143] Using a micromanipulator, the gel stamp was pressed against a pre-patterned substrate with holes (Si/SiO_2 wafer), transferring MoS_2 sheets onto the substrate. The gel stamp was then removed very slowly using the same micromanipulator. Nevertheless, mechanical exfoliation remains useful only in laboratory-scale investigations and not real-life applications.

1.3.2.2 Liquid/solution-phase production

In liquid-phase exfoliation, the key concept is to select a solvent that interacts with the sheets of the 2D material, such that this interaction can overcome van der Waals forces holding the sheets. At times, co-solvent and surfactants may be introduced to the mixture to improve solubility of the sheets. Mechanical stimulations, such as sonication mentioned in Section 1.2.2.2, may also be applied concurrently. Jawaid et al. provided a thorough discussion on the mechanism of liquid-phase exfoliation of MoS_2, testing out different types of solvent combinations, as well as how the yield is affected by the supplier, age, and storage conditions of the solvent.[32] It was found that NMP is an excellent solvent for MoS_2, as it has a surface energy of about 40 mJ/m^2, similar to the surface energy of MoS_2, which is about 46.5 mJ/m^2.[32]

To obtain aqueous dispersions of MoS_2, Yao et al. utilized a combination of grinding and sonication.[144] Bulk MoS_2 was first mixed with solvent and grounded with a mortar to produce a gel-like mixture, after which the solvent was removed through evaporation in a vacuum oven. The dried powder was next dispersed in a 45-vol% ethanol/water mixture and sonicated for varying durations. The concentrations of MoS_2 suspensions obtained using this method were up to 27 mg/mL, one of the highest reported. However, the thickness of the sheets ranged from 3.5 to 17.4 nm, indicating that most of the sheets comprised more than five layers of MoS_2.

Very recently, Li et al. attempted the use of a high-speed dispersive homogenizer to exfoliate MoS_2 into NMP.[145] The homogenizer used was relatively easy to operate and produced high sheer forces to aid the dispersion of MoS_2 into NMP. The authors reported that 65% of MoS_2 nanosheets obtained were less than four layers and 9% were monolayers. Yet, the highest concentration of MoS_2/NMP suspension was only 1.4 mg/mL. Tian, Wu and co-workers also used a sheer-assisted exfoliation, but with a supercritical liquid: CO_2. They found that the majority of exfoliated MoS_2 were less than 10 layers, more than 95%, among which about 50% were 1–4 layers. The highest concentration obtained was around 2.2 mg/mL, dropping to 0.97 mg/mL after 7 days. As seen in the aforementioned studies, the yield and quality of the products are dependent on many factors, which include, but are not limited to, type of solvent, strength of mechanical forces, and time. Thus, each of these factors has to be optimized to achieve the best outcome.

Solution or liquid-phase exfoliation usually results in the loss of pristine properties of the material, due to the formation of defects or structural changes. One way to mitigate this effect is through annealing. Eda et al. reported that chemically exfoliated MoS_2 comprises mostly the phase with metallic properties; however, by annealing the product to temperatures >300°C, the semiconducting properties can be restored.[146] Apart from placing the exfoliated compounds on heat-tolerant substrates during annealing, Chou et al. proposed the use of inert organic solvents with high boiling points, to reduce possible degradation of the nanosheets during annealing.[147] Another advantage of using solvents is that the annealed sheets are easily assembled as free-standing films for device fabrication.

Lately, Giri et al. reported the solution-phase synthesis of large-area, atomically thin $MoTe_2$ and WTe_2 films.[52] The authors designed a new Te precursor that allowed in situ chemical transformation from Te nanoparticles into thin films ($MoTe_2$ or WTe_2). In addition, the number of atomic layers can be controlled, which means greater precision in device fabrication. This study represents an important development in the large-scale production of TMD films.

1.3.2.3 Chemical vapor deposition

For device fabrication, CVD is a one of the most effective methods, using which large areas of thin films can be grown. In the case of MoS_2, MoO_3 is a suitable starting material, being commercially available. However, sulphurization of MoO_3 usually results in nanoparticles or nanorods instead of 2D sheets. In 2012, Lee et al. reported a breakthrough by demonstrating highly crystalline growth of atomically thin layers of MoS_2 using CVD.[33] The growth of MoS_2 was highly sensitive to the pretreatment of the substrate, and treating the substrate with graphene-like molecules promoted the growth of films in a 2D manner. Other pretreatment solutions were also tested, but it was found that the use of reduced graphene oxide solution resulted in the most homogenous growth. In a follow-up study, Lee et al. later showed that electronic transport and optical properties of these CVD-grown MoS_2 films were comparable to those of exfoliated MoS_2 films.[34] In addition, the films may be transferred to a variety of surfaces (HOPG and PET), thus allowing the fabrication of multilayered heterodevices.

In a recent study, Chen et al. reported a low-cost method for producing high-quality MoS_2 and $MoSe_2$ films.[46] $MoS_2/MoSe_2$ crystals were synthesized on molten glass using ambient-pressure CVD. Glass is cheap, chemically inert, and transparent, and is thus an excellent substrate for photonic and/or optoelectronic applications. The monolayer crystals reached sizes of up to 2.5 mm in 5 min, with a carrier mobility of up to 95 $cm^2/(V\ s)$.

Sulphurization may also be applied to other transition metal oxides, for example, tungsten oxide (WO_3), to produce WS_2.[59–61] In this two-step method, thin films of WO_3 are first thermally evaporated on a Si/SiO_2 substrate under high vacuum conditions. The films are then placed in a reaction chamber filled with inert Ar gas at atmospheric pressure. Next, the samples are heated to 800°C under an Ar flow. Ar acts as a carrier gas for S vapors generated by heating S powder placed upstream. WS_2 typically first forms islands at the earlier stages of growth; these islands later expand laterally to become films as the growth progresses. It was found that WS_2 monolayers may be obtained from this method, although the film consists of crystalline platelets of different orientations.[61]

Instead of synthetizing TMD films and then transferring these to another substrate to form heterostructures, Shi et al. developed a one-step method to assemble MoS_2 directly on graphene.[148] The substrate used was a graphene-coated copper foil, in which graphene acted as a growth template. A solid precursor, $(NH_4)_2MoS_4$, was first dissolved in *N,N*-dimethylformamide (DMF).

This solution was bubbled with Ar gas and DMF/$(NH_4)_2MoS_4$ vapor was carried into the growth chamber containing the substrate. The growth chamber was then slowly heated up to 400°C, at which the precursor decomposed into MoS_2, which was deposited on the graphene/Cu substrate. The crystalline flakes produced have a lateral size ranging from a few hundred nanometers to several millimeters, with a film thickness of 2–5 nm, which is highly encouraging. The authors proposed that this method can be extended to create various heterostructures by using other substrates such as hexagonal boron nitride.

1.4 Black phosphorus (phosphorene)

Following the success of graphene, phosphorene (a single layer of BP) is touted as an even more promising material due to the presence of a direct bandgap. For the purpose of this book, phosphorene refers to the monolayer and flakes which are few-layers thick will be termed black phosphorus. BP rose to prominence after it was discovered as a material that can "bridge the gap" between graphene and TMDs. The gap here does not simply refer to the optical or electrical bandgap, but also properties that complement those of graphene and TMDs. As can be seen in Fig. 1.6A, the bandgap of BP ranges from 0.3 in the bulk form to 2.0 eV as a monolayer.[69, 164] BP sits nicely in between the zero-gap graphene and TMDs with gaps of 1.5–2.5 eV. This layer-dependent direct bandgap is useful for tuning the properties of BP to match requirements, and it is likely to show great potential in applications in IR frequencies. In addition, the high carrier mobility of BP and relatively good on/off ratio also lie between those properties of graphene and TMDs, suggesting potential in unique optical and electronic applications, and rendering BP as an exciting addition to the 2D family.[165]

This direct bandgap is dependent on the number of layers, ranging from 0.3 eV in the bulk form to 2.0 eV as a monolayer. Thus, phosphorene serves as a material that "bridges the gap" between the zero-bandgap graphene and large-bandgap TMDs. BP also exhibits high carrier mobility at room temperature.

1.4.1 Structure and properties

Like carbon, phosphorus comes in many allotropes. The most common ones are white and red phosphorus. BP, however, is thermodynamically stable and the allotrope that we are going to discuss in greater detail. The structure of BP is shown in Fig. 1.7. Each phosphorus atom is sp^3 hybridized, with

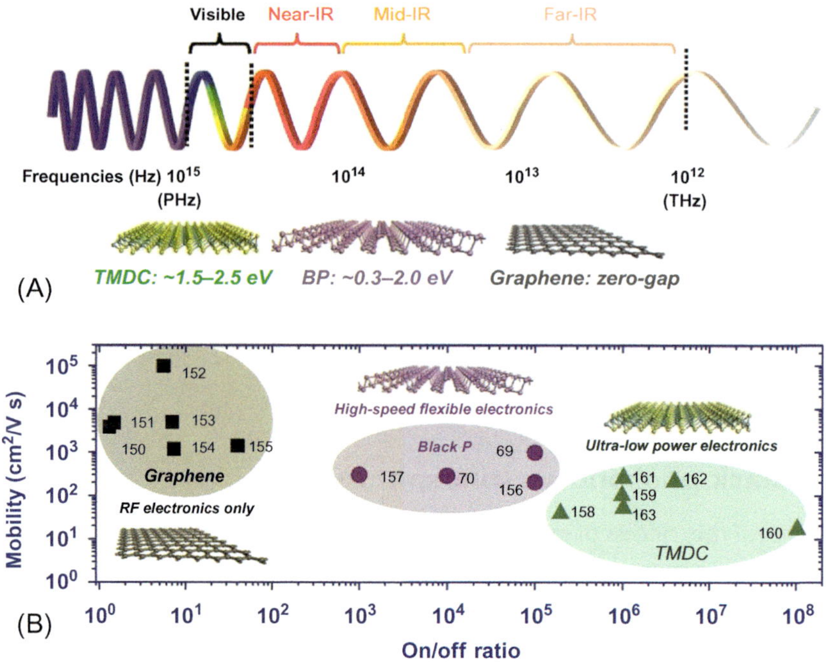

Fig. 1.6 (A) The electromagnetic wave spectrum and the bandgap ranges of graphene, BP, and TMDs. The frequency ranges corresponding to the bandgaps of 2D materials and their applications in optoelectronics are also indicated.[149] (B) The "electronics spectrum," i.e., the mobility/on-off ratio spectrum, of nanomaterials with corresponding performance regions indicated for graphene[150–155] (black squares and gray-shaded area), BP[69, 70, 156, 157] (*purple* dots and light purple-shaded area), and TMD (MoS_2,[158–160] WSe_2,[161, 162] and WS_2[163]) (*green* triangles and *light green*-shaded area) transistors. The dots correspond to data from specific references indicated next to them. The shaded regions are the approximate possible ranges of performance reported for the respective materials in the literature. *Adapted and reprinted from Ling, X.; Wang, H.; Huang, S.; Xia, F.; Dresselhaus, M. S., The Renaissance of Black Phosphorus. Proc. Natl. Acad. Sci.* ***2015****,* 112 *(15), 4523-4530, with permission from* PNAS.

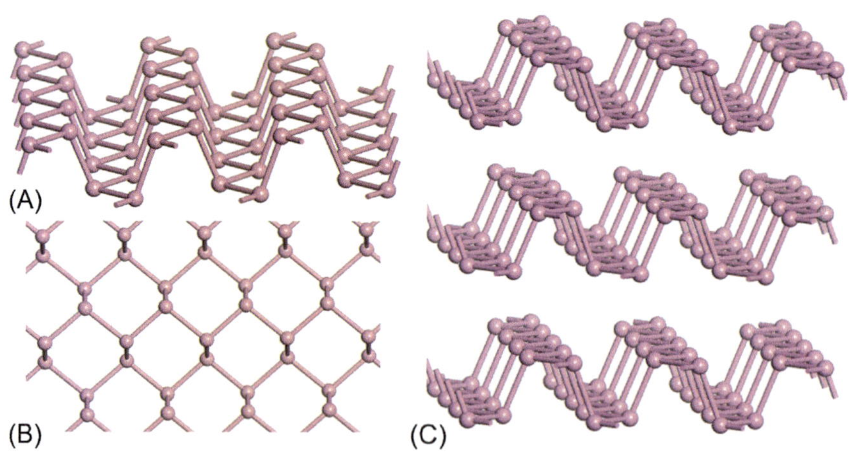

Fig. 1.7 Structure of (A) phosphorene (side view); (B) phosphorene (top view); (C) BP.

three valence electrons covalently bonded to other P atoms and leaving one lone pair. The result is a stacked, nonplanar honeycomb lattice. The layered structure renders BP somewhat similar to graphite in terms of appearance and properties.

Since BP is made up of stacked layers of phosphorene, the stacking order contributes significantly to its electronic and optical properties, as in the case of multilayer graphene. The most interesting property of phosphorene, however, is in-plane anisotropy, which arises from its puckered structure. In the BZ close to the Fermi level, the top of the valance bands and bottom of the conduction bands demonstrate high dispersion along the armchair direction, contrasting with the almost flat bands along the zigzag direction. As such, the effective masses of charge carriers, i.e., electrons and holes, also show anisotropy, and this is manifested in the anisotropic transport property, i.e., along different directions, the effective holes and electrons respond differently to strain and electric fields.[166] The anisotropic electrical conductance of few-layer BP may be tuned via strain engineering.[167] On the application on uniaxial strain ($<6\%$) or biaxial strain ($<4\%$), the preferred conduction direction can be rotated by 90 degrees. We will discuss the light-matter interactions in BP in greater detail in Chapter 6.

The major limitation in realizing devices with BP is its stability—or rather, the lack of it—in air. Therefore, when considering methods of production, we must also ensure that the products remain stable under ambient conditions.

1.4.2 Methods of production

1.4.2.1 Mechanical exfoliation

As in the case of graphene and TMDs, mechanical exfoliation is the go-to method for producing phosphorene or few-layer BP due to its layered structure. In early 2014, two independent groups announced that they had successfully obtained 2–3 layers of BP using the sticky tape method.[69, 70] Both groups also offered promising results from the BP FETs they fabricated. Li et al. reported carrier mobility of up to $\sim 1000\ cm^2/V\ s$,[69] while Liu et al. reported a high on-current of 194 mA/mm, a high hole field-effect mobility of $286\ cm^2/V\ s$, and an on/off ratio of up to 10^4.[70] Following those two studies, many other accounts on mechanically exfoliated BP soon appeared.

One method to suppress chemical degradation in ambient conditions following exfoliation is through chemical modification, as demonstrated by Ryder and co-workers.[168] After covalent aryl diazonium functionalization, the exfoliated BP remained stable after 3 weeks. The introduced functionalities also modified the electronic properties of BP, resulting in a strong p-type doping that is tunable. In addition, the FET with the functionalized BP showed improved carrier mobility and on/off current ratio.

1.4.2.2 Liquid-phase exfoliation

Liquid-phase exfoliation is a popular method for producing BP as it is the most feasible method for large-scale production. More importantly, the exfoliated flakes dispersed in a solvent are somewhat protected against degradation due to exposure to air. As there are numerous reports on the production of few-layer BP via liquid-phase exfoliation, we shall concentrate on a few significant studies. Readers are invited to refer to the review by Lin et al. for more information. In most circumstances, few-layer BP were prepared from bulk by dispersing the flakes in a solvent. The mixture was then sonicated at low temperature (ice bath) for long periods. The resulting brown solution was centrifuged to remove large particles that had not been exfoliated, and the supernatant was collected.[78]

As mentioned in Section 1.3.2.2, the choice of solvent is critical in liquid-phase exfoliation. Yasaei et al. considered several solvents and found that aprotic and polar solvents such as DMF and dimethyl sulfoxide present as suitable solvents for BP, and uniform and stable dispersions can be obtained.[169] For some solvents such as NMP, pretreatment of the solvent must be performed to remove water and oxygen that will cause the BP to degrade.[170, 171] Kang et al. also prepared a custom-tip sonicator setup to reduce exposure to water and oxygen, by sealing possible water and oxygen entry points with PMMA, Parafilm, and Teflon tapes.[170] Using the setup, stable BP dispersions with concentrations of up to 0.4 mg/mL were prepared. In addition, the exfoliated BP FET exhibited ambipolar behavior with high on/off current ratios (10^4) and mobilities (50 $cm^2/V\ s$). Guo, Zhang and co-workers also used NMP, but redispersed the exfoliated BP in water.[171] They also demonstrated nonlinear optical absorption by the exfoliatied BP and suggested that it could be exploited in a saturable absorber and Q-switcher in ultrafast photonics.

1.4.2.3 Other methods

CVD, which has been applied successfully to graphene and TMDs, unfortunately does not work as well for BP, due to stability issues. The CVD growth of BP on silicon was achieved by using a thin film of red phosphorus, and films as thin as four layers were obtained. However, the grown BP films were very sensitive to air and would start to decompose even during characterization.[172] Plasma treatment of BP flakes could, in theory, etch away layers of BP to the desired thickness. Jia et al. found that the treatment not only thinned the sample but also removed defects arising from mechanical exfoliation prior to plasma treatment, thus improving the morphology of BP films. To overcome stability issues, thinned BP films were coated with a polymer (PMMA). A FET made with PMMA-coated BP exhibits an initial hole mobility of 150 cm^2/V s and an on/off ratio of 10^4. More importantly, it maintains good I_d–V_g transfer characteristics over several weeks, suggesting that BP is well protected by PMMA.

Yang et al. attempted a new technique of synthesizing BP: pulsed laser deposition.[74] The ultrathin films of BP were grown on graphene/copper of SiO_2/Si substrates. Although the BP films obtained were amorphous, the bandgaps of the films were also thickness-dependent, ranging from 0.80 eV for a 2-nm film to 0.21–0.26 eV for an 8-nm film. The team fabricated a FET using a 2-nm film, and found that it demonstrated a moderate on/off current ratio up to 10^2, and high field-effect mobility up to 14 cm^2/V s, which is not outstanding per se but yet excellent when compared to those found in conventional elemental amorphous materials (e.g., a-Si:H and a-carbon).

1.5 Other 2D materials

Material scientists are on a constant search for new materials to overcome existing limitations, while physicists would be excited to study many phenomena such as quantum spin Hall effect, quantum anomalous Hall effect, quantum valley Hall effect, and superconductivity experimentally. This section and the next will be a short introduction to some of the other materials in 2D form. We do not attempt to discuss these in depth since photonic and optoelectronic applications for these materials are rarely proposed.

1.5.1 Group 14 elements: Silicene, germanene, stanene

The 2D form of silicon—silicene—is one of the first material scientists would like to consider, since silicon is just below carbon in the periodic table and widely used in the electronics industry. However, the production of Si is not as straightforward as that of graphene. First, there exists no naturally occurring allotrope of silicon in a layered structure, so mechanical exfoliation or assisted exfoliation, such as sonication, is not viable. Second, silicon dioxide is a strong and hard insulator, with a high melting point of >1700°C, and thus chemical methods to obtain graphene from graphite oxides cannot be applied. Therefore, many early studies of Si are theoretical predictions. Following the success of Vogt et al in the synthesis of epitaxial Si sheets on a silver (111) substrate in 2012,[173] experimental investigations are mostly performed on silicon sheets grown epitaxially.

By virtue of the larger atomic number of silicon and germanium, Si and Ge are expected to demonstrate stronger spin-orbit coupling. In addition, the honeycomb structure of Si and Ge, unlike graphene, is buckled,[174] and this buckling is expected to increase the spin-orbit coupling. Si, Ge, and similar materials discussed in this section are expected to have applications utilizing the quantum spin Hall effect. Readers interested to know more about Si may wish to refer to the review by Zhao et al., which details the synthesis, functionalization, and potential applications of Si.[79]

Although Ge is mostly synthesized through epitaxial growth, the hydrogenated form of Ge (GeH) has been obtained through chemical methods. Bianco et al. obtained GeH from calcium germanide ($CaGe_2$) by stirring $CaGe_2$ crystals in concentrated HCl for 5–10 days at low temperature (–40 to –20°C).[175] The produced GeH was stable for at least 5 months, and it had a direct bandgap of around 1.5 eV, a useful range for optoelectronic applications. However, the authors are not able to provide direct experimental evidence of this direct bandgap, and therefore more work needs to be done. In a review by Acun et al., Ge was examined with discussions on theoretical predictions, synthesis methods, and possible outlook.[83]

Like Si and Ge, Sn, tin in the 2D form, is synthesized through epitaxial growth. Zhu et al. first reported the successful growth of Sn on Bi_2Te_3(111) using molecular beam epitaxy.[85] They also showed that Sn has a large bandgap with 2D topological insulating behavior, suggesting improved thermoelectric performance, topological superconductivity, and quantum anomalous Hall effect (occurring near room temperature). Recently, superconductivity in Sn has been discovered. Liao et al. found that superconductivity exists for about 3–5 layers of Sn.[84] No superconductivity was observed in monolayers, while the growth became more three-dimensional in the form of islands beyond five layers. In addition, the underlying PbTe (111)/Bi_2Te_3 substrate has an effect on superconductivity properties. The superconducting transition temperature (T_c) varies depending on the thickness of the PbTe buffer layer. Such promising results will undoubtedly excite more interest in Sn.

1.5.2 Group 15 elements: Arsenene, antimonene, bismuthene

Apart from phosphorus, the other group 15 elements are just coming into the picture. As such, there are few experimental studies on As, Sb, and Bi. On the other hand, many theoretical studies have been performed.[87, 176–178] For As, calculations showed that there are two stable types: buckled and puckered.[177] In both structures, the bandgap is indirect; however, the application of strain causes transition to a direct bandgap. In the puckered structure, a 1% strain is sufficient to cause this indirect-to-direct bandgap transition. This idea is extended to Sb; calculations predicted a similar indirect-to-direct bandgap transition for Sb.[178] Therefore, strain engineering may be utilized for incorporating these materials in optoelectronic applications. For example, strain may be induced by selecting a suitable substrate. On the Ag (111) surface, deposition of Sb forms an $AgSb_2$ alloy after annealing.[179] Further deposition on this alloy leads to the formation of epitaxial layers of Sb with the buckled structure and a larger-than-usual lattice constant, corresponding to a tensile strain >20%.

Successful methods of producing 2D forms of group 15 elements are almost entirely limited to epitaxial growth. Tsai et al. developed a plasma-assisted method for the growth of graphene on 4H-SiC and later used it to prepare Ge on SiGe/Si and violet phosphorene on InP.[180–182] The authors attempted to expand this method to As.[86] Although multilayer As was successfully synthesized, its morphology was more like "a pile of multilayer nanoribbons" instead of a continuous 2D sheet. To the best of our knowledge, monolayer As has not been synthesized yet. There are, however, studies involving black arsenic-phosphorus since 2016. Arsenic is incorporated into BP in varying compositions to tune the bandgap from 0.15 to 0.3 eV, corresponding to the long-wavelength IR (LWIR) regime in the electromagnetic spectrum.[183] Thus, arsenic-phosphorus is viewed as a potential candidate for LWIR applications such as photodetection, sensing, and imaging. Theoretical calculations also found that As-based heterostructures, when paired with MoS_2, tetracyano-quinodimethane, or tetracyanonaphtho-quinodimethane, form type II band alignments, rendering such structures effective in photovoltaic and photocatalytic applications.[87] Nevertheless, technological challenges of material synthesis and device fabrication will have to be addressed first.

In the case of Sb, Ares et al. isolated few-layer Sb using mechanical exfoliation and found that the isolated flakes were stable under ambient conditions for months, even after immersion in water.[88] This is in contract with BP. The same group also reported that Sb may be produced via sonication-assisted liquid-phase exfoliation. Sb-isopropanol/water (4:1) suspensions with concentrations of up to 1.74 mg/L were also stable for weeks, which is highly promising. In terms of the bottom-up approach, Ji et al. first demonstrated van der Waals epitaxial growth of few-layer Sb on various substrates, including mica, silicon, and sapphire. The synthesized Sb polygens on mica showed good transparency in the visible range (>80). Electrical measurements of a flexible device made with Sb on mica found that the synthesized Sb sheet possessed good electrical conductivity, i.e., 1.6×10^4 S/m, usual for semi-metals. Wu et al. also later reported the epitaxial growth of high-quality monolayer on $PdTe_2$.[90] In both studies, Sb binds weakly to the substrates, which is excellent for studying the true properties of Sb. Moreover, the 2D Sb was shown to be stable in air. Therefore, Sb is likely to offer great prospects as a transparent conductor and in flexible electronics.

Bismuth, being heavier than the other group 15 elements, will display strong effects from spin-orbit coupling. In bulk bismuth, it was found that as little as an in-plane tensile strain of 0.3% is sufficient to cause a semimetal-to-semiconductor transition, while a compressive strain of only 0.4% can cause a trivial-to-topological phase transition.[184] In view of this, it is highly likely that Bi possesses many unique and attractive properties awaiting to be discovered. Epitaxial growth of Bi was achieved by Reis et al. on the SiC (0001) substrate.[94] The substrate not only stabilized Bi but is also a means to exploit the atomic (on-site) spin-orbit coupling to directly determine the topological energy gap. Experimental findings and theoretical predictions both portray Bi/SiC as a quantum spin Hall system. In another approach, Lu, Wang, and co-workers obtained Bi suspensions by first grinding bulk bismuth powder and then dispersing it in ethanol via sonication.[93] The suspension comprised few-layer Bi, with an average six layers. The authors also achieved all-optical switching induced by two-color continuous waves by utilizing the cross-phase modulation and strong nonlinear refraction effects in few-layer Bi. These promising results are a validation of the potential of Bi in photonic applications, and we expect to see more interesting work in this direction.

1.5.3 MXenes (2D metal carbides and nitrides)

In 2011, Gogotsi and his team at Drexel University wanted to improve the performance of lithium-ion battery anodes. They considered a family of layered, hexagonal carbides and nitrides that are conducting, known as

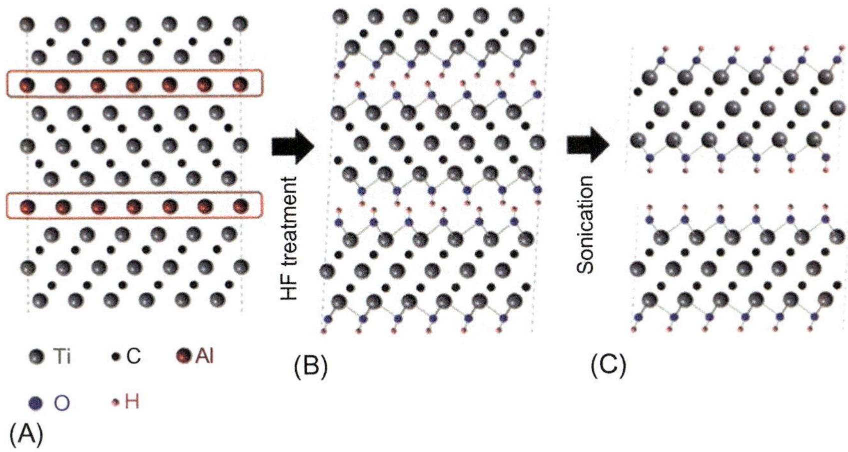

Fig. 1.8 Schematic of the exfoliation process for Ti_3AlC_2. (A) Ti_3AlC_2 structure. The rectangle highlights the layer of Al atoms that will be removed by HF. (B) Al atoms replaced by OH after reaction with HF. (C) Breakage of hydrogen bonds and separation of nanosheets after sonication in methanol.[95] *Reprinted with permission from Naguib, M.; Kurtoglu, M.; Presser, V.; Lu, J.; Niu, J.; Heon, M.; Hultman, L.; Gogotsi, Y.; Barsoum, M. W., Two-Dimensional Nanocrystals Produced by Exfoliation of Ti3AlC2. Adv. Mater.* ***2011****, 23 (37), 4248-4253. Copyright 2011 by John Wiley and Sons.*

MAX phases, where M is an early transition metal, A is a main-group element such as aluminum or silicon, and X refers to carbon or nitrogen. To increase the efficiency of Li ion movement in the anodes during charging and discharging, the team tried to remove aluminum ions from Ti_3AlC_2 by treating it with concentrated hydrofluoric acid (HF).[95] HF treatment worked and also created an extraordinary material. After all the aluminum atoms were removed from Ti_3AlC_2, the remaining Ti_3C_2 presented as exfoliated 2D sheets, similar to graphene (Fig. 1.8). Ti_3C_2 sheets are terminated with OH or OF groups, which can be chemically modified with ease.

More interestingly, because the MAX phases comprise a large family of materials, HF treatment can be used in a similar fashion to produce a new class of 2D materials, coined MXenes, a derivation from graphene. The team at Drexel University did just that, reporting in 2017 that more than 20 MXenes have been synthesized.[101] The review also provided a comprehensive table on the various applications of MXenes. Most of the proposed applications of MXenes are in the field of energy storage and conversion. Nevertheless, since MXenes are a large family of materials, numerous potential applications have been suggested, depending on the actual material. Apart from energy-related applications, other uses of MXenes include, but are not limited to, structural composite, shielding of electromagnetic interference, water treatment, chemical sensors, biosensors, and electronics.

Although most of the proposed applications of MXenes are not related to photonics or optoelectronics, that may change in the near future. Hantanasirisakul et al. produced $Ti_3C_2T_x$ films, where T_x refers to surface terminations (OH, O, or F), which are flexible, transparent, and conducting.[96] The optoelectronic properties of these films can be reversibly tuned by chemical or electrochemical intercalation of cations, suggesting potentials in flexible electronics or electrochromic applications. Recently, it was also shown that $Ti_3C_2T_x$ films can form mechanically responsive conductive coatings that maintain high conductivities of 2000 S/m after undergoing significant deformation.[102] The resistance changes of coating is reversible for bending of up to 2.5-mm bending radius and tensile strain stretching of up to 40%. Such flexible and stretchable conducting films will have much potential in electronics such as touchscreens or biometric sensing. As in the case of other 2D materials, MXenes may also be incorporated in van der Waals heterostructures to enhance the performance of electronic devices. For example, $Ti_3C_2T_X$ on n-type silicon (n-Si) of a vertical van der Waals heterostructure was utilized to form a self-driven vertical junction photodetector.[100] The measured response and recovery speed of the photodetector was relatively high, at 0.84 and 1.67 ms, respectively. It was found that the $Ti_3C_2T_X$ film not only functions as the transparent electrode but also aids in the separation and transport of photo-induced carriers, thus enhancing the performance of the photodetector.

In terms of photonic applications, it was found that metallic Ti_3CNT_x can serve as an excellent mode-locker, producing femtosecond laser pulses from fiber cavities.[98] The laser pulse width, when a Ti_3CNT_x saturable absorber was used, is the same order of magnitude as that obtained from the best TMDs (MoS_2, WS_2, and WTe_2) and one order of magnitude shorter than that of other TMDs ($MoSe_2$, $MoTe_2$, and WSe_2). The authors also believe that the operating range of the Ti_3CNT_x saturable absorber may be extended beyond the IR range (1500–2000 nm), fulfilling broadband applications. Another study revealed that $Ti_3C_2T_x$ films may also be used in photonic diodes.[99] The $Ti_3C_2T_x$ films exhibited a high modulation depth of up to ~50% and high threshold of ~70 mJ/cm^2 for optical damage. Moreover, saturation fluence can be easily tuned by adjusting the thickness of the $Ti_3C_2T_x$ films used. In view of these reports, we predict MXenes will become a significant player in photonic and optoelectronic technologies.

References

1. Novoselov, K. S.; Geim, A. K.; Morozov, S. V.; Jiang, D.; Zhang, Y.; Dubonos, S. V.; Grigorieva, I. V.; Firsov, A. A. Electric Field Effect in Atomically Thin Carbon Films. *Science* **2004**, *306*(5696), 666.
2. Feynman, R. There's Plenty of Room at the Bottom. *Eng. Sci.* **1960**, *23*(5), 22–36.
3. Wallace, P. R. The Band Theory of Graphite. *Phys. Rev.* **1947**, *71*(9), 622–634.
4. Kroto, H. W.; Heath, J. R.; O'Brien, S. C.; Curl, R. F.; Smalley, R. E. C60: Buckminsterfullerene. *Nature* **1985**, *318*, 162.
5. Iijima, S.; Ichihashi, T. Single-Shell Carbon Nanotubes of 1-nm Diameter. *Nature* **1993**, *363*, 603.
6. Cheon, G.; Duerloo, K.-A. N.; Sendek, A. D.; Porter, C.; Chen, Y.; Reed, E. J. Data Mining for New Two- and One-Dimensional Weakly Bonded Solids and Lattice-Commensurate Heterostructures. *Nano Lett.* **2017**, *17*(3), 1915–1923.
7. Mounet, N.; Gibertini, M.; Schwaller, P.; Campi, D.; Merkys, A.; Marrazzo, A.; Sohier, T.; Castelli, I. E.; Cepellotti, A.; Pizzi, G.; Marzari, N. Two-Dimensional Materials From High-Throughput Computational Exfoliation of Experimentally Known Compounds. *Nat. Nanotechnol.* **2018**, *13*(3), 246–252.
8. Yi, M.; Shen, Z. A Review on Mechanical Exfoliation for the Scalable Production of Graphene. *J. Mater. Chem. A* **2015**, *3*(22), 11700–11715.
9. Parvez, K.; Wu, Z.-S.; Li, R.; Liu, X.; Graf, R.; Feng, X.; Müllen, K. Exfoliation of Graphite into Graphene in Aqueous Solutions of Inorganic Salts. *J. Am. Chem. Soc.* **2014**, *136*(16), 6083–6091.
10. Park, S.; Ruoff, R. S. Chemical Methods for the Production of Graphenes. *Nat. Nanotechnol.* **2009**, *4*, 217.
11. Hernandez, Y.; Nicolosi, V.; Lotya, M.; Blighe, F. M.; Sun, Z.; De, S.; McGovern, I. T.; Holland, B.; Byrne, M.; Gun'Ko, Y. K.; Boland, J. J.; Niraj, P.; Duesberg, G.; Krishnamurthy, S.; Goodhue, R.; Hutchison, J.; Scardaci, V.; Ferrari, A. C.; Coleman, J. N. High-Yield Production of Graphene by Liquid-Phase Exfoliation of Graphite. *Nat. Nanotechnol.* **2008**, *3*, 563.
12. Lotya, M.; Hernandez, Y.; King, P. J.; Smith, R. J.; Nicolosi, V.; Karlsson, L. S.; Blighe, F. M.; De, S.; Wang, Z.; McGovern, I. T.; Duesberg, G. S.; Coleman, J. N. Liquid Phase Production of Graphene by Exfoliation of Graphite in Surfactant/Water Solutions. *J. Am. Chem. Soc.* **2009**, *131*(10), 3611–3620.
13. Ciesielski, A.; Samori, P. Graphene via Sonication Assisted Liquid-Phase Exfoliation. *Chem. Soc. Rev.* **2014**, *43*(1), 381–398.
14. Reina, A.; Jia, X.; Ho, J.; Nezich, D.; Son, H.; Bulovic, V.; Dresselhaus, M. S.; Kong, J. Large Area, Few-Layer Graphene Films on Arbitrary Substrates by Chemical Vapor Deposition. *Nano Lett.* **2009**, *9*(1), 30–35.
15. Yu, Q.; Lian, J.; Siriponglert, S.; Li, H.; Chen, Y. P.; Pei, S.-S. Graphene Segregated on Ni Surfaces and Transferred to Insulators. *Appl. Phys. Lett.* **2008**, *93*(11), 113103.
16. Bae, S.; Kim, H.; Lee, Y.; Xu, X.; Park, J.-S.; Zheng, Y.; Balakrishnan, J.; Lei, T.; Ri Kim, H.; Song, Y. I.; Kim, Y.-J.; Kim, K. S.; Özyilmaz, B.; Ahn, J.-H.; Hong, B. H.; Iijima, S. Roll-to-Roll Production of 30-Inch Graphene Films for Transparent Electrodes. *Nat. Nanotechnol.* **2010**, *5*, 574.
17. Berger, C.; Song, Z.; Li, X.; Wu, X.; Brown, N.; Naud, C.; Mayou, D.; Li, T.; Hass, J.; Marchenkov, A. N.; Conrad, E. H.; First, P. N.; de Heer, W. A. Electronic Confinement and Coherence in Patterned Epitaxial Graphene. *Science* **2006**, *312*(5777), 1191.
18. Hass, J.; Heer, W. A. D.; Conrad, E. H. The Growth and Morphology of Epitaxial Multilayer Graphene. *J. Phys. Condens. Matter* **2008**, *20*(32), 323202.
19. Jiao, L.; Zhang, L.; Wang, X.; Diankov, G.; Dai, H. Narrow Graphene Nanoribbons From Carbon Nanotubes. *Nature* **2009**, *458*, 877.

20. Liu, M.; Yin, X.; Ulin-Avila, E.; Geng, B.; Zentgraf, T.; Ju, L.; Wang, F.; Zhang, X. A Graphene-Based Broadband Optical Modulator. *Nature* **2011**, *474*, 64.
21. Maysonnave, J.; Huppert, S.; Wang, F.; Maero, S.; Berger, C.; de Heer, W.; Norris, T. B.; De Vaulchier, L. A.; Dhillon, S.; Tignon, J.; Ferreira, R.; Mangeney, J. Terahertz Generation by Dynamical Photon Drag Effect in Graphene Excited by Femtosecond Optical Pulses. *Nano Lett.* **2014**, *14*(10), 5797–5802.
22. Singh, R. S.; Nalla, V.; Chen, W.; Wee, A. T. S.; Ji, W. Laser Patterning of Epitaxial Graphene for Schottky Junction Photodetectors. *ACS Nano* **2011**, *5*(7), 5969–5975.
23. Wang, X.; Tian, H.; Mohammad, M. A.; Li, C.; Wu, C.; Yang, Y.; Ren, T.-L. A Spectrally Tunable All-Graphene-Based Flexible Field-Effect Light-Emitting Device. *Nat. Commun.* **2015**, *6*, 7767.
24. SERA Technologies. www.seratechnologies.com/what-is-a-graphene-led-bulb/.
25. Graphene Lighting PLC. http://graphenelighting.co/.
26. Fei, Z.; Rodin, A. S.; Andreev, G. O.; Bao, W.; McLeod, A. S.; Wagner, M.; Zhang, L. M.; Zhao, Z.; Thiemens, M.; Dominguez, G.; Fogler, M. M.; Neto, A. H. C.; Lau, C. N.; Keilmann, F.; Basov, D. N. Gate-Tuning of Graphene Plasmons Revealed by Infrared Nano-Imaging. *Nature* **2012**, *487*, 82.
27. Chen, J.; Badioli, M.; Alonso-González, P.; Thongrattanasiri, S.; Huth, F.; Osmond, J.; Spasenović, M.; Centeno, A.; Pesquera, A.; Godignon, P.; Zurutuza Elorza, A.; Camara, N.; de Abajo, F. J. G.; Hillenbrand, R.; Koppens, F. H. L. Optical Nano-Imaging of Gate-Tunable Graphene Plasmons. *Nature* **2012**, *487*, 77.
28. Sun, S.; Zhuang, X.; Wang, L.; Wang, C.; Zhang, B.; Chen, Y. Graphene-Based Functional Materials for Information Storage: Materials, Devices and Performance. *Prog. Chem.* **2016**, *28*(1), 18–39.
29. Dimitrios, B.; Hanene, A. B.; Hee, H. B.; Dal-Hee, M.; Kostas, K. Prospects and Challenges of Graphene in Biomedical Applications. *Adv. Mater.* **2013**, *25*(16), 2258–2268.
30. Novoselov, K. S.; Jiang, D.; Schedin, F.; Booth, T. J.; Khotkevich, V. V.; Morozov, S. V.; Geim, A. K. Two-Dimensional Atomic Crystals. *Proc. Natl. Acad. Sci. U. S. A.* **2005**, *102*(30), 10451.
31. Tonndorf, P.; Schmidt, R.; Böttger, P.; Zhang, X.; Börner, J.; Liebig, A.; Albrecht, M.; Kloc, C.; Gordan, O.; Zahn, D. R. T.; Michaelis de Vasconcellos, S.; Bratschitsch, R. Photoluminescence Emission and Raman Response of Monolayer MoS_2, $MoSe_2$, and WSe_2. *Opt. Express* **2013**, *21*(4), 4908–4916.
32. Jawaid, A.; Nepal, D.; Park, K.; Jespersen, M.; Qualley, A.; Mirau, P.; Drummy, L. F.; Vaia, R. A. Mechanism for Liquid Phase Exfoliation of MoS_2. *Chem. Mater.* **2016**, *28*(1), 337–348.
33. Yi-Hsien, L.; Xin-Quan, Z.; Wenjing, Z.; Mu-Tung, C.; Cheng-Te, L.; Kai-Di, C.; Ya-Chu, Y.; Tse-Wei, W. J.; Chia-Seng, C.; Lain-Jong, L.; Tsung-Wu, L. Synthesis of Large-Area MoS_2 Atomic Layers With Chemical Vapor Deposition. *Adv. Mater.* **2012**, *24*(17), 2320–2325.
34. Lee, Y.-H.; Yu, L.; Wang, H.; Fang, W.; Ling, X.; Shi, Y.; Lin, C.-T.; Huang, J.-K.; Chang, M.-T.; Chang, C.-S.; Dresselhaus, M.; Palacios, T.; Li, L.-J.; Kong, J. Synthesis and Transfer of Single-Layer Transition Metal Disulfides on Diverse Surfaces. *Nano Lett.* **2013**, *13*(4), 1852–1857.
35. Wang, G.; Zhang, S.; Zhang, X.; Zhang, L.; Cheng, Y.; Fox, D.; Zhang, H.; Coleman, J. N.; Blau, W. J.; Wang, J. Tunable Nonlinear Refractive Index of Two-Dimensional MoS_2, WS_2, and $MoSe_2$ Nanosheet Dispersions [Invited]. *Photon. Res.* **2015**, *3*(2), A51–A55.
36. Woodward, R. I.; Kelleher, E. J. R.; Howe, R. C. T.; Hu, G.; Torrisi, F.; Hasan, T.; Popov, S. V.; Taylor, J. R. Tunable Q-Switched Fiber Laser Based on Saturable Edge-State Absorption in Few-Layer Molybdenum Disulfide (MoS_2). *Opt. Express* **2014**, *22*(25), 31113–31122.

37. Lopez-Sanchez, O.; Lembke, D.; Kayci, M.; Radenovic, A.; Kis, A. Ultrasensitive Photodetectors Based on Monolayer MoS_2. *Nat. Nanotechnol.* **2013**, *8*, 497.
38. Tsai, M.-L.; Su, S.-H.; Chang, J.-K.; Tsai, D.-S.; Chen, C.-H.; Wu, C.-I.; Li, L.-J.; Chen, L.-J.; He, J.-H. Monolayer MoS_2 Heterojunction Solar Cells. *ACS Nano* **2014**, *8*(8), 8317–8322.
39. Bernardi, M.; Palummo, M.; Grossman, J. C. Extraordinary Sunlight Absorption and One Nanometer Thick Photovoltaics Using Two-Dimensional Monolayer Materials. *Nano Lett.* **2013**, *13*(8), 3664–3670.
40. Radisavljevic, B.; Radenovic, A.; Brivio, J.; Giacometti, V.; Kis, A. Single-Layer MoS_2 Transistors. *Nat. Nanotechnol.* **2011**, *6*, 147.
41. Wang, Q.; Ge, S.; Li, X.; Qiu, J.; Ji, Y.; Feng, J.; Sun, D. Valley Carrier Dynamics in Monolayer Molybdenum Disulfide From Helicity-Resolved Ultrafast Pump–Probe Spectroscopy. *ACS Nano* **2013**, 7(12), 11087–11093.
42. Zeng, H.; Dai, J.; Yao, W.; Xiao, D.; Cui, X. Valley Polarization in MoS_2 Monolayers by Optical Pumping. *Nat. Nanotechnol.* **2012**, 7, 490.
43. Mak, K. F.; He, K.; Shan, J.; Heinz, T. F. Control of Valley Polarization in Monolayer MoS_2 by Optical Helicity. *Nat. Nanotechnol.* **2012**, 7, 494.
44. Luo, Z.; Li, Y.; Zhong, M.; Huang, Y.; Wan, X.; Peng, J.; Weng, J. Nonlinear Optical Absorption of Few-Layer Molybdenum Diselenide ($MoSe_2$) for Passively Mode-Locked Soliton Fiber Laser [Invited]. *Photon. Res.* **2015**, *3*(3), A79–A86.
45. Wang, X.; Gong, Y.; Shi, G.; Chow, W. L.; Keyshar, K.; Ye, G.; Vajtai, R.; Lou, J.; Liu, Z.; Ringe, E.; Tay, B. K.; Ajayan, P. M. Chemical Vapor Deposition Growth of Crystalline Monolayer MoSe2. *ACS Nano* **2014**, *8*(5), 5125–5131.
46. Chen, J.; Zhao, X.; Tan, S. J. R.; Xu, H.; Wu, B.; Liu, B.; Fu, D.; Fu, W.; Geng, D.; Liu, Y.; Liu, W.; Tang, W.; Li, L.; Zhou, W.; Sum, T. C.; Loh, K. P. Chemical Vapor Deposition of Large-Size Monolayer $MoSe_2$ Crystals on Molten Glass. *J. Am. Chem. Soc.* **2017**, *139*(3), 1073–1076.
47. Zenghui, W.; Guoan, T.; Xufeng, W.; Tingsong, H.; Rui, W.; Wanlin, G. Large-Area Synthesis and Photoelectric Properties of Few-Layer $MoSe_2$ on Molybdenum Foils. *Nanotechnology* **2018**, *29*(12), 125605.
48. Chen, H.; Corboliou, V.; Solntsev, A. S.; Choi, D.-Y.; Vincenti, M. A.; de Ceglia, D.; de Angelis, C.; Lu, Y.; Neshev, D. N. Enhanced second-harmonic generation from two-dimensional $MoSe_2$ on a silicon waveguide. *Light Sci. Appl.* **2017**, *6e17060*.
49. Woodward, R. I.; Howe, R. C. T.; Runcorn, T. H.; Hu, G.; Torrisi, F.; Kelleher, E. J. R.; Hasan, T. Wideband Saturable Absorption in Few-Layer Molybdenum Diselenide ($MoSe_2$) for Q-Switching Yb-, Er- and Tm-Doped Fiber Lasers. *Opt. Express* **2015**, *23*(15), 20051–20061.
50. Kumar, N.; He, J.; He, D.; Wang, Y.; Zhao, H. Valley and Spin Dynamics in MoSe2 Two-Dimensional Crystals. *Nanoscale* **2014**, *6*(21), 12690–12695.
51. Ruppert, C.; Aslan, O. B.; Heinz, T. F. Optical Properties and Band Gap of Single- and Few-Layer $MoTe_2$ Crystals. *Nano Lett.* **2014**, *14*(11), 6231–6236.
52. Giri, A.; Yang, H.; Jang, W.; Kwak, J.; Thiyagarajan, K.; Pal, M.; Lee, D.; Singh, R.; Kim, C.; Cho, K.; Soon, A.; Jeong, U. Synthesis of Atomically Thin Transition Metal Ditelluride Films by Rapid Chemical Transformation in Solution Phase. *Chem. Mater.* **2018**, *30*(7), 2463–2473.
53. Jiadong, Z.; Fucai, L.; Junhao, L.; Xiangwei, H.; Juan, X.; Bowei, Z.; Qingsheng, Z.; Hong, W.; Chao, Z.; Lin, N.; Xuewen, W.; Wei, F.; Peng, Y.; Tay-Rong, C.; Chuang-Han, H.; Di, W.; Horng-Tay, J.; Yizhong, H.; Hsin, L.; Zexiang, S.; Changli, Y.; Li, L.; Kazu, S.; Wu, Z.; Pantelides, S. T.; Guangtong, L.; Zheng, L. Large-Area and High-Quality 2D Transition Metal Telluride. *Adv. Mater.* **2017**, *29*(3), 1603471.

54. Ma, P.; Flöry, N.; Salamin, Y.; Baeuerle, B.; Emboras, A.; Josten, A.; Taniguchi, T.; Watanabe, K.; Novotny, L.; Leuthold, J. Fast $MoTe_2$ Waveguide Photodetector With High Sensitivity at Telecommunication Wavelengths. *ACS Photonics* **2018**, *5*(5), 1846–1852.
55. Cho, S.; Kim, S.; Kim, J. H.; Zhao, J.; Seok, J.; Keum, D. H.; Baik, J.; Choe, D.-H.; Chang, K. J.; Suenaga, K.; Kim, S. W.; Lee, Y. H.; Yang, H. Phase Patterning for Ohmic Homojunction Contact in $MoTe_2$. *Science* **2015**, *349*(6248), 625.
56. Yang, J.; Lü, T.; Myint, Y. W.; Pei, J.; Macdonald, D.; Zheng, J.-C.; Lu, Y. Robust Excitons and Trions in Monolayer $MoTe_2$. *ACS Nano* **2015**, *9*(6), 6603–6609.
57. Mitioglu, A. A.; Plochocka, P.; Jadczak, J. N.; Escoffier, W.; Rikken, G. L. J. A.; Kulyuk, L.; Maude, D. K. Optical Manipulation of the Exciton Charge State in Single-Layer Tungsten Disulfide. *Phys. Rev. B* **2013**, *88*(24), 245403.
58. Zhao, W.; Ghorannevis, Z.; Chu, L.; Toh, M.; Kloc, C.; Tan, P.-H.; Eda, G. Evolution of Electronic Structure in Atomically Thin Sheets of WS_2 and WSe_2. *ACS Nano* **2013**, 7(1), 791–797.
59. Lin, X.; Liu, Y.; Wang, K.; Wei, C.; Zhang, W.; Yan, Y.; Li, Y. J.; Yao, J.; Zhao, Y. S. Two-Dimensional Pyramid-like WS_2 Layered Structures for Highly Efficient Edge Second-Harmonic Generation. *ACS Nano* **2018**, *12*(1), 689–696.
60. Janisch, C.; Wang, Y.; Ma, D.; Mehta, N.; Elías, A. L.; Perea-López, N.; Terrones, M.; Crespi, V.; Liu, Z. Extraordinary Second Harmonic Generation in Tungsten Disulfide Monolayers. *Sci. Rep.* **2014**, *4*, 5530.
61. Gutiérrez, H. R.; Perea-López, N.; Elías, A. L.; Berkdemir, A.; Wang, B.; Lv, R.; López-Urías, F.; Crespi, V. H.; Terrones, H.; Terrones, M. Extraordinary Room-Temperature Photoluminescence in Triangular WS_2 Monolayers. *Nano Lett.* **2013**, *13*(8), 3447–3454.
62. Yu, X.; Sivula, K. Photogenerated Charge Harvesting and Recombination in Photocathodes of Solvent-Exfoliated WSe_2. *Chem. Mater.* **2017**, *29*(16), 6863–6875.
63. Eichfeld, S. M.; Hossain, L.; Lin, Y.-C.; Piasecki, A. F.; Kupp, B.; Birdwell, A. G.; Burke, R. A.; Lu, N.; Peng, X.; Li, J.; Azcatl, A.; McDonnell, S.; Wallace, R. M.; Kim, M. J.; Mayer, T. S.; Redwing, J. M.; Robinson, J. A. Highly Scalable, Atomically Thin WSe_2 Grown via Metal–Organic Chemical Vapor Deposition. *ACS Nano* **2015**, *9*(2), 2080–2087.
64. Seyler, K. L.; Schaibley, J. R.; Gong, P.; Rivera, P.; Jones, A. M.; Wu, S.; Yan, J.; Mandrus, D. G.; Yao, W.; Xu, X. Electrical Control of Second-Harmonic Generation in a WSe_2 Monolayer Transistor. *Nat. Nanotechnol.* **2015**, *10*, 407.
65. Langer, F.; Schmid, C. P.; Schlauderer, S.; Gmitra, M.; Fabian, J.; Nagler, P.; Schüller, C.; Korn, T.; Hawkins, P. G.; Steiner, J. T.; Huttner, U.; Koch, S. W.; Kira, M.; Huber, R. Lightwave Valleytronics in a Monolayer of Tungsten Diselenide. *Nature* **2018**, *557*(7703), 76–80.
66. Lee, C.-H.; Silva, E. C.; Calderin, L.; Nguyen, M. A. T.; Hollander, M. J.; Bersch, B.; Mallouk, T. E.; Robinson, J. A. Tungsten Ditelluride: A Layered Semimetal. *Sci. Rep.* **2015**, *5*, 10013.
67. Ali, M. N.; Xiong, J.; Flynn, S.; Tao, J.; Gibson, Q. D.; Schoop, L. M.; Liang, T.; Haldolaarachchige, N.; Hirschberger, M.; Ong, N. P.; Cava, R. J. Large, Non-Saturating Magnetoresistance in WTe_2. *Nature* **2014**, *514*, 205.
68. Chen, Y.; Jiang, G.; Chen, S.; Guo, Z.; Yu, X.; Zhao, C.; Zhang, H.; Bao, Q.; Wen, S.; Tang, D.; Fan, D. Mechanically Exfoliated Black Phosphorus as a New Saturable Absorber for Both Q-Switching and Mode-Locking Laser Operation. *Opt. Express* **2015**, *23*(10), 12823–12833.
69. Li, L.; Yu, Y.; Ye, G. J.; Ge, Q.; Ou, X.; Wu, H.; Feng, D.; Chen, X. H.; Zhang, Y. Black Phosphorus Field-Effect Transistors. *Nat. Nanotechnol.* **2014**, *9*, 372.

70. Liu, H.; Neal, A. T.; Zhu, Z.; Luo, Z.; Xu, X.; Tománek, D.; Ye, P. D. Phosphorene: An Unexplored 2D Semiconductor With a High Hole Mobility. *ACS Nano* **2014**, *8*(4), 4033–4041.
71. Lu, S. B.; Miao, L. L.; Guo, Z. N.; Qi, X.; Zhao, C. J.; Zhang, H.; Wen, S. C.; Tang, D. Y.; Fan, D. Y. Broadband Nonlinear Optical Response in Multi-Layer Black Phosphorus: An Emerging Infrared and Mid-Infrared Optical Material. *Opt. Express* **2015**, *23*(9), 11183–11194.
72. Lin, S.; Chui, Y.; Li, Y.; Lau, S. P. Liquid-Phase Exfoliation of Black Phosphorus and Its Applications. *FlatChem.* **2017**, *2*, 15–37.
73. Jia, J.; Jang, S. K.; Lai, S.; Xu, J.; Choi, Y. J.; Park, J.-H.; Lee, S. Plasma-Treated Thickness-Controlled Two-Dimensional Black Phosphorus and Its Electronic Transport Properties. *ACS Nano* **2015**, *9*(9), 8729–8736.
74. Yang, Z.; Hao, J.; Yuan, S.; Lin, S.; Yau Hei, M.; Dai, J.; Lau Shu, P. Field-Effect Transistors Based on Amorphous Black Phosphorus Ultrathin Films by Pulsed Laser Deposition. *Adv. Mater.* **2015**, *27*(25), 3748–3754.
75. Buscema, M.; Groenendijk, D. J.; Blanter, S. I.; Steele, G. A.; van der Zant, H. S. J.; Castellanos-Gomez, A. Fast and Broadband Photoresponse of Few-Layer Black Phosphorus Field-Effect Transistors. *Nano Lett.* **2014**, *14*(6), 3347–3352.
76. Yu, X.; Zhang, S.; Zeng, H.; Wang, Q. J. Lateral Black Phosphorene P-N Junctions Formed via Chemical Doping for High Performance Near-Infrared Photodetector. *Nano Energy* **2016**, *25*, 34–41.
77. Shenghuang, L.; Shenghua, L.; Zhibin, Y.; Yanyong, L.; Wai, N. T.; Zaiquan, X.; Qiaoliang, B.; Jianhua, H.; Chun-Sing, L.; Charles, S.; Feng, Y.; Ping, L. S. Solution-Processable Ultrathin Black Phosphorus as an Effective Electron Transport Layer in Organic Photovoltaics. *Adv. Funct. Mater.* **2016**, *26*(6), 864–871.
78. Wang, H.; Yang, X.; Shao, W.; Chen, S.; Xie, J.; Zhang, X.; Wang, J.; Xie, Y. Ultrathin Black Phosphorus Nanosheets for Efficient Singlet Oxygen Generation. *J. Am. Chem. Soc.* **2015**, *137*(35), 11376–11382.
79. Zhao, J.; Liu, H.; Yu, Z.; Quhe, R.; Zhou, S.; Wang, Y.; Liu, C. C.; Zhong, H.; Han, N.; Lu, J.; Yao, Y.; Wu, K. Rise of Silicene: A Competitive 2D Material. *Prog. Mater. Sci.* **2016**, *83*, 24–151.
80. Ezawa, M. Spin Valleytronics in Silicene: Quantum Spin Hall–Quantum Anomalous Hall Insulators and Single-Valley Semimetals. *Phys. Rev. B* **2013**, *87*(15), 155415.
81. Derivaz, M.; Dentel, D.; Stephan, R.; Hanf, M.-C.; Mehdaoui, A.; Sonnet, P.; Pirri, C. Continuous Germanene Layer on Al(111). *Nano Lett.* **2015**, *15*(4), 2510–2516.
82. Li, L.; Lu, S. Z.; Pan, J.; Qin, Z.; Wang, Y. Q.; Wang, Y.; Cao, G. Y.; Du, S.; Gao, H. J. Buckled Germanene Formation on Pt(111). *Adv. Mater.* **2014**, *26*(28), 4820–4824.
83. Acun, A.; Zhang, L.; Bampoulis, P.; Farmanbar, M.; Houselt, A. V.; Rudenko, A. N.; Lingenfelder, M.; Brocks, G.; Poelsema, B.; Katsnelson, M. I.; Zandvliet, H. J. W. Germanene: The Germanium Analogue of Graphene. *J. Phys. Condens. Matter* **2015**, *27*(44), 443002.
84. Liao, M.; Zang, Y.; Guan, Z.; Li, H.; Gong, Y.; Zhu, K.; Hu, X.-P.; Zhang, D.; Xu, Y.; Wang, Y.-Y.; He, K.; Ma, X.-C.; Zhang, S.-C.; Xue, Q.-K. Superconductivity in Few-Layer Stanene. *Nat. Phys.* **2018**, *14*(4), 344–348.
85. Zhu, F.-F.; Chen, W.-J.; Xu, Y.; Gao, C.-L.; Guan, D.-D.; Liu, C.-H.; Qian, D.; Zhang, S.-C.; Jia, J.-F. Epitaxial Growth of Two-Dimensional Stanene. *Nat. Mater.* **2015**, *14*, 1020.
86. Tsai, H.-S.; Wang, S.-W.; Hsiao, C.-H.; Chen, C.-W.; Ouyang, H.; Chueh, Y.-L.; Kuo, H.-C.; Liang, J.-H. Direct Synthesis and Practical Bandgap Estimation of Multilayer Arsenene Nanoribbons. *Chem. Mater.* **2016**, *28*(2), 425–429.

87. Niu, X.; Li, Y.; Zhou, Q.; Shu, H.; Wang, J. Arsenene-Based Heterostructures: Highly Efficient Bifunctional Materials for Photovoltaics and Photocatalytics. *ACS Appl. Mater. Interfaces* **2017**, *9*(49), 42856–42861.
88. Pablo, A.; Fernando, A. G.; David, R. S. M.; Aldave, D. A.; Sergio, D. T.; Manuel, A.; Fernando, M.; Julio, G. H.; Félix, Z. Mechanical Isolation of Highly Stable Antimonene Under Ambient Conditions. *Adv. Mater.* **2016**, *28*(30), 6332–6336.
89. Carlos, G.; David, R. S. M.; Pablo, A.; Julio, G. H.; Maria, V.; Roland, G.; Janina, M.; Frank, H.; Andreas, H.; Gonzalo, A.; Félix, Z. Few-Layer Antimonene by Liquid-Phase Exfoliation. *Angew. Chem. Int. Ed.* **2016**, *55*(46), 14345–14349.
90. Wu, X.; Shao, Y.; Liu, H.; Feng, Z.; Wang, Y. L.; Sun, J. T.; Liu, C.; Wang, J. O.; Liu, Z. L.; Zhu, S. Y.; Wang, Y. Q.; Du, S. X.; Shi, Y. G.; Ibrahim, K.; Gao, H. J. Epitaxial Growth and Air-Stability of Monolayer Antimonene on $PdTe_2$. *Adv. Mater.* **2017**, *29*(11), 1605407.
91. Ji, J.; Song, X.; Liu, J.; Yan, Z.; Huo, C.; Zhang, S.; Su, M.; Liao, L.; Wang, W.; Ni, Z.; Hao, Y.; Zeng, H. Two-Dimensional Antimonene Single Crystals Grown by van der Waals Epitaxy. *Nat. Commun.* **2016**, 7, 13352.
92. Martín-Sánchez, J.; Trotta, R.; Mariscal, A.; Serna, R.; Piredda, G.; Stroj, S.; Edlinger, J.; Schimpf, C.; Aberl, J.; Lettner, T.; Wildmann, J.; Huang, H.; Yuan, X.; Ziss, D.; Stangl, J.; Rastelli, A. Strain-Tuning of the Optical Properties of Semiconductor Nanomaterials by Integration Onto Piezoelectric Actuators. *Semicond. Sci. Technol.* **2018**, *33*(1).
93. Lu, L.; Wang, W.; Wu, L.; Jiang, X.; Xiang, Y.; Li, J.; Fan, D.; Zhang, H. All-Optical Switching of Two Continuous Waves in Few Layer Bismuthene Based on Spatial Cross-Phase Modulation. *ACS Photonics* **2017**, *4*(11), 2852–2861.
94. Reis, F.; Li, G.; Dudy, L.; Bauernfeind, M.; Glass, S.; Hanke, W.; Thomale, R.; Schäfer, J.; Claessen, R. Bismuthene on a SiC Substrate: A Candidate for a High-Temperature Quantum Spin Hall Material. *Science* **2017**, *357*, 287–290.
95. Naguib, M.; Kurtoglu, M.; Presser, V.; Lu, J.; Niu, J.; Heon, M.; Hultman, L.; Gogotsi, Y.; Barsoum, M. W. Two-Dimensional Nanocrystals Produced by Exfoliation of Ti_3AlC_2. *Adv. Mater.* **2011**, *23*(37), 4248–4253.
96. Hantanasirisakul, K.; Zhao, M. Q.; Urbankowski, P.; Joseph, H.; Anasori, B.; Kota, S.; Chang, R. E.; Barsoum, M. W.; Yury, G. Fabrication of $Ti_3C_2T_x$ MXene Transparent Thin Films with Tunable Optoelectronic Properties. *Adv. Electron. Mater.* **2016**, *2*(6), 1600050.
97. Naguib, M.; Mashtalir, O.; Carle, J.; Presser, V.; Lu, J.; Hultman, L.; Gogotsi, Y.; Barsoum, M. W. Two-Dimensional Transition Metal Carbides. *ACS Nano* **2012**, *6*(2), 1322–1331.
98. Jhon, Y. I.; Koo, J.; Anasori, B.; Seo, M.; Lee, J. H.; Gogotsi, Y.; Jhon, Y. M. Metallic MXene Saturable Absorber for Femtosecond Mode-Locked Lasers. *Adv. Mater.* **2017**, *29*(40).
99. Dong, Y.; Chertopalov, S.; Maleski, K.; Anasori, B.; Hu, L.; Bhattacharya, S.; Rao, A. M.; Gogotsi, Y.; Mochalin, V. N.; Podila, R. Saturable Absorption in 2D Ti3C2 MXene Thin Films for Passive Photonic Diodes. *Adv. Mater.* **2018**, *30*(10).
100. Kang, Z.; Ma, Y.; Tan, X.; Zhu, M.; Zheng, Z.; Liu, N.; Li, L.; Zou, Z.; Jiang, X.; Zhai, T.; Gao, Y. MXene–Silicon Van Der Waals Heterostructures for High-Speed Self-Driven Photodetectors. *Adv. Electron. Mater.* **2017**, *3*(9), 1700165.
101. Anasori, B.; Lukatskaya, M. R.; Gogotsi, Y. 2D Metal Carbides and Nitrides (MXenes) for Energy Storage. *Nat. Rev. Mater.* **2017**, *2*, 16098.
102. An, H.; Habib, T.; Shah, S.; Gao, H.; Radovic, M.; Green, M. J.; Lutkenhaus, J. L. Surface-Agnostic Highly Stretchable and Bendable Conductive MXene Multilayers. *Sci. Adv.* **2018**, *4*(3).

103. Nag, A.; Raidongia, K.; Hembram, K. P. S. S.; Datta, R.; Waghmare, U. V.; Rao, C. N. R. Graphene Analogues of BN: Novel Synthesis and Properties. *ACS Nano* **2010**, *4*(3), 1539–1544.
104. Han, W.-Q.; Wu, L.; Zhu, Y.; Watanabe, K.; Taniguchi, T. Structure of Chemically Derived Mono- and Few-Atomic-Layer Boron Nitride Sheets. *Appl. Phys. Lett.* **2008**, *93*(22), 223103.
105. Wang, H.; Zhao, Y.; Xie, Y.; Ma, X.; Zhang, X. Recent Progress in Synthesis of Two-Dimensional Hexagonal Boron Nitride. *J. Semicond.* **2017**, *38*(3)031003.
106. Watanabe, K.; Taniguchi, T.; Kanda, H. Direct-Bandgap Properties and Evidence for Ultraviolet Lasing of Hexagonal Boron Nitride Single Crystal. *Nat. Mater.* **2004**, *3*, 404.
107. Watanabe, K.; Taniguchi, T.; Niiyama, T.; Miya, K.; Taniguchi, M. Far-Ultraviolet Plane-Emission Handheld Device Based on Hexagonal Boron Nitride. *Nat. Photonics* **2009**, *3*, 591.
108. Elbadawi, C.; Tran, T. T.; Shimoni, O.; Totonjian, D.; Lobo, C. J.; Grosso, G.; Moon, H.; Englund, D. R.; Ford, M. J.; Aharonovich, I.; Toth, M. *Ultra-Bright Emission From Hexagonal Boron Nitride Defects as a New Platform for Bio-Imaging and Bio-Labelling;* SPIE BioPhotonics Australasia, 2016; p 5.
109. Fahmida, R.; Taimur, A.; Sumeet, W.; Edwin, M.; Sharath, S.; Madhu, B.; Sivacarendran, B. Two-Dimensional MoO_3 via a Top-Down Chemical Thinning Route. *2D Mater.* **2017**, 4(3)035008.
110. Dagar, J.; Tyagi, P.; Ahmad, R.; Singh, R.; Sinha, O. P.; Suman, C. K.; Srivastava, R. Application of 2D-MoO_3 Nano-Flakes in Organic Light Emitting Diodes: Effect of Semiconductor to Metal Transition With Irradiation. *RSC Adv.* **2015**, *5*(11), 8397–8403.
111. Ji, F.; Ren, X.; Zheng, X.; Liu, Y.; Pang, L.; Jiang, J.; Liu, S. 2D-MoO_3 Nanosheets for Superior Gas Sensors. *Nanoscale* **2016**, *8*(16), 8696–8703.
112. Balendhran, S.; Walia, S.; Alsaif, M.; Nguyen, E. P.; Ou, J. Z.; Zhuiykov, S.; Sriram, S.; Bhaskaran, M.; Kalantar-zadeh, K. Field Effect Biosensing Platform Based on 2D α-MoO_3. *ACS Nano* **2013**, 7(11), 9753–9760.
113. Jacoby, M. 2-D Materials Go Beyond Graphene. *Chem. Eng. News* **2017**, *95*(22), 36–40.
114. Zhen, Z.; Zhu, H. 1—Structure and Properties of Graphene. In: *Graphene,* Academic Press, 2018, pp 1–12.
115. Bao, Q.; Hoh, H. Y.; Zhang, Y. *Graphene Photonics, Optoelectronics, and Plasmonics*. Pan Stanford Pte Ltd, 2017; p 204.
116. Avouris, P. Graphene: Electronic and Photonic Properties and Devices. *Nano Lett.* **2010**, *10*(11), 4285–4294.
117. Du, X.; Skachko, I.; Barker, A.; Andrei, E. Y. Approaching Ballistic Transport in Suspended Graphene. *Nat. Nanotechnol.* **2008**, *3*, 491.
118. Morozov, S. V.; Novoselov, K. S.; Katsnelson, M. I.; Schedin, F.; Elias, D. C.; Jaszczak, J. A.; Geim, A. K. Giant Intrinsic Carrier Mobilities in Graphene and Its Bilayer. *Phys. Rev. Lett.* **2008**, *100*(1) 016602.
119. Balandin, A. A.; Ghosh, S.; Bao, W.; Calizo, I.; Teweldebrhan, D.; Miao, F.; Lau, C. N. Superior Thermal Conductivity of Single-Layer Graphene. *Nano Lett.* **2008**, *8*(3), 902–907.
120. Lee, C.; Wei, X.; Kysar, J. W.; Hone, J. Measurement of the Elastic Properties and Intrinsic Strength of Monolayer Graphene. *Science* **2008**, *321*(5887), 385.
121. Jiang, J.-W.; Wang, J.-S.; Li, B. Young's Modulus of Graphene: A Molecular Dynamics Study. *Phys. Rev. B* **2009**, *80*(11), 113405.
122. Kuzmenko, A. B.; van Heumen, E.; Carbone, F.; van der Marel, D. Universal Optical Conductance of Graphite. *Phys. Rev. Lett.* **2008**, *100*(11), 117401.

123. Mak, K. F.; Ju, L.; Wang, F.; Heinz, T. F. Optical Spectroscopy of Graphene: From the Far Infrared to the Ultraviolet. *Solid State Commun.* **2012**, *152*(15), 1341–1349.
124. Nair, R. R.; Blake, P.; Grigorenko, A. N.; Novoselov, K. S.; Booth, T. J.; Stauber, T.; Peres, N. M. R.; Geim, A. K. Fine Structure Constant Defines Visual Transparency of Graphene. *Science* **2008**, *320*(5881), 1308.
125. Bao, Q.; Loh, K. P. Graphene Photonics, Plasmonics, and Broadband Optoelectronic Devices. *ACS Nano* **2012**, *6*(5), 3677–3694.
126. Bonaccorso, F.; Sun, Z.; Hasan, T.; Ferrari, A. C. Graphene Photonics and Optoelectronics. *Nat. Photonics* **2010**, *4*, 611.
127. Choi, W.; Lahiri, I.; Seelaboyina, R.; Kang, Y. S. Synthesis of Graphene and Its Applications: A Review. *Crit. Rev. Solid State Mater. Sci.* **2010**, *35*(1), 52–71.
128. Avouris, P.; Dimitrakopoulos, C. Graphene: Synthesis and Applications. *Mater. Today* **2012**, *15*(3), 86–97.
129. Jayasena, B.; Subbiah, S. A Novel Mechanical Cleavage Method for Synthesizing Few-Layer Graphenes. *Nanoscale Res. Lett.* **2011**, *6*(1), 95.
130. Chen, J.; Duan, M.; Chen, G. Continuous Mechanical Exfoliation of Graphene Sheets via Three-Roll Mill. *J. Mater. Chem.* **2012**, *22*(37), 19625–19628.
131. Zhong, Y. L.; Tian, Z.; Simon, G. P.; Li, D. Scalable Production of Graphene via Wet Chemistry: Progress and Challenges. *Mater. Today* **2015**, *18*(2), 73–78.
132. Abdelkader, A. M.; Cooper, A. J.; Dryfe, R. A. W.; Kinloch, I. A. How to Get Between the Sheets: A Review of Recent Works on the Electrochemical Exfoliation of Graphene Materials From Bulk Graphite. *Nanoscale* **2015**, 7(16), 6944–6956.
133. Wang, Q. H.; Kalantar-Zadeh, K.; Kis, A.; Coleman, J. N.; Strano, M. S. Electronics and Optoelectronics of Two-Dimensional Transition Metal Dichalcogenides. *Nat. Nanotechnol.* **2012**, 7, 699.
134. Li, X.; Zhu, H. Two-Dimensional MoS_2: Properties, Preparation, and Applications. *J. Mater.* **2015**, *1*(1), 33–44.
135. Lv, R.; Robinson, J. A.; Schaak, R. E.; Sun, D.; Sun, Y.; Mallouk, T. E.; Terrones, M. Transition Metal Dichalcogenides and Beyond: Synthesis, Properties, and Applications of Single- and Few-Layer Nanosheets. *Acc. Chem. Res.* **2015**, *48*(1), 56–64.
136. Chhowalla, M.; Shin, H. S.; Eda, G.; Li, L.-J.; Loh, K. P.; Zhang, H. The Chemistry of Two-Dimensional Layered Transition Metal Dichalcogenide Nanosheets. *Nat. Chem.* **2013**, *5*, 263.
137. Splendiani, A.; Sun, L.; Zhang, Y.; Li, T.; Kim, J.; Chim, C.-Y.; Galli, G.; Wang, F. Emerging Photoluminescence in Monolayer MoS_2. *Nano Lett.* **2010**, *10*(4), 1271–1275.
138. Mak, K. F.; Lee, C.; Hone, J.; Shan, J.; Heinz, T. F. Atomically Thin MoS_2: A New Direct-Gap Semiconductor. *Phys. Rev. Lett.* **2010**, *105*(13), 136805.
139. Xu, X.; Yao, W.; Xiao, D.; Heinz, T. F. Spin and Pseudospins in Layered Transition Metal Dichalcogenides. *Nat. Phys.* **2014**, *10*, 343.
140. Voiry, D.; Yamaguchi, H.; Li, J.; Silva, R.; Alves, D. C. B.; Fujita, T.; Chen, M.; Asefa, T.; Shenoy, V. B.; Eda, G.; Chhowalla, M. Enhanced Catalytic Activity in Strained Chemically Exfoliated WS_2 Nanosheets for Hydrogen Evolution. *Nat. Mater.* **2013**, *12*, 850.
141. Duerloo, K.-A. N.; Li, Y.; Reed, E. J. Structural Phase Transitions in Two-Dimensional Mo- and W-Dichalcogenide Monolayers. *Nat. Commun.* **2014**, *5*, 4214.
142. Song, S.; Keum, D. H.; Cho, S.; Perello, D.; Kim, Y.; Lee, Y. H. Room Temperature Semiconductor–Metal Transition of $MoTe_2$ Thin Films Engineered by Strain. *Nano Lett.* **2016**, *16*(1), 188–193.
143. Andres, C. G.; van Leeuwen, R.; Buscema, M.; van der Zant, H. S.; Steele, G. A.; Venstra, W. J. Single-Layer MoS_2 Mechanical Resonators. *Adv. Mater.* **2013**, *25*(46), 6719–6723.

144. Yagang, Y.; Lorenzo, T.; Zhongzheng, Y.; Xiaojuan, S.; Wen, Z.; Yongsheng, C.; Ching-ping, W. High-Concentration Aqueous Dispersions of MoS_2. *Adv. Funct. Mater.* **2013**, *23*(28), 3577–3583.
145. Li, Y.; Yin, X.; Wu, W. Preparation of Few-Layer MoS_2 Nanosheets via an Efficient Shearing Exfoliation Method. *Ind. Eng. Chem. Res.* **2018**, *57*(8), 2838–2846.
146. Eda, G.; Yamaguchi, H.; Voiry, D.; Fujita, T.; Chen, M.; Chhowalla, M. Photoluminescence From Chemically Exfoliated MoS_2. *Nano Lett.* **2011**, *11*(12), 5111–5116.
147. Chou, S. S.; Huang, Y.-K.; Kim, J.; Kaehr, B.; Foley, B. M.; Lu, P.; Dykstra, C.; Hopkins, P. E.; Brinker, C. J.; Huang, J.; Dravid, V. P. Controlling the Metal to Semiconductor Transition of MoS_2 and WS_2 in Solution. *J. Am. Chem. Soc.* **2015**, *137*(5), 1742–1745.
148. Shi, Y.; Zhou, W.; Lu, A.-Y.; Fang, W.; Lee, Y.-H.; Hsu, A. L.; Kim, S. M.; Kim, K. K.; Yang, H. Y.; Li, L.-J.; Idrobo, J.-C.; Kong, J. van der Waals Epitaxy of MoS_2 Layers Using Graphene As Growth Templates. *Nano Lett.* **2012**, *12*(6), 2784–2791.
149. Xia, F.; Wang, H.; Xiao, D.; Dubey, M.; Ramasubramaniam, A. Two-Dimensional Material Nanophotonics. *Nat. Photonics* **2014**, *8*, 899.
150. Liang, X.; Fu, Z.; Chou, S. Y. Graphene Transistors Fabricated via Transfer-Printing in Device Active-Areas on Large Wafer. *Nano Lett.* **2007**, 7(12), 3840–3844.
151. Lemme, M. C.; Echtermeyer, T. J.; Baus, M.; Kurz, H. A Graphene Field-Effect Device. *IEEE Electron Device Letters* **2007**, *28*(4), 282–284.
152. Das, A.; Pisana, S.; Chakraborty, B.; Piscanec, S.; Saha, S. K.; Waghmare, U. V.; Novoselov, K. S.; Krishnamurthy, H. R.; Geim, A. K.; Ferrari, A. C.; Sood, A. K. Monitoring Dopants by Raman Scattering in an Electrochemically Top-Gated Graphene Transistor. *Nat. Nanotechnol.* **2008**, *3*, 210.
153. Kedzierski, J.; Hsu, P. L.; Healey, P.; Wyatt, P. W.; Keast, C. L.; Sprinkle, M.; Berger, C.; Heer, W. A. D. Epitaxial Graphene Transistors on SiC Substrates. *IEEE Trans. Electron Devices* **2008**, *55*(8), 2078–2085.
154. Jaesung, P.; Byeok, J. S.; Young-Jun, Y.; Youngsoo, K.; Won, Y. J.; Hyoung, L. W.; Ho, K. H.; Hee, H. B.; Philip, K.; Kilwon, C.; Kim, K. S. Single-Gate Bandgap Opening of Bilayer Graphene by Dual Molecular Doping. *Adv. Mater.* **2012**, *24*(3), 407–411.
155. Szafranek, B. N.; Schall, D.; Otto, M.; Neumaier, D.; Kurz, H. High On/Off Ratios in Bilayer Graphene Field Effect Transistors Realized by Surface Dopants. *Nano Lett.* **2011**, *11*(7), 2640–2643.
156. Xia, F.; Wang, H.; Jia, Y. Rediscovering Black Phosphorus as an Anisotropic Layered Material for Optoelectronics and Electronics. *Nat. Commun.* **2014**, *5*, 4458.
157. Koenig, S. P.; Doganov, R. A.; Schmidt, H.; Neto, A. H. C.; Özyilmaz, B. Electric Field Effect in Ultrathin Black Phosphorus. *Appl. Phys. Lett.* **2014**, *104*(10), 103106.
158. Ayari, A.; Cobas, E.; Ogundadegbe, O.; Fuhrer, M. S. Realization and Electrical Characterization of Ultrathin Crystals of Layered Transition-Metal Dichalcogenides. *J. Appl. Phys.* **2007**, *101*(1) 014507.
159. Kim, S.; Konar, A.; Hwang, W.-S.; Lee, J. H.; Lee, J.; Yang, J.; Jung, C.; Kim, H.; Yoo, J.-B.; Choi, J.-Y.; Jin, Y. W.; Lee, S. Y.; Jena, D.; Choi, W.; Kim, K. High-Mobility and Low-Power Thin-Film Transistors Based on Multilayer MoS_2 Crystals. *Nat. Commun.* **2012**, *3*, 1011.
160. Wu, W.; De, D.; Chang, S.-C.; Wang, Y.; Peng, H.; Bao, J.; Pei, S.-S. High Mobility and High On/Off Ratio Field-Effect Transistors Based on Chemical Vapor Deposited Single-Crystal MoS_2 Grains. *Appl. Phys. Lett.* **2013**, *102*(14), 142106.
161. Fang, H.; Chuang, S.; Chang, T. C.; Takei, K.; Takahashi, T.; Javey, A. High-Performance Single Layered WSe_2 p-FETs With Chemically Doped Contacts. *Nano Lett.* **2012**, *12*(7), 3788–3792.

162. Liu, W.; Kang, J.; Sarkar, D.; Khatami, Y.; Jena, D.; Banerjee, K. Role of Metal Contacts in Designing High-Performance Monolayer n-Type WSe_2 Field Effect Transistors. *Nano Lett.* **2013**, *13*(5), 1983–1990.
163. Ovchinnikov, D.; Allain, A.; Huang, Y.-S.; Dumcenco, D.; Kis, A. Electrical Transport Properties of Single-Layer WS_2. *ACS Nano* **2014**, *8*(8), 8174–8181.
164. Ling, X.; Wang, H.; Huang, S.; Xia, F.; Dresselhaus, M. S. The Renaissance of Black Phosphorus. *Proc. Natl. Acad. Sci.* **2015**, *112*(15), 4523–4530.
165. Jain, A.; McGaughey, A. J. H. Strongly Anisotropic In-Plane Thermal Transport in Single-Layer Black Phosphorene. *Sci. Rep.* **2015**, *5*, 8501.
166. He, J.; He, D.; Wang, Y.; Cui, Q.; Bellus, M. Z.; Chiu, H.-Y.; Zhao, H. Exceptional and Anisotropic Transport Properties of Photocarriers in Black Phosphorus. *ACS Nano* **2015**, *9*(6), 6436–6442.
167. Fei, R.; Yang, L. Strain-Engineering the Anisotropic Electrical Conductance of Few-Layer Black Phosphorus. *Nano Lett.* **2014**, *14*(5), 2884–2889.
168. Ryder, C. R.; Wood, J. D.; Wells, S. A.; Yang, Y.; Jariwala, D.; Marks, T. J.; Schatz, G. C.; Hersam, M. C. Covalent Functionalization and Passivation of Exfoliated Black Phosphorus via Aryl Diazonium Chemistry. *Nat. Chem.* **2016**, *8*, 597.
169. Yasaei, P.; Kumar, B.; Foroozan, T.; Wang, C.; Asadi, M.; Tuschel, D.; Indacochea, J. E.; Klie Robert, F.; Salehi-Khojin, A. High-Quality Black Phosphorus Atomic Layers by Liquid-Phase Exfoliation. *Adv. Mater.* **2015**, *27*(11), 1887–1892.
170. Kang, J.; Wood, J. D.; Wells, S. A.; Lee, J.-H.; Liu, X.; Chen, K.-S.; Hersam, M. C. Solvent Exfoliation of Electronic-Grade, Two-Dimensional Black Phosphorus. *ACS Nano* **2015**, *9*(4), 3596–3604.
171. Guo, Z.; Zhang, H.; Lu, S.; Wang, Z.; Tang, S.; Shao, J.; Sun, Z.; Xie, H.; Wang, H.; Yu, X. F.; Chu Paul, K. From Black Phosphorus to Phosphorene: Basic Solvent Exfoliation, Evolution of Raman Scattering, and Applications to Ultrafast Photonics. *Adv. Funct. Mater.* **2015**, *25*(45), 6996–7002.
172. Smith, J. B.; Hagaman, D.; Ji, H.-F. Growth of 2D Black Phosphorus Film From Chemical Vapor Deposition. *Nanotechnology* **2016**, *27*(21), 215602.
173. Vogt, P.; De Padova, P.; Quaresima, C.; Avila, J.; Frantzeskakis, E.; Asensio, M. C.; Resta, A.; Ealet, B.; Le Lay, G. Silicene: Compelling Experimental Evidence for Graphenelike Two-Dimensional Silicon. *Phys. Rev. Lett.* **2012**, *108*(15), 155501.
174. Takeda, K.; Shiraishi, K. Theoretical Possibility of Stage Corrugation in Si and Ge Analogs of Graphite. *Phys. Rev. B* **1994**, *50*(20), 14916–14922.
175. Bianco, E.; Butler, S.; Jiang, S.; Restrepo, O. D.; Windl, W.; Goldberger, J. E. Stability and Exfoliation of Germanane: A Germanium Graphane Analogue. *ACS Nano* **2013**, 7(5), 4414–4421.
176. Shengli, Z.; Meiqiu, X.; Fengyu, L.; Zhong, Y.; Yafei, L.; Erjun, K.; Wei, L.; Zhongfang, C.; Haibo, Z. Semiconducting Group 15 Monolayers: A Broad Range of Band Gaps and High Carrier Mobilities. *Angew. Chem. Int. Ed.* **2016**, *55*(5), 1666–1669.
177. Kamal, C.; Ezawa, M. Arsenene: Two-Dimensional Buckled and Puckered Honeycomb Arsenic Systems. *Phys. Rev. B* **2015**, *91*(8)085423.
178. Shengli, Z.; Zhong, Y.; Yafei, L.; Zhongfang, C.; Haibo, Z. Atomically Thin Arsenene and Antimonene: Semimetal–Semiconductor and Indirect–Direct Band-Gap Transitions. *Angew. Chem. Int. Ed.* **2015**, *54*(10), 3112–3115.
179. Mao, Y.-H.; Zhang, L.-F.; Wang, H.-L.; Shan, H.; Zhai, X.-F.; Hu, Z.-P.; Zhao, A.-D.; Wang, B. Epitaxial Growth of Highly Strained Antimonene on Ag(111). *Front. Phys.* **2018**, *13*(3), 138106.
180. Tsai, H.-S.; Lai, C.-C.; Medina, H.; Lin, S.-M.; Shih, Y.-C.; Chen, Y.-Z.; Liang, J.-H.; Chueh, Y.-L. Scalable Graphene Synthesised by Plasma-Assisted Selective Reaction on Silicon Carbide for Device Applications. *Nanoscale* **2014**, *6*(22), 13861–13869.

181. Tsai, H.-S.; Chen, Y.-Z.; Medina, H.; Su, T.-Y.; Chou, T.-S.; Chen, Y.-H.; Chueh, Y.-L.; Liang, J.-H. Direct Formation of Large-Scale Multi-Layered Germanene on Si Substrate. *Phys. Chem. Chem. Phys.* **2015**, *17*(33), 21389–21393.
182. Tsai, H.-S.; Lai, C.-C.; Hsiao, C.-H.; Medina, H.; Su, T.-Y.; Ouyang, H.; Chen, T.-H.; Liang, J.-H.; Chueh, Y.-L. Plasma-Assisted Synthesis of High-Mobility Atomically Layered Violet Phosphorus. *ACS Appl. Mater. Interfaces* **2015**, 7(25), 13723–13727.
183. Bilu, L.; Marianne, K.; Abbas, A. N.; Xiaomu, W.; Qiushi, G.; Yichen, J.; Fengnian, X.; Richard, W.; Frederik, B.; Florian, P.; Han, W.; Rohan, D.; Cronin, S. B.; Mingyuan, G.; Xin, F.; Tom, N.; Chongwu, Z. Black Arsenic–Phosphorus: Layered Anisotropic Infrared Semiconductors With Highly Tunable Compositions and Properties. *Adv. Mater.* **2015**, *27*(30), 4423–4429.
184. Aguilera, I.; Friedrich, C.; Blügel, S. Electronic Phase Transitions of Bismuth Under Strain From Relativistic Self-Consistent GW Calculations. *Phys. Rev. B* **2015**, *91*(12), 125129.
185. Castro Neto, A. H.; Guinea, F.; Peres, N. M. R.; Novoselov, K. S.; Geim, A. K. The Electronic Properties of Graphene. *Rev. Mod. Phys.* **2009**, *81*(1), 109–162.

CHAPTER 2

Optical modulators based on 2D materials

Jianji Dong, Xinliang Zhang
Wuhan National Laboratory for Optoelectronics, Huazhong University of Science and Technology, Wuhan, China

2.1 Introduction

Optical modulation[1] is a process to load information onto the light beam. Generally, modulation can be achieved by altering several attributes of the light beam, including the direction, amplitude, phase, frequency, and polarization state, as shown in Fig. 2.1. Depending on the physical principle to achieve modulation, optical modulators can be classified into all-optical,[2] electro-optical,[3] thermo-optical,[4] magneto-optical,[5] and acousto-optical modulators.[6] Among these, all-optical modulation, electro-optic modulation, and thermo-optic modulation are very common in practical applications, since these can be easily achieved within the present optical materials and technologies.[1] In all-optical modulation, one light beam is utilized to control the parameters of a certain light beam, including the amplitude and the phase. Because the all-optical modulation is completed entirely in the optical domain, this type of modulation can avoid electrical-optical-electrical conversions, resulting in a faster and less noisy modulation than other modulation schemes. An electro-optical modulator is the one based on the electro-optic effect. The electro-optic effect is the change in the refractive index of a material resulting from a direct current (DC) or an alternating current (AC) electric field, which can be acquired in nonlinear optic materials through the Pockels effect,[7] the Franz-Keldysh effect,[7] or in silicon through the plasma dispersion effect.[8] To date, electro-optical modulators have been widely applied in optical interconnections and communications due to their ability to connect the electrical domain with the optical domain. Thermo-optical modulators rely on the variation of the material's refractive index when a temperature change occurs. Therefore, thermo-optic modulation typically employs heating to change phase of the light beam. Thermo-optical modulators have rather slow response time due to the intrinsically slow thermal diffusivity. As a result, these are often applied in

2D Materials for Photonic and Optoelectronic Applications
https://doi.org/10.1016/B978-0-08-102637-3.00002-4

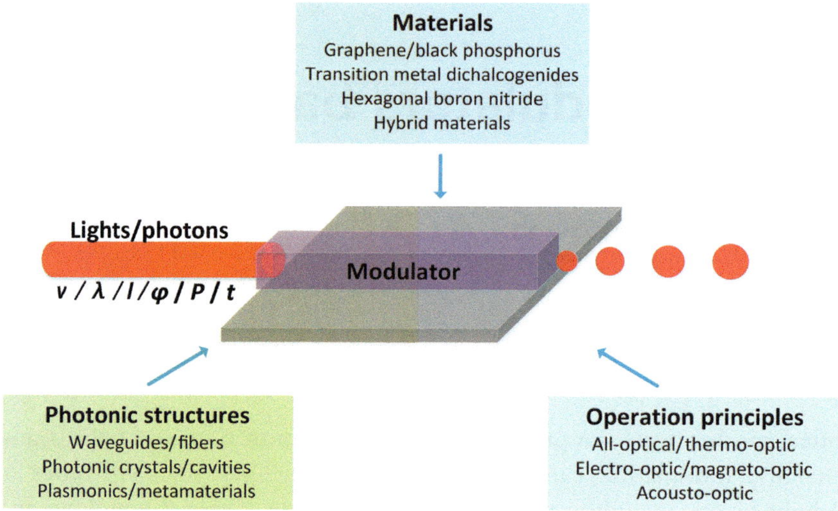

Fig. 2.1 The concept of optical modulator with 2D materials.

areas where high speed is not required such as optical switching and routing. Apart from these three kinds of typical modulators mentioned before, magneto-optical and acousto-optical modulators that employ magnetic fields and acoustic waves to modulate light, respectively, can also be found in certain applications due to their unique properties. The operation theory of these modulators will be discussed in detail in the following sections.

To characterize the performance of a modulator, several figures of merit are critical, including modulation bandwidth, modulation depth, optical bandwidth, insertion loss, and power consumption. Modulation bandwidth or speed is regarded one of the most important figures of merit for optical modulators since it characterizes the modulator's ability to carry data at a certain rate. It is usually defined by the frequency at which modulation intensity is reduced to 50% of its maximum value. A larger modulation bandwidth indicates a higher modulation speed, which is normally given in the form of bit rate. A higher modulation speed is always required in data transmission applications. Modulation depth, which is also known as the extinction ratio, is defined as the ratio between the maximum and minimum transmittance in the modulator. For simplicity, modulation depth is usually expressed in decibel unit by $10 \cdot \log_{10}\left(\frac{T_{max}}{T_{min}}\right)$. A modulation depth >7 dB is preferable for most optical modulation schemes. The optical bandwidth characterizes the useful operational wavelength range of a modulator. It depends on the structure of the modulators, with resonant structure-based modulators tending to function over a relatively narrow band compared with Mach-Zehnder interferometer-based devices. As a result of the broad absorption bandwidth of graphene,[9] it may contribute to widen the optical bandwidth of the modulator. Insertion loss is defined as the difference between the output and the input optical power of the modulator, which corresponds to the overall power loss of the system employing optical modulators. Normally, the power consumption of an optical modulator is defined as the energy expended in generating each bit of data.[10] This metric has become particularly important since energy consumption has attracted significant focus in recent decades.

To realize optical modulation, various photonic structures have been employed ranging from discrete fiber optic devices to nanophotonic structures. The lithium niobate ($LiNbO_3$) electro-optical modulator has been widely used in fiber optic communications as a well-developed commercial product. Compared to optical fibers, integrated optics can provide a more attractive platform to manipulate light beam. Numerous structures have been utilized in optical modulation, including the straight waveguide, the microring cavity, or the photonic crystal structure, most of which have demonstrated impressive modulation performance. In recent years, we have also witnessed several subwavelength photonic structures such as the metamaterial and plasmonic waveguide being employed in optical modulation, owing to the rapid development of nanophotonics.

As a typical 2D material, graphene has proved its potential possibility to realize light modulation with superior performance, thanks to its unusual electrical and optical properties. Optical modulators based on graphene have shown competitive performance, such as extremely broad operation bandwidth covering spectral range from the visible to microwave regions, ultrafast modulation speed, and ultralow power consumption. Apart from graphene, other 2D monolayer materials such as transition metal dichalcogenides (TMDs), black phosphorus, and heterostructures based on monolayer 2D materials also have a great potential to realize optical modulation in the future. However, up to now, most optical modulators with 2D materials are graphene-based.

2.2 All-optical modulators with 2D materials

2.2.1 Theory of all-optical modulators

An all-optical modulator employing 2D materials is regarded a promising candidate of ultra-high-speed modulation. All-optical modulation can be accomplished in the optical fiber or integrated photonic chips. Since it allows signal processing to be realized fully in the optical domain and avoids electrical bottleneck, all-optical modulation can realize ultrafast, low-loss, and broadband optical signal processing in relatively simple configurations. In most all-optical modulation cases, light transmission through 2D materials at the signal's wavelength is modulated (or switched) by another light beam. Up to now, most all-optical modulators based on 2D materials are realized with graphene.[11–13] In general, there are three different physical mechanisms to achieve all-optical modulation in graphene, including Pauli blocking,[11, 12] nonlinear Kerr effect,[13] and optical doping.[14, 15]

Fig. 2.2 illustrates the theory of Pauli blocking in graphene.[16] Electrons from the valence band (orange) are excited into the conduction band (yellow) when there is an incident light with energy $\hbar\omega$ absorbed by graphene. These hot electrons thermalize and cool down to form a hot Fermi-Dirac distribution with electronic temperature T_e after excitation within 10–150 fs, which can block some of the originally possible interband optical transitions in a range of $k_B T_e$ (k_B is the Boltzmann constant) around the Fermi energy E_F and decrease the absorption of photons. Afterward, intraband phonon scattering further cools the thermalized carriers in the following 1 ps. Then, electron-hole recombination will dominate the process until the equilibrium electron and hole distribution is restored. The process only describes a linear optical transition under low excitation intensity. When excitation intensity is high, the concentration of photogenerated carriers will increase significantly and become much larger than the intrinsic electron and

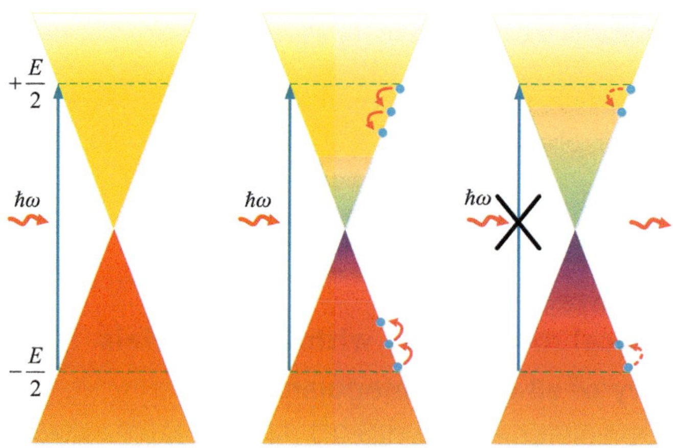

Fig. 2.2 The process of Pauli blocking in graphene.

hole carrier densities in graphene at room temperature. Thus, the photogenerated carriers fill the states near the edge of the conduction and valence band, blocking further absorption. As a result, the transmission of light will increase due to lower absorption. This phenomenon is called the Pauli blocking process in graphene. Based on the Pauli blocking theory, one can use a high-frequency pump light with high power to shift the Fermi level of graphene to a higher position, thus inducing a lower attenuation of a low-frequency signal light. This mechanism has been exploited to realize all-optical intensity modulation in graphene.

Under a strong laser illumination, graphene exhibits not only a transmittance increase due to the Pauli blocking effect but also a nonlinear phase shift of the transmitted light due to the Kerr effect.[13] The nonlinear phase shift is caused by graphene's giant nonlinear refractive index n_2 of around 10^{-7} cm^2/W, which is almost nine orders of magnitude larger than bulk dielectrics.[17] The large nonlinear refractive index indicates that one can realize all-optical phase modulation by employing optical field-induced refractive index change. Moreover, phase modulation can be converted into intensity modulation by utilizing phase modulation into the Mach-Zehnder interferometer (MZI).

Another mechanism to achieve all-optical modulation in graphene is optical doping, which is mainly applied in the terahertz (THz) wave modulation. In this case, graphene is usually integrated on silicon (Si)[14] or germanium (Ge).[15] A typical scheme is shown in Fig. 2.3. Both the modulation infrared beam and the modulated THz beam are incident from the graphene side and is partially reflected from the top and bottom surfaces of the graphene film. The THz beam propagates through the graphene/silicon or

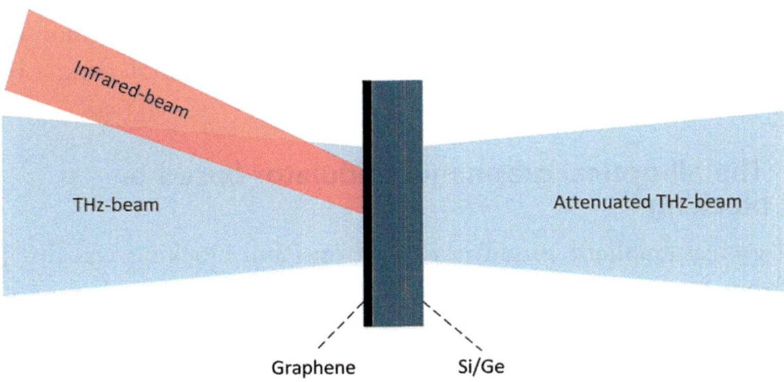

Fig. 2.3 The schematic of the THz modulator based on graphene.

germanium substrate and experiences attenuation, where transmittance depends on the electrical conductivity of the system. A small fraction of the modulation infrared beam is absorbed in graphene, whereas the remaining fraction of the beam is absorbed by silicon and free carriers are generated. These photogenerated free carriers diffuse into the graphene layer, resulting in a strong change of electrical conductivity. As a result, the modulation beam changes the doping level of the graphene/silicon or germanium system as well as the transmittance of the THz wave propagating through the system, realizing all-optical modulation. The mechanism is often employed in the THz all-optical modulators due to the dependence of the THz wave on electrical conductivity. Besides, graphene can also give rise to an enhancement of attenuation of THz waves in graphene on silicon or germanium in comparison with transmission through pure silicon or germanium due to its ultra-high carrier mobility. The enhancement effect becomes even more dominant when the thickness of the semiconductor substrate is reduced.

In conclusion, both Pauli blocking and the nonlinear Kerr effect employed in all-optical modulation can be regarded as a result of the large complex nonlinear refractive index of graphene. Under strong illumination, the transmittance of light in fiber with graphene exhibits an increase due to Pauli blocking, which can be summarized mathematically as the result of the imaginary part of the complex nonlinear refractive index of graphene, whereas the Kerr effect is the result of the real part of the complex nonlinear refractive index. It should be noted that, on the one hand, both the Pauli blocking effect and the nonlinear Kerr effect exist simultaneously in most all-optical modulators based on graphene. On the other hand, the optical doping effect indicates the system's electrical conductivity change induced by the modulation light, which is usually utilized in the modulators of the THz domain. For all-optical modulators based on optical doping effect, the existence of graphene contributes to an enhancement of attenuation of the THz beam propagating through the spatial modulator, indicating a better modulation depth.

2.2.2 The all-optical graphene modulator based on Pauli blocking

An all-optical graphene modulator based on Pauli blocking was first proposed by Liu et al. in 2013 using a graphene-covered microfiber (GCMF) device.[12] As shown in Fig. 2.4, the microfiber was sandwiched between a low-refractive-index magnesium fluoride (MgF_2) substrate and a polydimethylsiloxane (PDMS)-supported graphene film. The microfiber

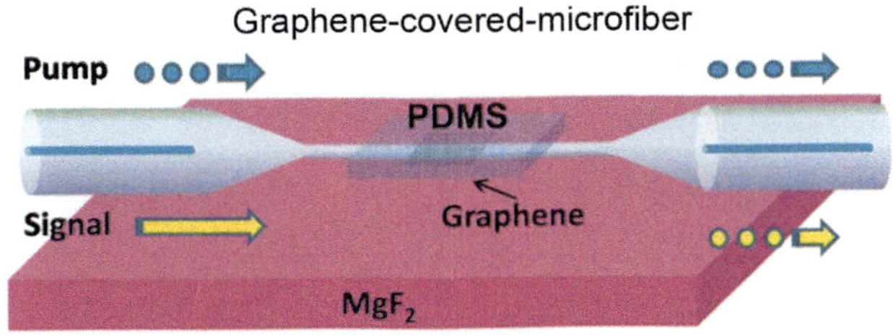

Fig. 2.4 Schematic of the graphene-covered microfiber structure.

in this work has a diameter of 8 μm, a length of 1 cm, and a low insertion loss down to 0.1 dB. Compared with the standard single-mode fiber, microfibers have many advantages such as flexible configurability, strong light confinement, and large evanescent fields, so it may be easily coupled to graphene material with low loss. PDMS was chosen as the support for graphene because of its flexibility, high adsorption capacity, and low refractive index of 1.413. When the microfiber is tightly covered by the PDMS-supported graphene film, the interaction between the propagating light and graphene can be significantly enhanced due to the strong evanescent field of the microfiber. In this work, a continuous wave (CW) laser with a wavelength of 1060 nm was used as the pump light to induce the Pauli blocking effect in graphene. When the pump light power changed from 0 to 2.2 mW, the loss of GCMF decreased from 13 to 8 dB due to the Pauli blocking effect, indicating a modulation depth of 5 dB. The modulation speed was characterized by measuring the frequency response of the probe light to the pump light. The results show that the modulator can reach a modulation bandwidth of 1 MHz.

To enhance the interaction between light and graphene, Li et al. reduced the diameter of the microfiber to the subwavelength scale of around 1 μm.[11] A thin layer of graphene was wrapped around a single-mode microfiber to form the graphene-clad microfiber (GCM), as shown in Fig. 2.5. The operation principle was also based on Pauli blocking. A weak infrared signal wave coupled into the GCM experienced significant attenuation due to absorption in graphene as it propagated along the microfiber. When a switch light was introduced, the carriers in graphene were excited and the absorption threshold of graphene was shifted to a higher level through the Pauli blocking of interband transitions, resulting in a much lower attenuation of the signal wave. The switch light led to the modulation of the output signal, while its response time was limited by the relaxation of excited carriers. Due to the tightly confined evanescent field guided along the surface of the microfiber,

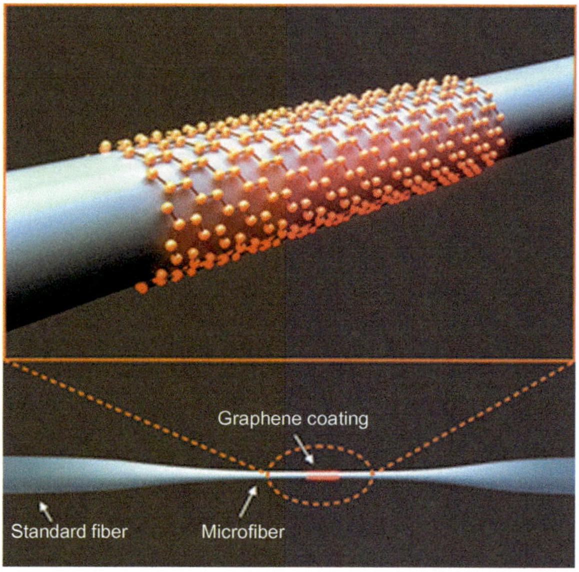

Fig. 2.5 Schematic of a GCM all-optical modulator.

the light-graphene interaction was significantly enhanced. According to experimental results, the transmittance of GCM increased from 15.5% to 24% when the power density in GCM was increased from 0.2 to 3 GW/cm^2. Thus, a strong pump light-dependent absorption of GCM can be directly employed for all-optical modulation. When a pump laser with a wavelength of 1064 nm pulses and a repetition frequency of 2.4 kHz propagated through GCM with an average power of 300 μW, the intensity of the 1550-nm probe light was modulated with a modulation depth of 30%. To further explore the response time of all-optical modulation, they replaced the previously used pump laser with a femtosecond titanium/sapphire laser with 789 nm wavelength, 35 fs duration, 1 kHz repetition rate. A modulation depth of 38% and a decay time of 2.2 ps was obtained, which was corresponding to a maximum modulation bandwidth of 200 GHz for Gaussian pulses.

To further enhance the light-graphene interaction, Chen et al. proposed and experimentally demonstrated a stereo graphene-microfiber (GMF) structure,[18] which is shown in Fig. 2.6A. Since the stereo GMF has a larger light-graphene interaction area than the previously proposed structure, it may provide a better modulation performance. The fabrication process of the stereo GMF includes several steps as depicted in Fig. 2.6A. First, to prevent the loss induced by the relatively high index of the PMMA rod with a

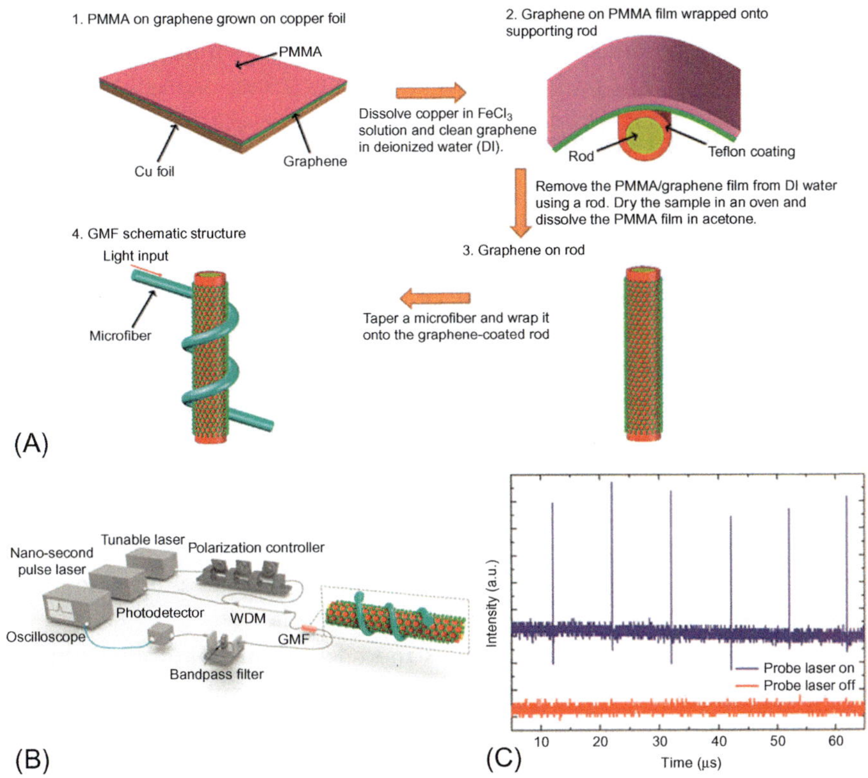

Fig. 2.6 (A) Fabrication process of the stereo GMF structure. (B) Experiment setup. (C) Performance of the stereo GMF structure.

diameter of 2 mm, they dip-coated a thin low-index Teflon layer with tens of micrometers in thickness and a refractive index of 1.31 on the rod's surface. Graphene on copper (Cu) foil was first spin-coated with a 4% PMMA-anisole solution. After dissolving the Cu foil using aqueous iron(III) chloride ($FeCl_3$), the PMMA/graphene film was transferred onto a rod. The surface tension led to the film tightly encapsulating the rod when the film left de-ionized water. Then, the rod with the graphene/PMMA film was heated and the PMMA film was removed with acetone. Finally, a tapered microfiber was wrapped onto the graphene-functionalized rod to finish the fabrication of the stereo GMF structure.

To measure the time-domain response of the stereo GMF structure, a nanosecond laser at 1064 nm with a pulse width of 6 ns and a repetition rate of 100 kHz acted as the pump light, while a CW light at 1550 nm acted as the probe light. Owing to a stronger light-matter interaction of the stereo

GMF, one can find that a high modulation depth of 7.5 dB was successfully achieved with a modulation speed of 100 kHz, as shown in Fig. 2.6C.

2.2.3 The all-optical graphene modulator based on Kerr effect

The modulation depth of all-optical modulator based on Pauli blocking is relatively low, owing to the weak light-matter interaction. An all-optical graphene modulator based on the nonlinear Kerr effect may have the potential to increase the modulation depth of the all-optical modulator. Considering graphene's giant nonlinear refractive index, a GCM was proposed as a phase modulator utilizing the optical field-induced refractive index change under strong illumination.[13] Furthermore, the phase modulator was placed in one arm of a fiber MZI structure to covert phase change to intensity change at the output of the MZI, as shown in Fig. 2.7A. GCM had a diameter of 1 μm and a graphene cladding length of 15 μm. A CW laser beam with 100 kHz line width at 1550 nm was used as signal light, and 1064-nm nanosecond pulses with a pulse width of 8 ns and a repetition frequency of 4.8 kHz were used as the switching light.

To compare optically induced loss modulation and optically induced phase modulation, GCM was directly employed as the intensity modulator and as a phase modulator in MZI structure, respectively. According to Fig. 2.7B, the modulation depth in optically induced phase modulation was 52.5% when the peak power of the pump light was 1.18 W, which was nearly 4.6 times better than the case of optically induced loss modulation with a modulation depth of 11.5%. The modulation depth of 52.5% corresponded to a phase shift of about 0.18π in the GCM arm. The overall

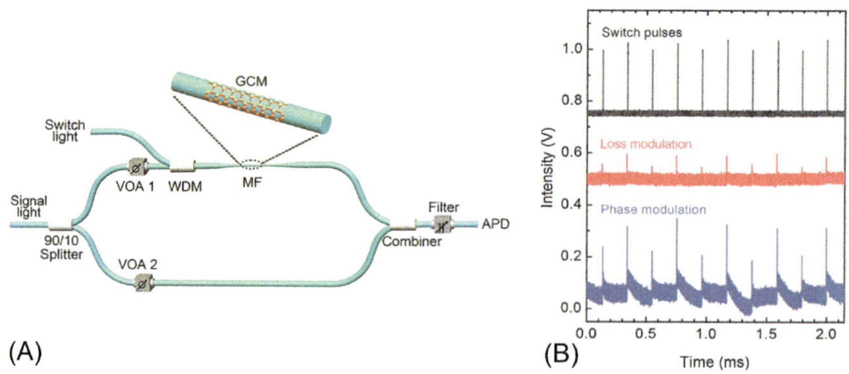

Fig. 2.7 (A) Schematic of all-optical modulators based on the Kerr effect. (B) Performance of all-optical modulators based on the Kerr effect.

transmittance was also higher than the optically induced loss modulation. Besides, the pulse-modulated signal had a long tail with a decay time of 100 μs. It should be noted that the nanosecond pulse comes from the refractive index change induced by an optical excitation of carriers in graphene, which has a picosecond time response. Therefore, the long tail was mainly attributed to the thermally induced refractive index change by laser heating of graphene. Note that although almost every all-optical modulator fabricated with 2D materials was realized with graphene, the all-optical phase modulation scheme can also be applied using other 2D materials with a high nonlinear refractive index such as MoS_2,[19] WS_2,[20] and WSe_2.[21]

2.2.4 The all-optical graphene modulator based on optical doping

As we described before, most all-optical graphene modulators realized in fiber optic devices are based on Pauli blocking and the nonlinear Kerr effect. Besides, graphene can be also employed in free-space all-optical THz modulators based on the optical doping effect. An all-optical THz modulator based on the optical doping effect usually relies on the pump light-induced electrical conductivity change, which leads to the transmission change of the THz wave. Weis et al. introduced a monolayer graphene to the silicon modulator to form a graphene-on-silicon (GOS) structure to enhance the modulation depth in optically driven pure silicon modulators.[14] To induce the optical doping effect, a pulsed laser with a wavelength of 780 nm and a pulse duration of 100 fs was aligned to overlap with the modulated THz beam. The amplitude spectra of the transmitted electric THz field through silicon and GOS samples are shown in Fig. 2.8. We can find that when there is no infrared light, silicon provided a high transmission up to 100%. When electron-hole pairs were induced due to the infrared light, the increase of conductivity of the structure caused a broadband attenuation of the transmitted THz wave under an optical doping power of 40 mW. The transmission of the THz wave decreased over a wide frequency range from 0.2 to 2 THz. Besides, compared to pure silicon, THz beam attenuation was significantly enhanced at equivalent power levels of optical doping due to the existence of graphene with a maximum difference of 18%.

To obtain a high-speed THz all-optical modulator driven by a 1.55-μm laser, Wen et al. replaced silicon with Ge to fabricate a graphene-on-Ge (GOG) system.[15] Since Ge has a small bandgap of 0.66 eV, it allows the pump light with 1.3–1.55 μm wavelength. Meanwhile, compared to silicon, Ge has a higher bulk mobility for both electrons and holes, which should

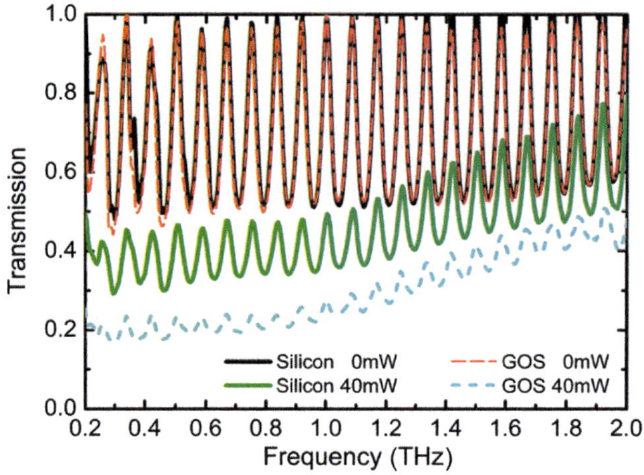

Fig. 2.8 Transmission spectra of silicon and GOS with and without infrared light illumination.

ideally correspond to an increase in surface mobility and ultimately an increase in device performance such as modulation depth and speed. The spatial configuration of GOG is shown in Fig. 2.9A. The THz wave was overlapped by a 1550-nm modulation laser beam, and both beams were incident from the graphene side. The 1550-nm modulation laser was generated by an Er-fiber laser of 100 fs pulse duration with 100 MHz repetition frequency. The frequency-domain spectra with and without the modulation laser are shown in Fig. 2.9B. Upon pumping by 1.55 μm laser with a power of 400 mW, a modulation depth of 83% was obtained within the frequency from 0.25 to 1 THz for GOG, while that for pure Ge was 68%. As depicted in Fig. 2.9C, the 3-dB modulation bandwidth of the GOG modulator was measured to be 200 kHz, which is still relatively narrow. The reason can be attributed to the long carrier recombination time of Ge around 2 μs. Fortunately, by optimizing the Ge materials and optical lasers, the modulation bandwidth of GOG has the potential of reaching tens of megahertz according to theoretical predictions.

In conclusion, an all-optical modulator with 2D materials has a great potential to reach ultrafast modulation speed since it can complete modulation totally in the optical domain. Meanwhile, the employment of graphene significantly enhances the modulation depth of the THz all-optical modulator. However, even though the fastest decay time of 2.2 ps has already been realized, the high-power pump light may inevitably induce the thermo-optic effect, which is a negative factor in terms of the theoretical ultrafast

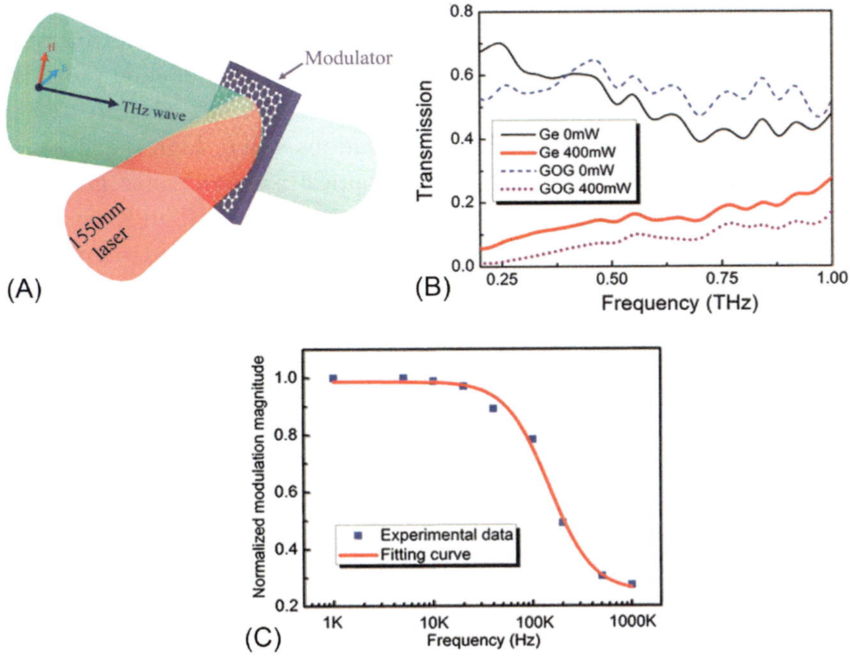

Fig. 2.9 (A) Schematic of the GOG all-optical THz modulator. (B, C) Performance of the GOG all-optical THz modulator.

modulation speed. Therefore, to fully realize the inherent ultrafast modulation speed in the future, one must avoid thermo-optic modulation in all-optic modulation. Besides, until now, the all-optical modulators have yet to be experimentally proposed on the integrated photonic platform such as silicon and indium phosphide chip. This may be another valuable path to improve the relatively poor modulation depth of existing all-optical modulators.

2.3 Thermo-optical modulators with 2D materials

2.3.1 Theory of thermo-optical modulators

The thermo-optic effect is defined as the variation of the refractive index of the material with temperature. Since silicon has a large thermo-optic coefficient[22] of 1.8×10^{-4} K^{-1}, thermo-optical modulators have been extensively employed in silicon photonics.[23, 24] The temperature increase is often induced by Ohmic heating of a metallic heater when the external voltage is applied to the heater. The change of refractive index leads to a phase change of the signal, which can also be transferred to intensity change using MZI, microring resonator (MRR), or plasmonic structure. Unfortunately, the modulation speed of thermo-optic modulators is rather slow (~MHz) due to the intrinsic slow thermal diffusivity. Therefore, thermo-optical modulators are often considered for applications where high speed is not necessary, such as optical switching and optical routing. Since graphene has a high intrinsic thermal conductivity up to 5300 W/m K and low absorption of light at telecommunication wavelength compared to a metallic heater,[7] it is recognized as an ideal candidate to act as transparent heaters and conductors with a fast response speed. Due to easy fabrication and compatibility with conventional optic devices, the graphene microheater has been comprehensively studied in recent years in both fiber optics and integrated optics, which will be discussed in detail later in this chapter. Apart from graphene, other 2D layered materials and their heterostructures might also be potentially employed for thermo-optic refractive modulation. However, up to now, the thermo-optic modulator based on other 2D layer materials has not been proposed yet. These novel thermo-optic modulators might spring up in the future after a full thermal stability assessment of these materials.

2.3.2 Thermo-optical modulators with 2D materials in optical fibers

Thermo-optical modulators based on graphene were first employed in optical fiber systems. In 2015, Gan et al. proposed a thermo-optical modulation scheme based on fiber.[25] To implement the modulation, they used a GMF to generate Joule heating through the interaction with the microfiber evanescent field, which subsequently heated the microfiber. As Fig. 2.10A shows, a microfiber with a uniform diameter of ~10 μm over a length of ~5 mm was employed to propagate the light. The graphene film used in their work consisted of five graphene layers to provide stronger interaction with the evanescent field than a single-layer graphene. Due to the zero bandgap in graphene, a low-power pump light did not induce the radioactive process in the electronic relaxation. Hence, in the GMF, graphene could generate Ohmic heating effectively by absorbing the evanescent field and heating the microfiber. The GMF was inserted into one arm of the MZI as shown in Fig. 2.10B to realize intensity modulation. The pump light was supplied by a 980-nm laser diode and the signal light source was a telecom tunable laser with a linewidth of 400 kHz. When the 980-nm pump light with a power of 5.3 mW was injected into the system, a 0.024-nm blueshift of the interference fringe without any distortion can be observed (Fig. 2.10C),

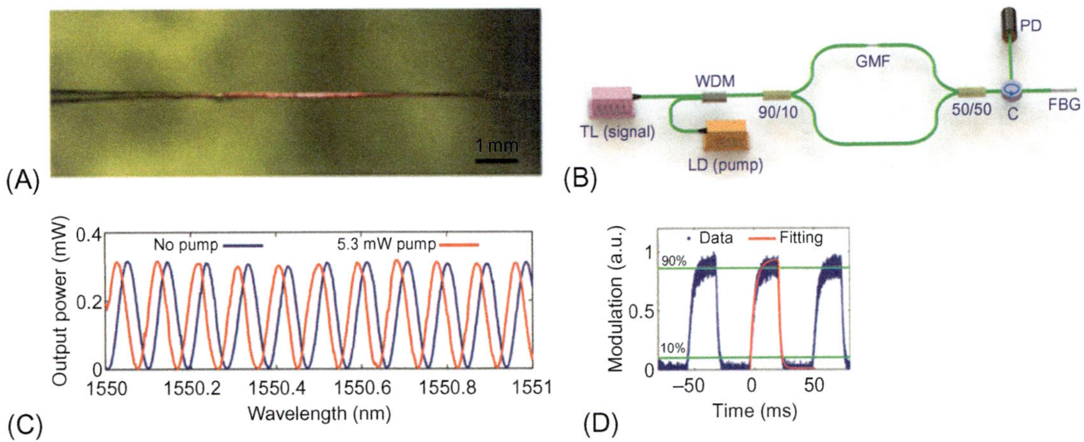

Fig. 2.10 (A) Schematic of the GMF thermo-optical modulator. (B) Experiment setup. (C, D) Performance of the GMF thermo-optical modulator.

indicating that the pump light induced a 0.51π phase shift over the 5-mm-long GMF. The maximum phase change of 21π could be induced when the incident pump power increased to 230 mW, corresponding to a temperature increase of 95 K in the GMF. The temporal response was also examined with the rise/fall time of 4/1.4 ms, respectively, as displayed in Fig. 2.10D. Compared to the standard single-mode fiber, the measured results of the GMF exhibited a significant improvement, which can be attributed to the small volume of the GMF and fast heat dissipation.

2.3.3 Thermo-optical modulators with 2D materials in silicon photonics

Compared to the fiber optic modulators, the integrated silicon modulator holds several distinctive advantages, including compact size, low energy consumption, and the possibility to be monolithically integrated with other on-chip devices. A number of works on thermo-optical modulators with graphene have been reported in silicon photonics. In 2013, Kim et al. developed a thermo-optic-mode extinction modulator based on graphene plasmonic waveguide.[26] As Fig. 2.11 shows, the modulator consisted of two graphene microribbons arranged in a cross and a polymer dielectric with a high thermo-optic coefficient. The long graphene strip supported a low-loss symmetric surface plasmon polaritons (SPP) mode, i.e., long-range

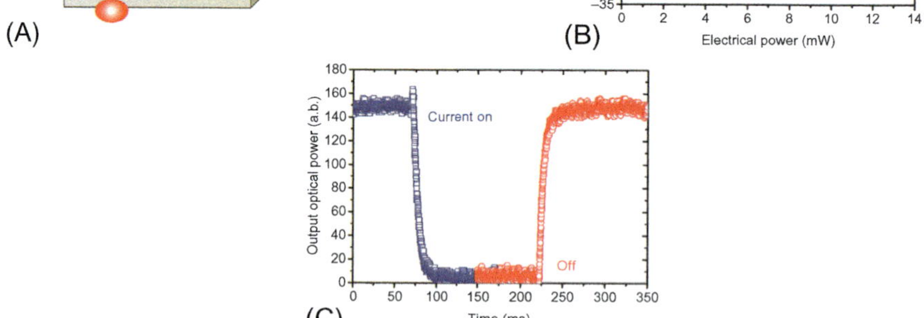

Fig. 2.11 (A) Schematic of the mode extinction modulator. (B, C) Performance of the mode extinction modulator.

SPP (LRSPP) mode, while the short graphene strip acted as a microheater when an external voltage was applied through the metal pad. Since the transmission loss of the LRSPP heavily relied on the refractive index and material loss of the surrounding dielectric, the refractive index change induced by the graphene heater led to an LRSPP stripe mode to be cut off, thus enabling an intensity modulation through the thermo-optic effect.

The static and dynamic measurement results of the mode extinction modulator are shown in Fig. 2.11B and C. We can find a significant attenuation increase of 30 dB when the input electrical power was 12.5 mW. Besides, the falling and rising times were measured to be 15 and 10 ms, respectively. Although the modulation depth of 30 dB was high, the response time remained relatively slow. In addition, the fabrication of an extinction modulator was somehow complicated, which might limit its practical applications in the future.

In order to lower fabrication complexity, a thermo-optic microring modulator assisted by graphene was proposed by Gan et al.,[27] with a simple fabrication flow. The schematic of the graphene microring modulator is shown in Fig. 2.12. A single straight waveguide was coupled to a race-track ring resonator, which was covered by a monolayer graphene. MRR was fabricated on a silicon-on-insulator (SOI) wafer with a 340-nm-thick top silicon layer and a 1-μm-thick buried oxide layer. The widths of both the straight and ring waveguides were 400 nm with a height of 220 nm. The gap between the straight and ring waveguide was set as 100 nm to approach the critical coupling condition. The bending radius of the ring was 3.4 μm and the length of straight section was 1.96 μm. Graphene could produce Joule heating when electrical signals were applied to two electrodes. Heat was transferred onto the microring and therefore caused a change in the real and imaginary parts of

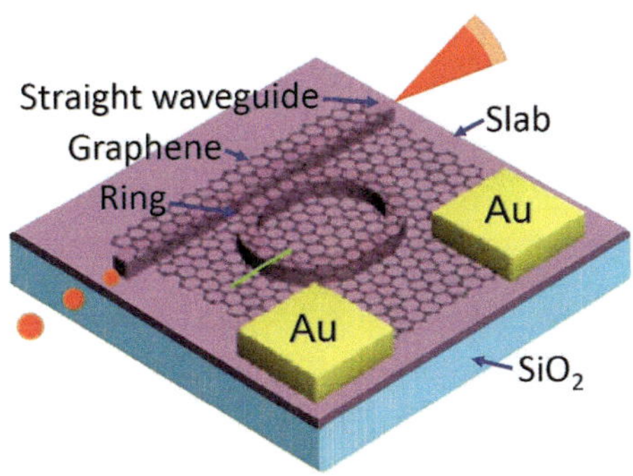

Fig. 2.12 Schematic of the graphene MRR thermo-optic modulator.

refractive index due to the thermo-optic effect. Consequently, the resonance wavelength and the transmission of the MRR could be tuned.

The measurement of the static response of the modulator was conducted by applying direct current signal to the metal pad. The resonance wavelength of the MRR experienced a red shift of 2.9 nm when the external voltage was increased from 0 to 7 V. The resonance wavelength shift resulted in a decrease of the quality factor. An external voltage of 7 V corresponded to a heating power of 28 mW, indicating a thermal tuning efficiency of 0.1 nm/mW. A maximum modulation depth of 7 dB could be obtained. Besides, the wavelength range with a modulation depth >3 dB was 6.2 nm. The rise and fall time were measured as 750 and 800 ns, respectively. Such a high modulation speed is attributed to superior thermal conductivity and fast heat generation of graphene. However, the thermal tuning efficiency remains low in this MRR thermo-optic modulator.

In order to boost tuning efficiency, Yu et al. proposed a thermo-optic silicon microdisk resonator with graphene microheater.[28] Fig. 2.13A shows

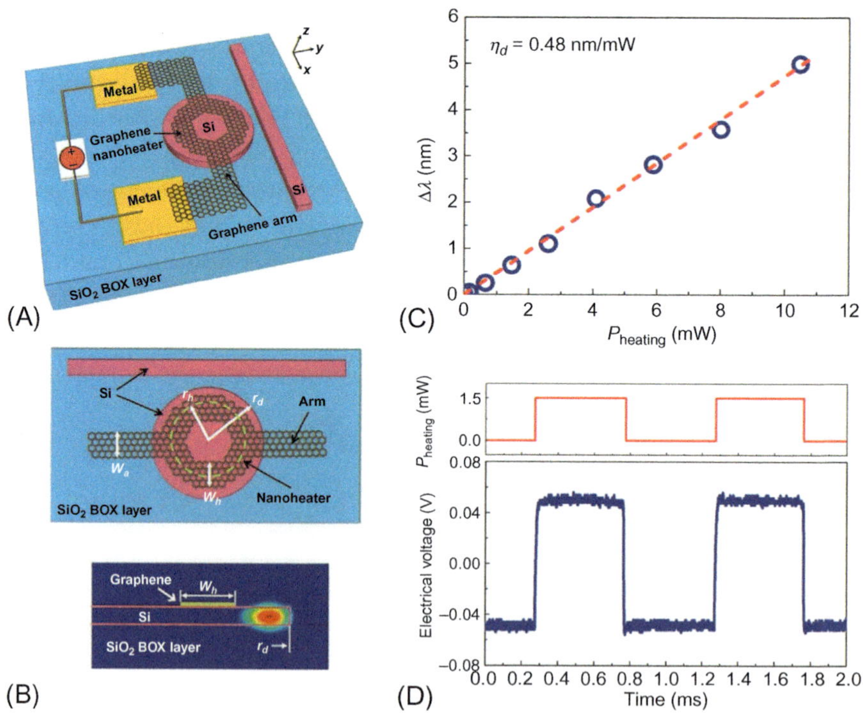

Fig. 2.13 (A, B) Schematic of the microdisk thermo-optic modulator. (C, D) Performance of the microdisk thermo-optic modulator.

a schematic of the proposed modulator, which was fabricated on a commercial SOI wafer with a 250-nm-thick top silicon and an SiO_2 layer of 3 μm. The radius of the microdisk was set as 5 μm with the shape and the position of the graphene microheater optimized to avoid excess loss induced by graphene, as shown in Fig. 2.13B. Fig. 2.13C shows the thermal tuning efficiency of 0.48 nm/mW, and Fig. 2.13D shows that the 90% rising time and the decaying time of thermal tuning were about 12.8 and 8.8 μs, respectively. Besides, they also proved that tuning efficiency could be further improved to 1.67 nm/mW by reducing the size of the radius.

Photonic crystal structures have also been utilized to improve the performance of the thermo-optic modulator. In 2017, Yan et al. proposed a slow-light-enhanced graphene microheater to obtain high tuning efficiency and fast response simultaneously.[29] An optimized silicon photonic crystal waveguide with a group index >20 within 10 nm bandwidth replaced the conventional strip waveguide to provide stronger light-graphene interaction, as shown in Fig. 2.14A. Meanwhile, the photonic crystal waveguide with a graphene microheater was integrated in an MZI structure to convert the phase change into a resonance shift of the transmission spectrum. Thanks to both the slow light effect and the structure of photonic crystal slab; they experimentally demonstrated an energy-efficient graphene microheater with a tuning efficiency of 1.07 nm/mW and power consumption per free spectral range of 3.99 mW. The rise and decay times (10%–90%) were only 750 and 525 ns, which are the fastest reported response times for microheaters in silicon photonics. Recently, Qiu et al. combined a silicon photonic crystal nanobeam (PCN) cavity with an energy-efficient graphene

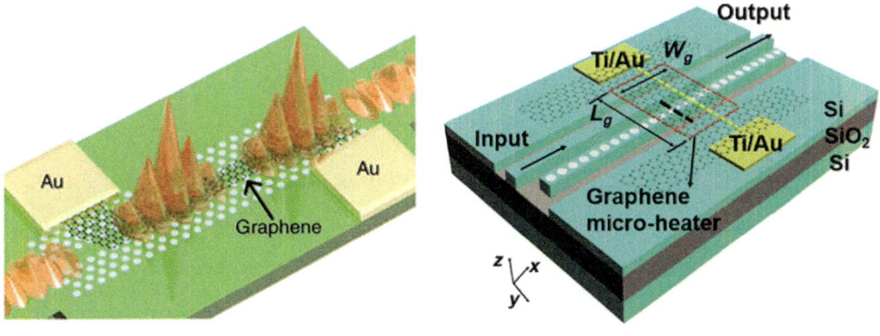

(A) (B)

Fig. 2.14 (A) Schematic of thermo-optic modulator based on the photonic crystal waveguide. (B) Schematic of thermo-optic modulator based on the photonic crystal nanocavity.

microheater to achieve a high-performance thermo-optical modulator.[30] The structure is shown in Fig. 2.14B. A tuning efficiency as high as 1.5 nm/mW was experimentally demonstrated due to the ultra-small optical mode volume of the PCN cavity. The rise and decay times were measured to be 1.11 and 1.47 μs, respectively.

Apart from the microheater, graphene can also act as a thermal conductor[31] in the thermo-optic modulator as Yu et al. proposed in 2014. As shown in Fig. 2.15, this graphene heat conductor covered the metal and a part of the heated MZI arm. The SOI waveguide was covered partially in the lateral direction to reduce excess loss due to graphene absorption. Heat generated by the non-local metal heater was delivered toward the waveguide to influence the temperature of the waveguide, thus shifting the resonance wavelength of the MZI transfer spectrum. However, in this scheme, power consumption is relatively high with 110 mW heating power only inducing a redshift of 7 nm, while the response time is about 20 μs.

Owing to its ultra-high thermal conductivity and low absorption loss toward the light of the communication wavelength, graphene has become a promising candidate as a replacement of the traditional metal microheater. Impressive thermo-optical modulation performance has been demonstrated with the photonic crystal structure because of its strong light-matter interaction. One should note that although the modulation speed of thermo-optical modulation is inherently limited by its operation principle compared to the all-optical and electro-optical modulations, the thermo-optical modulator can still be regarded as an economic manner of optical modulation where ultra-high modulation speed is not required, because of its easy fabrication. Besides, power consumption in all-optical modulation has attracted more attention than other modulation schemes since large heating power

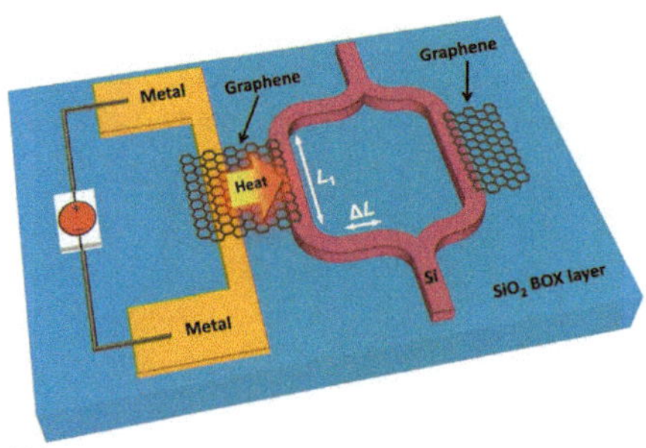

(A)

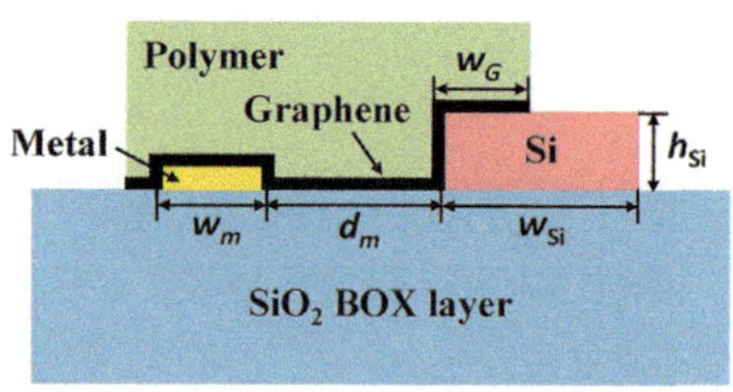

(B)

Fig. 2.15 (A) Schematic of the SOI thermo-optic modulator with a graphene heat conductor. (B) Cross-section view of the modulator.

may deteriorate the performance of the chip due to thermal crosstalk. Therefore, achieving a fast modulation speed and low power consumption simultaneously for the thermo-optical modulator is a long-pursuit goal. Moreover, the thermo-optic modulation performance can be further enhanced by boosting the light-graphene interaction with specific optical structures, including photonic crystal nanocavity and plasmonic waveguides.

2.4 Electro-optical modulators with 2D materials

2.4.1 Theory of electro-optical modulators

The electro-optical modulators exploit electro-optic effects to electrically control the light properties, including refractive index and propagation loss. Since graphene holds an ultra-high electron mobility, the electro-optical modulators may reach an ultra-high modulation speed of a hundred gigahertz, which is particularly attractive for data communication. Meanwhile, the absorption bandwidth of 2D materials, such as graphene, can cover a spectral range from the ultraviolet to the THz, including the visible, infrared, and THz wavelength, which is significantly broader than conventional semiconductor materials. Besides, 2D material-based electro-optic modulators also have distinctive advantages of compactness, low operation voltage, and CMOS compatibility. Therefore, electro-optical modulators with 2D materials have been a research focus since it was proposed. To date, most electro-optical modulators with 2D materials utilize the gate-tunable electro-absorption effect[32, 33] and gate-tunable refractive index effect[34] in graphene, which corresponds to intensity modulation and phase modulation, respectively. Although other electro-optic effects in 2D materials, such as the Franz-Keldysh effect[7] and the quantum-confined Stark[7] effect, are also possible for light modulation, no experimental results have been demonstrated yet, to the best of our knowledge.

Modulation speed and modulation depth are two crucial figures of merit for electro-optical modulators. Even though graphene has a carrier mobility exceeding 200,000 cm^2 V^2/s at room temperature, the modulation speed of the electro-optical modulator is still limited by the RC constant of the modulator with a typical value of a few gigahertz, which might be significantly enhanced with structural optimization of electro-optical modulators. The modulation depth is also intrinsically limited by a low absorption rate of a monolayer graphene. Since a monolayer graphene absorbs only 2.3% of white vertically incident light,[35, 36] intrinsic modulation of monolayer graphene-based free-space modulators can only be up to 2.3% (~0.1 dB), which is obviously insufficient for most practical applications. Therefore, overcoming the low absorption in 2D materials is another crucial issue for electro-optical modulators.

2.4.2 Electro-optical modulators in communication wavelength

The first graphene-based silicon waveguide electro-optical modulator was reported by Liu and colleagues in the University of California, Berkeley.[9] As Fig. 2.16A displays, a 50-nm-thick Si layer was used to connect the 250-nm-thick Si bus waveguide and one of the gold electrodes. Seven-millimeter-thick Al_2O_3 was then uniformly deposited on the surface of the waveguide by atom layer deposition. A monolayer graphene sheet grown by chemical vapor deposition was then mechanically transferred onto the Si waveguide. Another electrode with platinum was deposited 500 nm away from the waveguide to avoid disturbing the optical mode in the waveguide. Only the regions on top of the waveguide and between the waveguide and the platinum electrode was left, while the excess part was removed by oxygen plasma. Graphene, Al_2O_3, and silicon together formed a capacitor structure, which was the basic block of the graphene electro-optical modulator. The cross-sectional view of the modulator and the

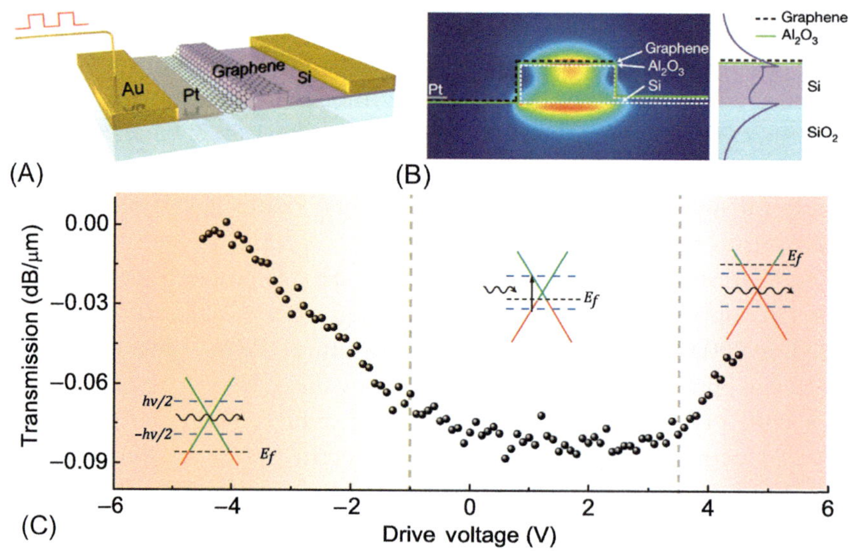

Fig. 2.16 (A) Schematic of the graphene electro-optical modulator. (B) Optical field distribution of the guided mode. (C) Transmission with different applied voltages.

optical field distribution of the guided mode is illustrated in Fig. 2.16B. The silicon waveguide was optimized to have an electric field maximized at its top and bottom surfaces in order to maximize the light-graphene interaction.

Fig. 2.16C shows the static electro-optical response of the device at different driving voltages where three different regions can be seen with their band structures shown as insets. In the middle region, the interband transition of a photon of 1.53 μm was allowed since the Fermi level was close to the Dirac point. Thus, the graphene sheet was absorptive to incident photons, resulting in a modulation depth of 0.1 dB/μm. In the left-hand region, the Fermi level was lower than half the photon energy. Therefore, there were no electrons available for the interband transition, and graphene exhibited non-absorption to incident light. In the right-hand region, all electron states in resonance with incident photons were occupied and the transition was forbidden, also resulting in the low absorption rate of graphene. Thus, in the last two cases, light transmission through the waveguide was increased when external voltage was increased. The measured 3-dB modulation bandwidth was 1.2 GHz, which was relatively low due to the large *RC* constant of the circuit. The modulation wavelength range covered from 1.35 to 1.6 μm as a result of the broad absorption wavelength range of graphene. Although the modulation performance of the proposed modulator is yet to be optimized, it is the first experimental verification of the integrated graphene electro-optical modulator in communication wavelength.

Since both modulation depth and modulation speed of the aforementioned modulator can hardly meet the requirements of the telecommunication application, huge efforts have been carried out to boost these two crucial figures of merit. In order to increase the modulation depth of graphene modulators, one has to enhance the inherently low light-graphene interaction in the current modulation scheme. For this purpose, various methods have been proposed, including using nonresonant waveguides, heterostructure, and cavity enhancement methods.

For nonresonant waveguides, the double-layer graphene structure was frequently employed to obtain a high modulation depth,[37] which was first proposed by Liu et al. in 2012. As Fig. 2.17 displays, they used two graphene layers and a thick oxide layer to form a p-oxide-n-like junction, which had a similar structure as a forward/reverse-biased silicon modulator. The major difference between the graphene modulator and the conventional silicon modulator was that the doped silicon was replaced by graphene, which could effectively lower the *RC* constant of the modulator, thus increasing

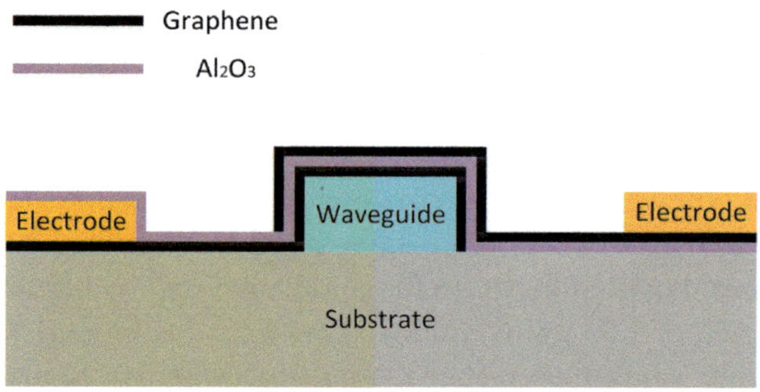

Fig. 2.17 Schematic of the double-layer graphene electro-optical modulator.

the modulation bandwidth. Meanwhile, due to the use of two graphene layers as the active medium, optical absorption could be significantly increased, thus leading to a larger modulation depth. When no driving voltage was applied, both graphene layers were undoped, indicating that the Fermi levels were close to the Dirac point and both graphene sheets exhibited absorption to light. When a voltage was added between them, two graphene layers formed a simple parallel capacitor model with one graphene layer doped by holes and the other by electrons at the same doping level. When the Fermi level shift in both graphene layers reached half photon energy of incident light, both graphene layers became transparent simultaneously. Therefore, a modulation depth as high as 6.5 dB was reached for the 40-μm-length waveguide with double-layer graphene. The 3-dB modulation bandwidth was 1.2 GHz, which was still relatively low. A similar structure was proposed by Mohsin et al. in 2014, where a hydrogen silsesquioxane (HSQ) layer with a thickness of 30 nm was deposited on the waveguide.[38] According to experimental results, the modulation depth was increased to 16 dB with an insertion loss of 3.3 dB operating at a 1550 nm wavelength.

For cavity enhancement, Gan et al. reported that the photonic crystal cavity could contribute to enhancing the graphene modulator's performance.[39] A modulation depth of 10 dB could be reached due to its strong light localization effect. In 2015, Gao et al. proposed a graphene-BN heterostructure integrated with a silicon photonic crystal nanocavity to enhance the light-graphene interaction,[40] as shown in Fig. 2.18. Along with the silicon photonic crystal nanocavity, a BN/graphene/BN/graphene/BN five-layer stack was built by the van der Waals assembly technique transferred onto a quartz substrate to enhance the light-graphene interaction. The two graphene sheets were positioned across each other to act as a gate

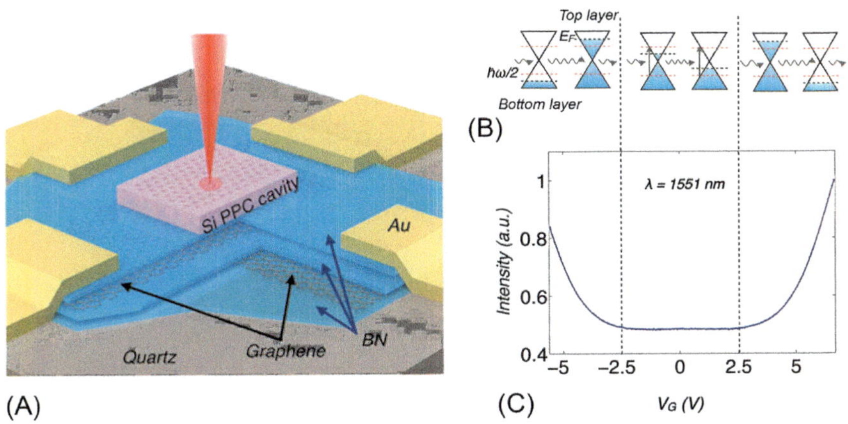

Fig. 2.18 (A) Schematic of the graphene-boron nitride heterostructure modulator with silicon photonic crystal nanocavity. (B, C) Performance of the graphene-boron nitride heterostructure with silicon photonic crystal nanocavity.

and supply voltage. The modulation theory of this structure is explained in Fig. 2.18B. An increase in gate voltage led to Pauli blocking and a reduced optical absorption in graphene, thus a larger reflection from the cavity. A modulation depth of 3.2 dB was reached when the gate voltage was increased to 6.7 V. A high modulation depth mainly resulted from the employment of the graphene-BN heterostructure. Owing to its enhancement of light-matter interaction, the graphene-BN heterostructure has also been applied in straight waveguides[41] and SPP waveguides[42] in order to reach a high modulation depth.

Apart from the photonic crystal cavity, the MRR is also considered a promising candidate for cavity enhancement. Compared to the photonic crystal nanocavity, the MRR has less fabrication complexity and more fabrication tolerance. A silicon MRR with graphene was proposed by Qiu et al.[43] to act as an electro-optical modulator, as illustrated in Fig. 2.19. This structure employs evanescent mode coupling between graphene and silicon to enhance the light-matter interaction. When an external voltage was applied to the metal contact, the change of the Fermi level of graphene resulted in changes in both resonance wavelength and quality factor of the MRR. Thus, light transmission at the resonance wavelength experienced a significant increase. According to experimental results, a modulation of 40% was demonstrated at 1555.97 nm with a gate voltage of 6 V. However, the modulation speed was not experimentally measured in their work. The modulation mechanism in the electro-optical MRR modulator was also not explained clearly.

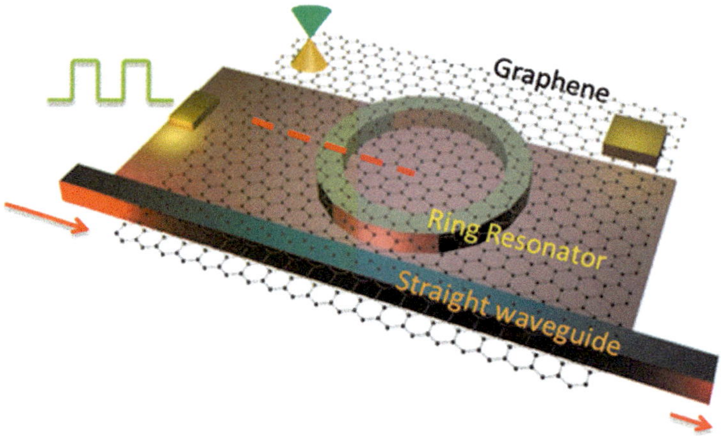

Fig. 2.19 Schematic of the silicon MRR with graphene.

To enhance modulation performance and provide a thorough theoretical analysis of electro-optic modulation in the microring with graphene, Ding et al. proposed an electro-optical graphene modulator with a high modulation depth and moderate modulation speed.[44] A false-color scanning electron microscope (SEM) image of the structure is shown in Fig. 2.20. We find that in their work, graphene did not fully cover the all-pass-type MRR, but only covered 25% of the ring waveguide. The MRR with graphene was designed to work close to the critical coupling condition when no external voltage was applied. When an external voltage was applied to graphene, the propagation loss of the graphene-silicon bend waveguide was tuned. Thus, the roundtrip transmission coefficient of the MRR was tuned correspondingly. The alteration of the roundtrip transmission coefficient indicated a change of the coupling condition from critical coupling to under-coupling.

Fig. 2.20 A false-color SEM image of a highly efficient silicon MRR.

As a result, the extinction ratio of the MRR was decreased, leading to a significant transmission change at the output. In their experiments, a modulation depth of 12.5 dB was reached with a bias voltage of 8.8 V. Besides, a modulation bandwidth of 100 kHz was obtained with an extinction ratio of 3.8 dB, which can be further increased by reducing the associated *RC* time constant of the electric circuit. The value of Ding's work lies in not only the experimental demonstration of an efficient electro-optical modulator with graphene but also the systematic interpretation of how MRR can enhance the modulation depth by altering the coupling condition.

By altering the coupling condition, the MRR can convert the loss of waveguide to the transmission increases, which induces a significant improvement in modulation depth. Meanwhile, as we mentioned before, the double-layer graphene can also contribute to a better modulation performance. Therefore, by combining the double-layer graphene structure with the silicon nitride (Si_3N_4) MRR, Phare et al. demonstrated a high-performance graphene electro-optic modulator.[45] A schematic of the proposed modulator is shown in Fig. 2.21. An Si_3N_4 waveguide was employed to guide light with transverse electric (TE) mode. The MRR was designed to operate in under-coupled condition. A graphene/graphene capacitor consisting of two sheets of monolayer graphene was fabricated on top of a portion of the ring resonator. Between graphene, approximately a 65-nm layer of atomic layer-deposited Al_2O_3 formed the interlayer dielectric. Since the thickness of the dielectric was five times larger than the previous work, a lower capacitance and a higher modulation speed can be obtained by reducing the *RC* constant of the circuit.

The experimental results are shown in Fig. 2.21B and C. At 0 V bias, both graphene sheets in the capacitor were lightly doped and exhibited a large absorption rate to the light; thus the ring was under-coupled to the bus waveguide. When an external voltage was applied to the capacitor, the MRR was critically coupled to the bus waveguide, decreasing the system's transmission. According to experimental results, the modulator had a modulation depth of 15 dB when the driven voltage was 10 V. Besides, the device also exhibited a modulation bandwidth of 30 GHz, which, to the best of our knowledge, is the largest experimentally verified modulation bandwidth of the electro-optical modulator with graphene, to the best of our knowledge. Meanwhile, the power consumption of the modulator can be as low as ~100 fJ/bit. Such an impressive performance is almost already comparable with current semiconductor modulation technologies. The employment of double-layer graphene capacitor can not only enhance

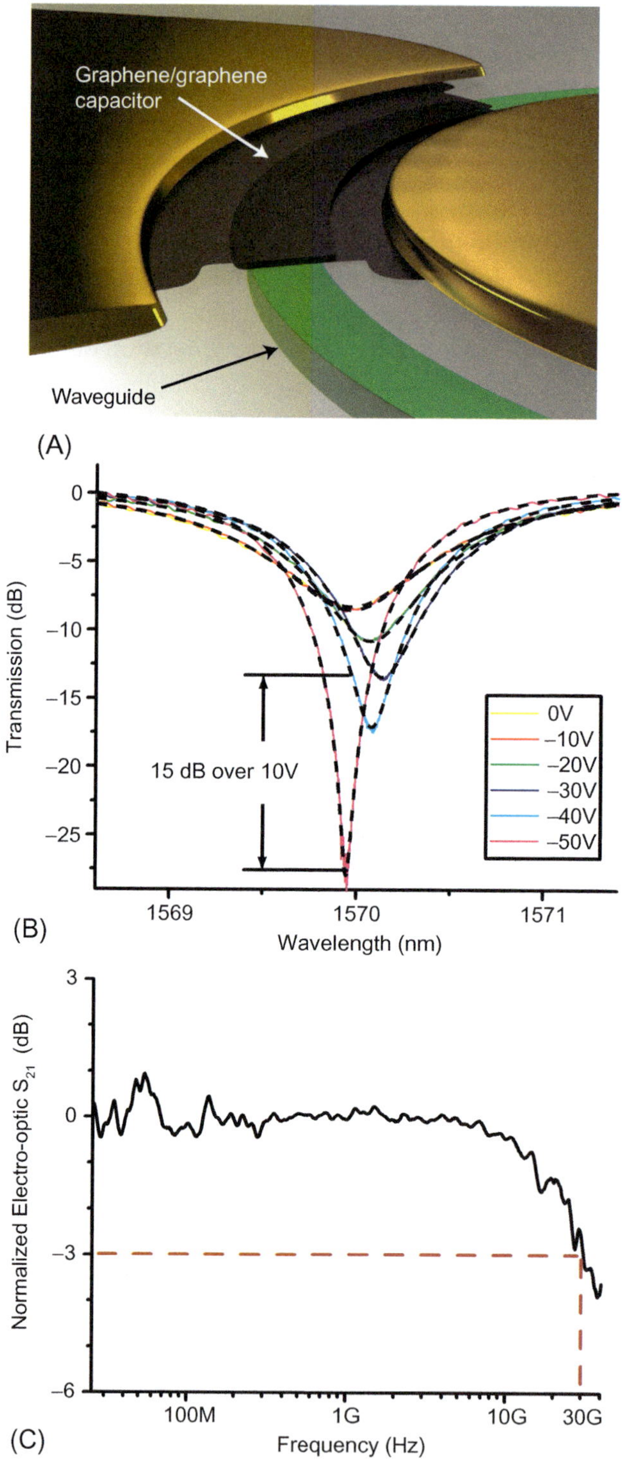

Fig. 2.21 (A) Schematic of the Si_3N_4 MRR. (B, C) Performance of the Si_3N_4 MRR.

the light-matter interaction but also increase the modulation bandwidth by reducing the *RC* constant. Therefore, the double-layer graphene structure has also been widely applied in other photonic structures such as D-shaped microfiber,[46] amorphous silicon waveguides[47] (35 GHz in experimental results), hybrid waveguides with polycrystalline silicon on the top and single-crystalline silicon on the bottom[48] (55 GHz in theoretical calculation), silicon ridge waveguide[49] (THz scale in theoretical calculation), and dual-graphene-on-graphene configuration[50] (100 GHz in theoretical calculation).

To compare the performance of electro-optical modulators with graphene in communication wavelength, several key parameters verified with experiments are listed in Table 2.1. We can find that among these works, the Si_3N_4 MRR with double-layer graphene exhibits the most impressive performance with a 30-GHz modulation bandwidth and a 15-dB modulation depth. To the best of our knowledge, this is the electro-optical modulator with best performance among the reported works. Since their work, the double-layer graphene structure has become a common method to obtain a better modulation performance, which can not only enhance the modulation bandwidth by reducing the *RC* constant but also increase the modulation depth by strengthening the light-graphene interaction. More importantly, we believe that the combination of double-layer graphene and resonance structure represents the future trend to obtain superior

Table 2.1 Key parameters of the electro-optical modulators in communication wavelength.

Ref.	Structure	Modulation bandwidth	Modulation depth
9	Straight Si waveguide	1.2 GHz	0.1 dB/μm
37	Straight Si waveguide	1.2 GHz	6.5 dB
38	Straight Si waveguide	670 MHz	16 dB
39	Photonic crystal cavity	Not mentioned	10 dB
40	Si photonic crystal nanocavity	1.2 GHz	3.2 dB
41	Straight Si waveguide	Not mentioned	0.09 dB/μm
43	Si microring resonator	Not mentioned	40%
44	Si microring resonator	100 kHz	12.5 dB
45	Si_3N_4 microring resonator	30 GHz	15 dB
46	D-shaped microfiber	Estimated 97.26 GHz	Not mentioned
47	Amorphous Si waveguides	35 GHz	2 dB

modulation performance in electro-optical modulators with 2D materials. Moreover, modulation bandwidth can be further optimized by increasing the thickness of the dielectric layer or optimizing the contact resistance. Although it may increase fabrication complexity significantly, extensive work is inevitable to optimize the fabrication process of electro-optical modulators in the future. Therefore, despite the difficulty to implement the theoretical modulation bandwidth limit of graphene electro-optical modulators, we can still expect graphene as well as other 2D materials to complement the intrinsic drawbacks of bulk materials in electro-optical modulation.

2.4.3 Electro-optical modulators in THz wavelength

In the past decades, THz modulators have been attracting significant attention since there is a great demand for components in this spectral range for various applications, including health, environmental, and security applications. However, conventional modulation manners have been challenged in this spectral range due to the lack of efficient devices to manipulate THz waves. Although the optical doping effect in graphene can be employed to realize all-optical THz modulation, the most mature and convenient of modulation way remains electro-optic modulation, owing to the development of the electrical information industry.

The first THz electro-optical modulator with graphene was proposed by Berardi et al.,[51] with a structure as depicted in Fig. 2.22A. A single-layer graphene was deposited on an SiO_2/p-Si substrate. The top metal contacts were used to monitor graphene conductivity, and the bottom ring-shaped gate was used to tune the conductivity of graphene. At zero bias, the Fermi level was at the Dirac point of all graphene layers, thus exhibiting minimum

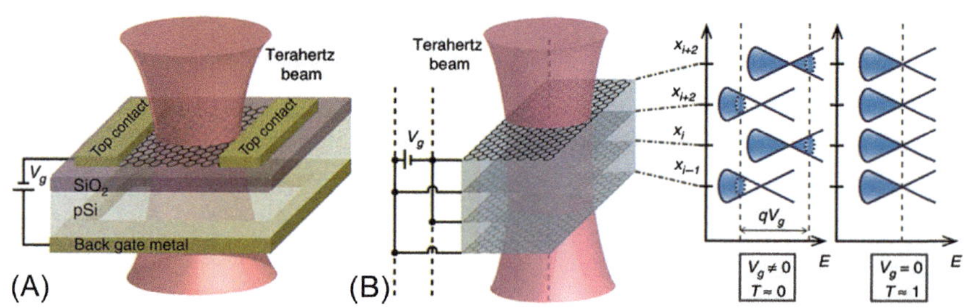

Fig. 2.22 (A) Schematic of the graphene electro-optical modulator in THz wavelength. (B) Operation principle of the modulator.

signal attenuation. According to a simple Drude model,[52, 53] when a bias voltage was applied, the accumulation of carriers led to the shift of Fermi level and electrical conductivity, thus modulating THz absorption in graphene. Besides, stacked graphene-semiconductor or graphene-graphene pairs were employed to increase the potentially limited modulation effect in a single graphene layer, as Fig. 2.22B shows. After removing the substrate effect, the proposed device had a flat modulation depth of 16% at the gate voltage of 50 V in the 570–630 GHz frequency band. The 10%–90% rise time of the modulator was measured to be 38 μs; thus, the 3-dB bandwidth was calculated to be around 18 kHz, which was mainly limited by the large *RC* constant. Although several other experimental demonstrations of the electro-optical THz modulators have been proposed by other groups using similar structure[54] or patterned graphene,[42] the modulation depth remains relatively low and the modulation speed is inherently limited by the large size of the modulator, resulting in a large *RC* constant.

To obtain a larger modulation depth, Berardi et al. proposed a THz modulator consisting of a single layer of graphene on top of an SiO_2/Si substrate with a metal back gate.[55] The schematic of the modulator is shown in Fig. 2.23A. In the proposed structure, the THz wave intensity at the back metal interface is zero, while the THz wave intensity in the active graphene layer relied on the substrate optical thickness and the THz wavelength. When the substrate optical thickness was an odd-multiple of the THz wavelength, the field intensity in graphene was at its maxima. Consequently, very large absorption was expected at graphene, indicating an extraordinary modulation when graphene conductivity was tuned. By applying a voltage

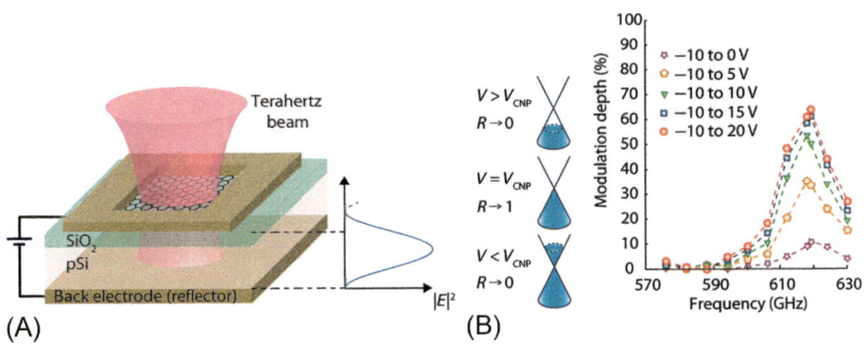

Fig. 2.23 (A) Schematic of the THz modulator with metal reflector. (B) Operation principle and performance of the modulator.

between the top graphene layer and the back metal, the absorption rate of THz waves in graphene could be significantly tuned. On the other hand, if the substrate optical thickness was an even-multiple of the THz wavelength, graphene did not absorb any incident light; thus, modulator reflectance became unity and independent of its conductivity, since the field intensity is zero in the active graphene layer. Therefore, selecting the optical thickness of the substrate according to the THz wavelength could effectively increase the light-graphene interaction. As shown in Fig. 2.23B, experimental results showed that a maximum modulation depth of 64% occurred near 620 GHz when the gating voltage varied from −10 to 20 V with an insertion loss <2 dB owing to the relatively high minimum conductivity achieved in this graphene sample. Compared to the previously reported work, the modulation depth was effectively enhanced, due to the optimization of the structure. The modulation bandwidth was measured to be around 4 kHz, which was limited by its *RC* constant.

Another highly efficient THz modulator was demonstrated by Liang et al. in 2015; in this work, a THz quantum laser was monolithically integrated with a graphene modulator.[56] This device could efficiently modulate the light intensity of THz radiation from the laser. The modulation depth was as high as 100% for the certain region of the pumping current as a result of the strongly enhanced interaction between the laser field and graphene. A schematic of the device is shown in Fig. 2.24A, which consists of an underlying surface-emitting concentric circular grating (CCG) THz quantum cascade laser (QCL) and a graphene modulator. The Fermi level of graphene could be dynamically tuned by a voltage applied to the gating electrode. The active region of the QCL was sandwiched between a bottom gold plate and the upper CCG metal grating. Single-mode THz radiation

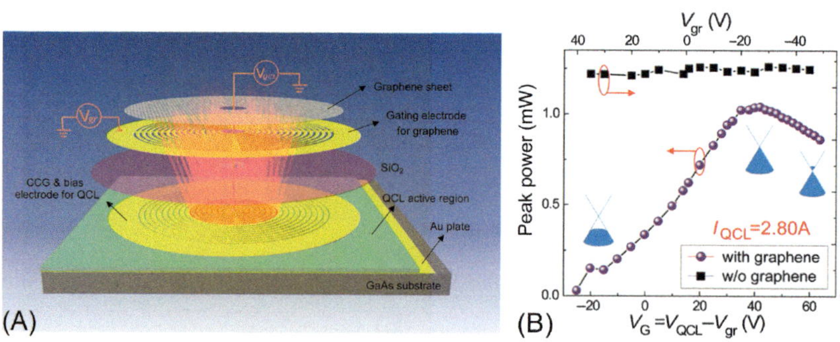

Fig. 2.24 (A) Schematic of the CCG THz QCLs with a graphene modulator. (B) Performance of the modulator.

was emitted vertically through the grating. Owing to electric field enhancement in graphene near the output aperture of the QCL, a larger modulation depth was expected when the THz CCG QCL was integrated with a graphene sheet. Theoretical analysis showed that an intensity enhancement factor of ~140 could be reached when the thickness of SiO_2 spacing was set as 450 nm. The static response of the device showed that the peak intensity of the THz laser peak varied from nearly 0 mW to a maximum of 1 mW. Such a high modulation depth can be attributed to the stronger interaction between the graphene sheet and evanescent waves of the cavity mode through the output apertures, as shown in Fig. 2.24B. Moreover, a 3-dB modulation bandwidth response was as high as 110 MHz, which was higher than most THz modulators with graphene. The reason was attributed to the fact that miniaturization reduced parasitic capacitance and resistance of the device, thus increasing modulation depth as well as modulation speed. The highlight of this work can be concluded as the discovery of the monolithic integration of THz laser and the electro-optical graphene modulator, which results in a striking modulation performance. Despite obvious fabrication complexity, the integration of the THz source remains the most promising path to greatly enhance the THz modulation performance in the future.

2.4.4 Electro-optical modulators in plasmonic optics and metamaterials

Due to its ultra-small modal volumes and strong field enhancement, graphene modulators based on plasmonic waveguides may provide a higher modulation depth. An efficient hybrid graphene plasmonic waveguide modulator was realized by Ansell et al. They used the wedge SPP waveguide supported by the edge of planar section of the waveguide[57] to guide the SPP mode, as shown in Fig. 2.25. This SPP mode could not only enhance the in-graphene-plane electric fields near the edge of the strip but also had superior field confinement characteristics. The plasmonic modes were excited by a laser with the power of ~3 mW at a wavelength of 1.5 μm. To realize the modulation of plasmonic waveguides, the Pauli blocking effect was induced by applying an external voltage to the graphene sheet. Experimental results showed that the wedge plasmons mode had a transmission response variation of 3.3×10^{-2} dB/μm when the gating voltage was set as 6 V. This indicated a modulation depth of 8.7% for a 12-μm modulation plasmonic waveguide.

Apart from the plasmonic waveguide, several other optical structures such as patterned graphene structures,[58, 59] hybrid structures,[60] and metamaterials[61, 62] have also been discussed to tune the device response with a

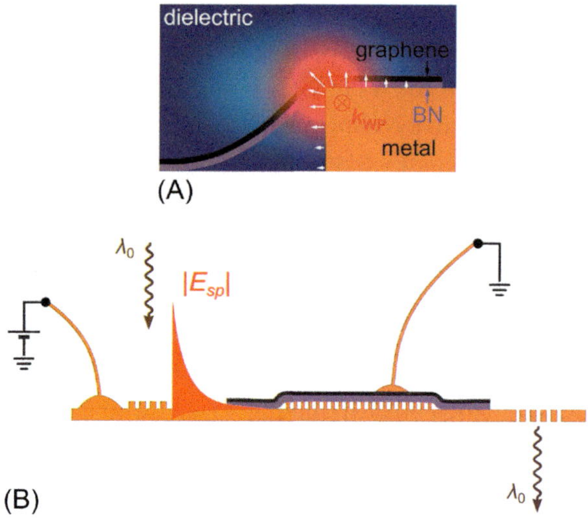

Fig. 2.25 (A) Schematic of the graphene plasmonic waveguide modulator. (B) Operation principle of the modulator.

large modulation depth. Apart from these structures, the metasurface with 2D materials has raised huge attention, thanks to its extraordinary properties. A typical work was proposed by Yao and colleagues in 2014. The proposed structure consisted of a widely tunable metasurface on graphene.[62] As Fig. 2.26 shows, an electrically tunable perfect absorber was formed by a metasurface on graphene, an aluminum oxide dielectric layer, and an aluminum substrate. The whole structure could be regarded as an asymmetric Fabry-Perot (FP) resonator with two mirrors, i.e., a tunable metasurface reflector as a partially reflecting mirror in the front

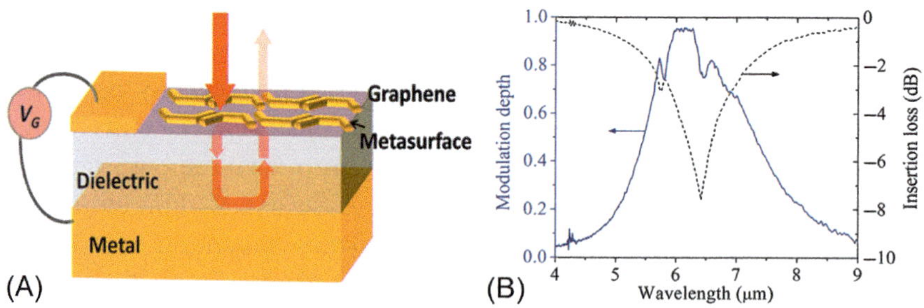

Fig. 2.26 (A) Schematic of the electro-optical modulator with graphene on metasurface. (B) Performance of the modulator.

and a metallic fully reflecting mirror in the back. When the critical coupling condition of the FP resonator was satisfied, all the incident light was absorbed in the resonator and the reflection coefficient was zero. Due to electrostatic doping of the graphene sheet underneath the optical metasurface, one can modulate the coupling condition by employing an external voltage, thus affecting the reflecting ratio toward the incident light. According to experimental results in Fig. 2.26B, the proposed device achieved a modulation depth >95% at 6 μm and >50% over a broad wavelength range of 5.4–7.3 μm. Similar to the MRR graphene modulator in communication wavelength, the large modulation depth in this work was realized by altering the coupling condition of the FP resonator. Besides, owing to the high carrier conductivity of graphene, the proposed modulator based on metasurface was experimentally verified to have a fast response time <10 ns, indicating a 3-dB modulation bandwidth of 20 GHz. Along with this work, various other modulation schemes based on metamaterial structures with 2D materials have also been proposed, including wavelength modulation,[59] phase modulation,[63] and polarization modulation.[58] It should be noted that in these plasmonic and metasurface structures, other 2D materials can also be effectively utilized for light modulation in a similar way.

2.5 Magneto-optic and acousto-optic modulators with 2D materials

Magneto-optic modulators refer to those employing magneto-optic effects such as the Faraday effect[64, 65] or magneto-optic Kerr effect[64] for light modulation. Compared to the mature all-optical and electrical-optic modulation schemes, the magneto-optic modulator has not received so much attention due to fabrication and operation complexity. However, the unique nonreciprocal property of the magneto-optic modulator provides the possibility to obtain various special devices that are not easy to realize with other modulators, such as optical isolators, circulators, polarization controllers, and electric and magnetic field sensors.

Up to date, magneto-optic Faraday rotation and Kerr rotation are the two most commonly used physical mechanisms to implement magneto-optic modulation. Both effects have been observed in graphene[64] by Shimano et al. in 2013. As Fig. 2.27 displays, a large-area monolayer graphene epitaxially grown on the Si face of silicon carbide acted as a magneto-optic rotator. A Faraday rotation angle up to 0.1 rad at a magnetic

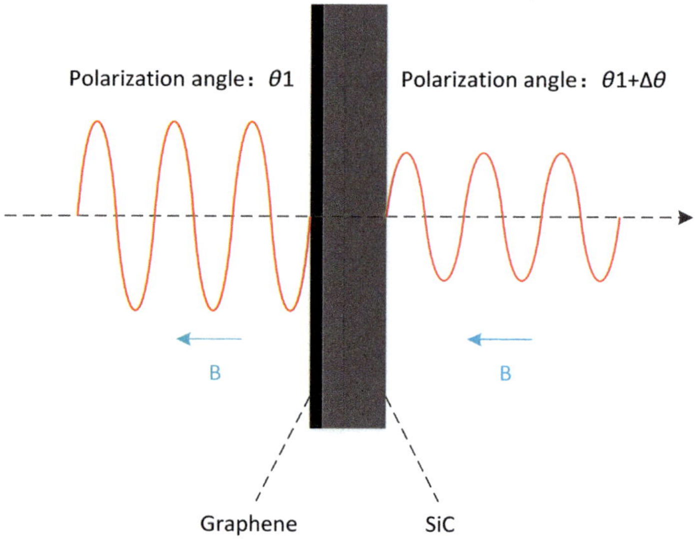

Fig. 2.27 Schematic of a graphene-based Faraday rotator.

field of 7 T was demonstrated for the incident THz pulse. This rotation angle was quite large considering the single-atom thickness of the monolayer graphene. Another magneto-optic effect, the Kerr rotation effect, had a relative small fixed ration angle of -15 mrad independent of the magnetic field >5.5 T, compared to the Faraday rotation. Although the magneto-optic effect has been successfully demonstrated in 2D materials, reducing the required magnetic field and moving its operation wavelength toward shorter wavelengths should be considered before it is practically employed in magneto-optic modulators with 2D materials.

Acousto-optic modulators refer to those using acoustic waves to modulate the refractive index of certain materials for light diffraction and frequency changing, which have been widely used for Q-switching and signal modulation in optical telecommunications and displays. Graphene has been exploited to generate, propagate, amplify, and detect surface acoustic waves, which indicates its potential in constructing 2D material-based acousto-optic modulators.[66] Other 2D materials such as MoS_2 have also been demonstrated to be technically feasible in the acousto-optic modulation scheme.[67] Owing to their large surface area and unique properties, 2D material-based acousto-optic modulators may have the potential to be widely used to miniaturize the current bulk acousto-optic modulators for specific applications in the future.

2.6 Conclusion

Optical modulators are heralded as one of the most crucial devices in various fields such as optical interconnection, environmental monitoring, biosensing, medicine, and security applications. 2D materials such as graphene, TMDs, and black phosphorus hold several distinct optoelectronic properties due to its unique 2D layer structure. Therefore, the exploitation of 2D materials in optical modulation may give rise to significant improvements of modulation performance, including modulation bandwidth, modulation depth, and energy consumption. In the past decades, modulators with 2D materials have been extensively explored in various approaches such as all-optic modulation, thermo-optic modulation, and electro-optic modulation. Impressive works have been reported such as thermo-optic modulation with nanosecond-scale response time and electro-optic modulators with 30 GHz modulation bandwidth. Nonetheless, a significant demand for performance improvement is still ongoing to compete with mature traditional technologies on modulation speed and energy consumption for optical interconnection applications. Stronger light-matter interaction and smaller RC constant of the modulators are two key points to realize better performances in the future.

References

1. Reed, G. T.; Mashanovich, G.; Gardes, F. Y.; Thomson, D. Silicon Optical Modulators. *Nat. Photonics* **2010**, *4*(8), 518.
2. Almeida, V. R.; Barrios, C. A.; Panepucci, R. R.; Lipson, M. All-Optical Control of Light on a Silicon Chip. *Nature* **2004**, *431*(7012), 1081.
3. Xu, Q.; Schmidt, B.; Pradhan, S.; Lipson, M. Micrometre-Scale Silicon Electro-Optic Modulator. *Nature* **2005**, *435*(7040), 325.
4. Cocorullo, G.; Rendina, I. Thermo-Optical Modulation at 1.5 μm in Silicon Etalon. *Electron. Lett.* **1992**, *28*(1), 83–85.
5. Petrosyan, D.; Malakyan, Y. P. Magneto-Optical Rotation and Cross-Phase Modulation via Coherently Driven Four-Level Atoms in a Tripod Configuration. *Phys. Rev. A* **2004**, *70*(2), 023822.
6. Berg, N. J.; Lee, J. N. Acousto-Optic Signal Processing: Theory and Implementation. *Optical Engineering* vol 2; Marcel Dekker, Inc.: New York, 1983, p 496.
7. Ferrari, A. C.; Bonaccorso, F.; Fal'Ko, V.; Novoselov, K. S.; Roche, S.; Bøggild, P.; Borini, S.; Koppens, F. H.; Palermo, V.; Pugno, N. Science and Technology Roadmap for Graphene, Related Two-Dimensional Crystals, and Hybrid Systems. *Nanoscale* **2015**, 7(11), 4598–4810.
8. Soref, R.; Bennett, B. Electrooptical Effects in Silicon. *IEEE J. Quantum Electron.* **1987**, *23*(1), 123–129.
9. Liu, M.; Yin, X.; Ulin-Avila, E.; Geng, B.; Zentgraf, T.; Ju, L.; Wang, F.; Zhang, X. A Graphene-Based Broadband Optical Modulator. *Nature* **2011**, *474*(7349), 64.

10. Miller, D. A. Energy Consumption in Optical Modulators for Interconnects. *Opt. Express* **2012**, *20*(102), A293–A308.
11. Li, W.; Chen, B.; Meng, C.; Fang, W.; Xiao, Y.; Li, X.; Hu, Z.; Xu, Y.; Tong, L.; Wang, H.; Liu, W.; Bao, J.; Shen, Y. R. Ultrafast All-Optical Graphene Modulator. *Nano Lett.* **2014**, *14*(2), 955–959. https://pubs.acs.org/doi/abs/10.1021/nl404356t.
12. Liu, Z.-B.; Feng, M.; Jiang, W.-S.; Xin, W.; Wang, P.; Sheng, Q.-W.; Liu, Y.-G.; Wang, D.; Zhou, W.-Y.; Tian, J.-G. Broadband All-Optical Modulation Using a Graphene-Covered-Microfiber. *Laser Phys. Lett.* **2013**, *10*(6), 065901.
13. Yu, S.; Wu, X.; Chen, K.; Chen, B.; Guo, X.; Dai, D.; Tong, L.; Liu, W.; Shen, Y. R. All-Optical Graphene Modulator Based on Optical Kerr Phase Shift. *Optica* **2016**, *3*(5), 541–544.
14. Weis, P.; Garcia-Pomar, J. L.; Höh, M.; Reinhard, B.; Brodyanski, A.; Rahm, M. Spectrally Wide-Band Terahertz Wave Modulator Based on Optically Tuned Graphene. *ACS Nano* **2012**, *6*(10), 9118–9124.
15. Wen, Q.-Y.; Tian, W.; Mao, Q.; Chen, Z.; Liu, W.-W.; Yang, Q.-H.; Sanderson, M.; Zhang, H.-W. Graphene Based All-Optical Spatial Terahertz Modulator. *Sci. Rep.* **2014**, *4*, 7409.
16. Bonaccorso, F.; Sun, Z.; Hasan, T.; Ferrari, A. Graphene Photonics and Optoelectronics. *Nat. Photonics* **2010**, *4*(9), 611.
17. Zhang, H.; Virally, S.; Bao, Q.; Kian Ping, L.; Massar, S.; Godbout, N.; Kockaert, P. Z-Scan Measurement of the Nonlinear Refractive Index of Graphene. *Opt. Lett.* **2012**, *37*(11), 1856–1858.
18. Chen, J.-H.; Zheng, B.-C.; Shao, G.-H.; Ge, S.-J.; Xu, F.; Lu, Y.-Q. An All-Optical Modulator Based on a Stereo Graphene–Microfiber Structure. *Light Sci. Appl.* **2015**, *4*(12), e360.
19. Malard, L. M.; Alencar, T. V.; Barboza, A. P. M.; Mak, K. F.; de Paula, A. M. Observation ofIntense Second Harmonic Generation From MoS_2 Atomic Crystals. *Phys. Rev. B* **2013**, *87*(20), 201401.
20. Zeng, H.; Liu, G.-B.; Dai, J.; Yan, Y.; Zhu, B.; He, R.; Xie, L.; Xu, S.; Chen, X.; Yao, W. Optical Signature of Symmetry Variations and Spin-Valley Coupling in Atomically Thin Tungsten Dichalcogenides. *Sci. Rep.* **2013**, *3*, 1608.
21. Seyler, K. L.; Schaibley, J. R.; Gong, P.; Rivera, P.; Jones, A. M.; Wu, S.; Yan, J.; Mandrus, D. G.; Yao, W.; Xu, X. Electrical Control of Second-Harmonic Generation in a WSe2 Monolayer Transistor. *Nat. Nanotechnol.* **2015**, *10*(5), 407.
22. Espinola, R.; Tsai, M.; Yardley, J. T.; Osgood, R. Fast and Low-Power Thermooptic Switch on Thin Silicon-On-Insulator. *IEEE Photon. Technol. Lett.* **2003**, *15*(10), 1366–1368.
23. Dong, P.; Qian, W.; Liang, H.; Shafiiha, R.; Feng, D.; Li, G.; Cunningham, J. E.; Krishnamoorthy, A. V.; Asghari, M. Thermally Tunable Silicon Racetrack Resonators With Ultralow Tuning Power. *Opt. Express* **2010**, *18*(19), 20298–20304.
24. Dong, P.; Qian, W.; Liang, H.; Shafiiha, R.; Feng, N.-N.; Feng, D.; Zheng, X.; Krishnamoorthy, A. V.; Asghari, M. Low Power and Compact Reconfigurable Multiplexing Devices Based on Silicon Microring Resonators. *Opt. Express* **2010**, *18*(10), 9852–9858.
25. Gan, X.; Zhao, C.; Wang, Y.; Mao, D.; Fang, L.; Han, L.; Zhao, J. Graphene-Assisted All-Fiber Phase Shifter and Switching. *Optica* **2015**, *2*(5), 468–471.
26. Kim, J. T.; Chung, K. H.; Choi, C.-G. Thermo-Optic Mode Extinction Modulator Based on Graphene Plasmonic Waveguide. *Opt. Express* **2013**, *21*(13), 15280–15286.
27. Gan, S.; Cheng, C.; Zhan, Y.; Huang, B.; Gan, X.; Li, S.; Lin, S.; Li, X.; Zhao, J.; Chen, H. A Highly Efficient Thermo-Optic Microring Modulator Assisted by Graphene. *Nanoscale* **2015**, 7(47), 20249–20255.

28. Yu, L.; Yin, Y.; Shi, Y.; Dai, D.; He, S. Thermally Tunable Silicon Photonic Microdisk Resonator With Transparent Graphene Nanoheaters. *Optica* **2016**, *3*(2), 159–166.
29. Yan, S.; Zhu, X.; Frandsen, L. H.; Xiao, S.; Mortensen, N. A.; Dong, J.; Ding, Y. Slow-Light-Enhanced Energy Efficiency for Graphene Microheaters on Silicon Photonic Crystal Waveguides. *Nat. Commun.* **2017**, *8,* 14411.
30. Xu, Z.; Qiu, C.; Yang, Y.; Zhu, Q.; Jiang, X.; Zhang, Y.; Gao, W.; Su, Y. Ultra-Compact Tunable Silicon Nanobeam Cavity With an Energy-Efficient Graphene Micro-Heater. *Opt. Express* **2017**, *25*(16), 19479–19486.
31. Yu, L.; Dai, D.; He, S. Graphene-Based Transparent Flexible Heat Conductor for Thermally Tuning Nanophotonic Integrated Devices. *Appl. Phys. Lett.* **2014**, *105*(25), 251104.
32. Li, Z.; Henriksen, E. A.; Jiang, Z.; Hao, Z.; Martin, M. C.; Kim, P.; Stormer, H.; Basov, D. N. Dirac Charge Dynamics in Graphene by Infrared Spectroscopy. *Nat. Phys.* **2008**, *4*(7), 532.
33. Wang, F.; Zhang, Y.; Tian, C.; Girit, C.; Zettl, A.; Crommie, M.; Shen, Y. R. Gate-Variable Optical Transitions in Graphene. *Science* **2008**, *320*(5873), 206–209.
34. Xu, F.; Das, S.; Gong, Y.; Liu, Q.; Chien, H.-C.; Chiu, H.-Y.; Wu, J.; Hui, R. Complex Refractive Index Tunability of Graphene at 1550 nm Wavelength. *Appl. Phys. Lett.* **2015**, *106*(3), 031109.
35. Nair, R. R.; Blake, P.; Grigorenko, A. N.; Novoselov, K. S.; Booth, T. J.; Stauber, T.; Peres, N. M.; Geim, A. K. Fine Structure Constant Defines Visual Transparency of Graphene. *Science* **2008**, *320*(5881), 1308.
36. Mak, K. F.; Sfeir, M. Y.; Wu, Y.; Lui, C. H.; Misewich, J. A.; Heinz, T. F. Measurement of the Optical Conductivity of Graphene. *Phys. Rev. Lett.* **2008**, *101*(19), 196405.
37. Liu, M.; Yin, X.; Zhang, X. Double-Layer Graphene Optical Modulator. *Nano Lett.* **2012**, *12*(3), 1482–1485.
38. Mohsin, M.; Neumaier, D.; Schall, D.; Otto, M.; Matheisen, C.; Giesecke, A. L.; Sagade, A. A.; Kurz, H. Experimental Verification of Electro-Refractive Phase Modulation in Graphene. *Sci. Rep.* **2015**, *5,* 10967.
39. Gan, X.; Shiue, R.-J.; Gao, Y.; Mak, K. F.; Yao, X.; Li, L.; Szep, A.; Walker, D., Jr.; Hone, J.; Heinz, T. F. High-Contrast Electrooptic Modulation of a Photonic Crystal Nanocavity by Electrical Gating of Graphene. *Nano Lett.* **2013**, *13*(2), 691–696.
40. Gao, Y.; Shiue, R.-J.; Gan, X.; Li, L.; Peng, C.; Meric, I.; Wang, L.; Szep, A.; Walker, D., Jr.; Hone, J. High-Speed Electro-Optic Modulator Integrated With Graphene-Boron Nitride Heterostructure and Photonic Crystal Nanocavity. *Nano Lett.* **2015**, *15*(3), 2001–2005.
41. Koester, S. J.; Li, M. High-Speed Waveguide-Coupled Graphene-On-Graphene Optical Modulators. *Appl. Phys. Lett.* **2012**, *100*(17), 171107.
42. Sensale-Rodriguez, B.; Yan, R.; Zhu, M.; Jena, D.; Liu, L.; Grace Xing, H. Efficient Terahertz Electro-Absorption Modulation Employing Graphene Plasmonic Structures. *Appl. Phys. Lett.* **2012**, *101*(26), 261115.
43. Qiu, C.; Gao, W.; Vajtai, R.; Ajayan, P. M.; Kono, J.; Xu, Q. Efficient Modulation of 1.55 μm Radiation With Gated Graphene on a Silicon Microring Resonator. *Nano Lett.* **2014**, *14*(12), 6811–6815.
44. Ding, Y.; Zhu, X.; Xiao, S.; Hu, H.; Frandsen, L. H.; Mortensen, N. A.; Yvind, K. Effective Electro-Optical Modulation With High Extinction Ratio by a Graphene–Silicon Microring Resonator. *Nano Lett.* **2015**, *15*(7), 4393–4400.
45. Phare, C. T.; Lee, Y.-H. D.; Cardenas, J.; Lipson, M. Graphene Electro-Optic Modulator With 30 GHz Bandwidth. *Nat. Photonics* **2015**, *9*(8), 511.
46. Shah, M. K.; Ye, S.-W.; Zou, X.-H.; Yuan, F.; Jha, A.; Zhang, Y.; Lu, R.-G.; Liu, Y. Graphene-Assisted Electroabsorption Optical Modulator Using D-Microfiber. *IEEE J. Sel. Top. Quantum Electron.* **2017**, *23*(1), 89–93.

47. Dalir, H.; Xia, Y.; Wang, Y.; Zhang, X. Athermal Broadband Graphene Optical Modulator With 35 GHz Speed. *ACS Photonics* **2016**, *3*(9), 1564–1568.
48. Kim, K.; Choi, J.-Y.; Kim, T.; Cho, S.-H.; Chung, H.-J. A Role for Graphene in Silicon-Based Semiconductor Devices. *Nature* **2011**, *479*(7373), 338.
49. Gosciniak, J.; Tan, D. T. Theoretical Investigation of Graphene-Based Photonic Modulators. *Sci. Rep.* **2013**, *3*, 1897.
50. Ye, S.; Wang, Z.; Tang, L.; Zhang, Y.; Lu, R.; Liu, Y. Electro-Absorption Optical Modulator Using Dual-Graphene-On-Graphene Configuration. *Opt. Express* **2014**, *22*(21), 26173–26180.
51. Sensale-Rodriguez, B.; Yan, R.; Kelly, M. M.; Fang, T.; Tahy, K.; Hwang, W. S.; Jena, D.; Liu, L.; Xing, H. G. Broadband Graphene Terahertz Modulators Enabled by Intraband Transitions. *Nat. Commun.* **2012**, *3*, 780.
52. Dawlaty, J. M.; Shivaraman, S.; Strait, J.; George, P.; Chandrashekhar, M.; Rana, F.; Spencer, M. G.; Veksler, D.; Chen, Y. Measurement of the Optical Absorption Spectra of Epitaxial Graphene From Terahertz to Visible. *Appl. Phys. Lett.* **2008**, *93*(13), 131905.
53. Winnerl, S.; Orlita, M.; Plochocka, P.; Kossacki, P.; Potemski, M.; Winzer, T.; Malic, E.; Knorr, A.; Sprinkle, M.; Berger, C. Carrier Relaxation in Epitaxial Graphene Photoexcited Near the Dirac Point. *Phys. Rev. Lett.* **2011**, *107*(23), 237401.
54. Maeng, I.; Lim, S.; Chae, S. J.; Lee, Y. H.; Choi, H.; Son, J.-H. Gate-Controlled Nonlinear Conductivity of Dirac Fermion in Graphene Field-Effect Transistors Measured by Terahertz Time-Domain Spectroscopy. *Nano Lett.* **2012**, *12*(2), 551–555.
55. Sensale-Rodriguez, B.; Yan, R.; Rafique, S.; Zhu, M.; Li, W.; Liang, X.; Gundlach, D.; Protasenko, V.; Kelly, M. M.; Jena, D. Extraordinary Control of Terahertz Beam Reflectance in Graphene Electro-Absorption Modulators. *Nano Lett.* **2012**, *12*(9), 4518–4522.
56. Liang, G.; Hu, X.; Yu, X.; Shen, Y.; Li, L. H.; Davies, A. G.; Linfield, E. H.; Liang, H. K.; Zhang, Y.; Yu, S. F. Integrated Terahertz Graphene Modulator With 100% Modulation Depth. *ACS Photonics* **2015**, *2*(11), 1559–1566.
57. Ansell, D.; Radko, I.; Han, Z.; Rodriguez, F.; Bozhevolnyi, S.; Grigorenko, A. Hybrid Graphene Plasmonic Waveguide Modulators. *Nat. Commun.* **2015**, *6*, 8846.
58. Ju, L.; Geng, B.; Horng, J.; Girit, C.; Martin, M.; Hao, Z.; Bechtel, H. A.; Liang, X.; Zettl, A.; Shen, Y. R. Graphene Plasmonics for Tunable Terahertz Metamaterials. *Nat. Nanotechnol.* **2011**, *6*(10), 630.
59. Yan, H.; Li, X.; Chandra, B.; Tulevski, G.; Wu, Y.; Freitag, M.; Zhu, W.; Avouris, P.; Xia, F. Tunable Infrared Plasmonic Devices Using Graphene/Insulator Stacks. *Nat. Nanotechnol.* **2012**, 7(5), 330.
60. Kim, J.; Son, H.; Cho, D. J.; Geng, B.; Regan, W.; Shi, S.; Kim, K.; Zettl, A.; Shen, Y.-R.; Wang, F. Electrical Control of Optical Plasmon Resonance With Graphene. *Nano Lett.* **2012**, *12*(11), 5598–5602.
61. Papasimakis, N.; Luo, Z.; Shen, Z. X.; De Angelis, F.; Di Fabrizio, E.; Nikolaenko, A. E.; Zheludev, N. I. Graphene in a Photonic Metamaterial. *Opt. Express* **2010**, *18*(8), 8353–8359.
62. Yao, Y.; Shankar, R.; Kats, M. A.; Song, Y.; Kong, J.; Loncar, M.; Capasso, F. Electrically Tunable Metasurface Perfect Absorbers for Ultrathin Mid-Infrared Optical Modulators. *Nano Lett.* **2014**, *14*(11), 6526–6532.
63. Lee, S. H.; Choi, M.; Kim, T.-T.; Lee, S.; Liu, M.; Yin, X.; Choi, H. K.; Lee, S. S.; Choi, C.-G.; Choi, S.-Y. Switching Terahertz Waves With Gate-Controlled Active Graphene Metamaterials. *Nat. Mater.* **2012**, *11*(11), 936.
64. Shimano, R.; Yumoto, G.; Yoo, J.; Matsunaga, R.; Tanabe, S.; Hibino, H.; Morimoto, T.; Aoki, H. Quantum Faraday and Kerr Rotations in Graphene. *Nat. Commun.* **2013**, *4*, 1841.

65. Crassee, I.; Levallois, J.; Walter, A. L.; Ostler, M.; Bostwick, A.; Rotenberg, E.; Seyller, T.; Van Der Marel, D.; Kuzmenko, A. B. Giant Faraday Rotation in Single- and Multilayer Graphene. *Nat. Phys.* **2011**, 7(1), 48.
66. Thalmeier, P.; Dóra, B.; Ziegler, K. Surface Acoustic Wave Propagation in Graphene. *Phys. Rev. B* **2010**, *81*(4), 041409.
67. Preciado, E.; Schülein, F. J.; Nguyen, A. E.; Barroso, D.; Isarraraz, M.; von Son, G.; Lu, I.-H.; Michailow, W.; Möller, B.; Klee, V. Scalable Fabrication of a Hybrid Field-Effect and Acousto-Electric Device by Direct Growth of Monolayer MoS_2/$LiNbO_3$. *Nat. Commun.* **2015**, *6*, 8593.

CHAPTER 3

2D Materials for laser applications

Xinxin Jin, Meng Zhang
School of Electronic and Information Engineering, Beihang University, Beijing, China

3.1 Background

Pulsed lasers have become an essential tool across a variety of scientific and engineering applications, ranging from ultrafast phenomena research to biomedical imaging, metrology, and laser micromachining. Of the many available schemes for short and stable pulse generation, passive Q-switching and mode-locking techniques are highly preferred as these avoid the employment of complex and costly active electrical modulators. While Q-switching modulates the quality factor of a laser cavity to produce low-repetition-rate (kHz) pulses with microsecond-nanosecond pulse duration, mode-locking could synchronize the phases of the longitudinal modes of the cavity and enable much higher repetition rates (MHz-GHz) with short pulse durations down to fs scale. Both Q-switching and mode-locking technology require an indispensable component called saturable absorber (SA), which exhibits a nonlinear decreased absorption with increasing light intensity and thus transforms a continuous wave into pulses. The saturable absorption mechanism is generally based on Pauli exclusion principles, according to which the electrons in the fully filled, upper excited states of materials would block further absorption of optoelectrons. An ideal SA should have these properties: fast response time, broad wavelength range, low optical loss, and high damage threshold. In the early years since the invention of lasers in 1960, organic dyes and colored glass were initially developed as SAs. However, these SAs suffered from inconvenient collimation, low optical damage, and high losses, restricting their wider application for stable pulse generation. With the advent of semiconductor saturable absorber mirror (SESAM) in the early 1990s, the SA field made an enormous progress. Since its invention, SESAM soon grew to be the most commercialized device for pulse generation in either fiber lasers or solid-state lasers. However, there are still some drawbacks for SESAM. First, the specific energy bandgap in semiconductors results in a narrow operation bandwidth at the designed wavelength. Second, costly fabrication and packing are inevitable in order to customize the key optical parameters of SESAM, e.g.,

2D Materials for Photonic and Optoelectronic Applications
https://doi.org/10.1016/B978-0-08-102637-3.00003-6

carrier relaxation times. These limitations propel continual research, with the intention to seek novel SA materials that outperform SESAMs. In recent years, intense research on two-dimensional (2D) materials greatly promoted the development of 2D SA. The 2D SA research interest was first triggered by the successful demonstration of a graphene mode locker in a fiber laser in 2009,[1] and then inspired by the subsequent 2D SA members, e.g., topological insulators, semiconducting transition metal dichalcogenides (TMDs), and black phosphorus (BP). 2D materials offer promising advantages such as strong intensity-dependent absorption, broadband working wavelength range, ultrafast carrier dynamics, in addition to the cost-efficient and convenient fabrication, making them comparable or even superior to SESAM.

In this chapter, we restrict the focus to the two classes of novel 2D materials that are most widely studied: TMDs and BP. Integration strategies of 2D SA devices and their nonlinear absorption characterizations are summarized. Significant applications in pulsed fiber lasers and solid-state lasers are reviewed. We also give an outlook on future directions of research.

3.2 2D SA incorporation and nonlinear absorption characterization

There are various methods to fabricate 2D materials such as mechanical exfoliation, liquid-phase exfoliation, and chemical vapor deposition (CVD), as mentioned in Chapter 1. For the produced 2D materials, further integration is required in order to be suitably incorporated into lasers. The most commonly exploited method for 2D materials to be integrated as practical SA devices is shown in Fig. 3.1.

For those applied in fiber lasers, 2D materials could be transferred on a fiber connector end facet (Fig. 3.1A), tapered fiber (Fig. 3.1B), side-polished

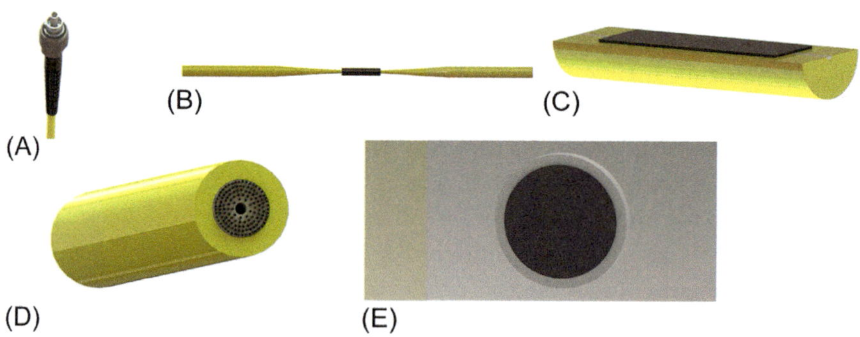

Fig. 3.1 Platform for 2D materials to be integrated as practical SA devices. (A) Fiber end facet. (B) Side-polished fiber. (C) Tapered fiber. (D) Photonic crystal fiber. (E) Substrate for free-space coupling.

fiber (Fig. 3.1C), or filled into the hollow channels of a photonic crystal fiber (PCF) (Fig. 3.1D). The fiber connector end facet scheme is superior in simple integration and flexible handle. Side-polished fiber and tapered fiber scheme have the advantage of high power tolerance as only part of the light (i.e., evanescent field) interacts with the 2D materials, in addition to the stronger nonlinearity as the interaction length is far increased. The PCF scheme also shows high power handling and long light-matter interaction length, but low coupling efficiency remains a problem. For those schemes applied in solid-state lasers, 2D materials could be transferred onto substrates such as quartz glass plate or mirrors (Fig. 3.1E), where the interaction between the 2D materials and light is enabled by free-space coupling through the transmission or reflection approach. Besides, different strategies have been developed to transfer 2D materials on the aforementioned platforms. The selection of a proper transfer strategy depends on the production method employed to exfoliate few-layer 2D materials.

For the mechanical exfoliation technique, few-layer flakes could be transferred on the fiber connector end facet by pushing down a fiber end to the scotch tape adhered with peeled few-layer flakes, and the fiber core would be covered by 2D materials due to the adhesive force between flakes and ceramic. This method utilizes pure 2D materials such as SA; however, it could be time-consuming. For the solution-based exfoliation technique (e.g., liquid-phase exfoliation and lithium-ion intercalation exfoliation), the resulting flakes could be embedded in polymer thin film with scalability, and then adhered to the fiber end or side-polished fiber using an index matching gel. Although the polymer thin film is compact and flexible for transfer, its low thermal damage threshold restricts their further applications where high pulse energy is required. In addition, inkjet printing is emerging as an attractive strategy based on solution exfoliation nanomaterials. Ink containing 2D material flakes could be printed onto suitable substrates, e.g., PET. This method is scalable and highly controllable, enabling stable and reproducible SA fabrication. Another common transfer strategy for solution exfoliated nanomaterial is optical-driven deposition. The fiber end facet or microfiber is first immersed in solutions and then injected with intense light, which introduces thermal gradient in the solution that facilitates few-layer flakes to be attracted to the substrate. For few-layer 2D materials fabricated by CVD, these could be transferred through few steps. First, the materials are coated with polymer. Second, the original substrate is removed by chemical etching. Then the 2D material-polymer thin film is transferred onto the target substrate. Last, the polymer is removed by acetone.

The nonlinear absorption of 2D SAs could be characterized experimentally by two types of measurements, i.e., Z-scan and balanced twin-detector

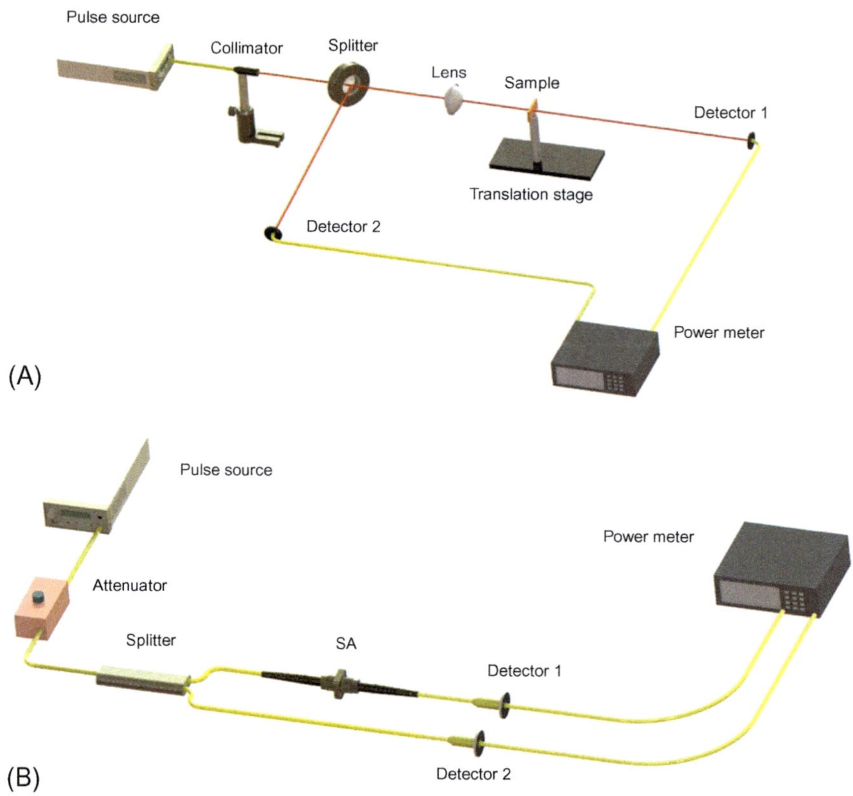

Fig. 3.2 Nonlinear absorption characterization setups. (A) Z-scan measurement. (B) Balanced twin-detector measurement.

measurements. The Z-scan measurement is implemented to characterize the free-space-type SA sample. As shown in Fig. 3.2A, the pulsed light from the pumping source is first split into two beams by the splitter. The beam propagating along the path incorporated with the SA sample functions as measurement beam; the other beam functions as reference beam. A lens focuses the measurement beam to the SA sample that is mounted on a *Z*-direction translation stage. By moving the sample in the *z*-axis along the measurement beam, light intensity per unit area of the sample could be changed due to different beam sizes. The power from both paths is collected by the dual-channel power meter. The nonlinear absorption of the sample can be obtained by comparing the power from the two detectors. For SAs integrated as the fiber-based device, balanced twin-detector measurement is preferred for nonlinear absorption characterization. The principle is quite similar to Z-scan, other than the setup is fully fiberized (Fig. 3.2B). The light from a pulsed laser is separated into measurement beam and reference beam by an optical coupler. Distinct from the Z-scan method, the variation of light intensity in twin-detector measurement is achieved by a variable optical attenuator placed before the optical coupler. Again, the power of both beams is measured by the dual-channel power meter and the absorption of the SA versus incident power could be obtained.

The critical parameters of SA deduced from the measurement are modulation depth (α_0), saturation intensity (I_{sat}), and nonsaturable losses (α_{ns}). The nonlinear absorption could thus be described by a two-level SA model:

$$\alpha(I) = \alpha_0/(1 + I/I_{sat}) + \alpha_{ns}$$

3.3 TMDs in pulsed lasers

TMDs possess a layer count-dependent optoelectronic property. The bandgap of molybdenum disulfide (MoS_2) varies from an indirect gap of ~1.29 eV (961 mm) in bulk to an increased direct gap of 1.80 eV (689 nm) in a monolayer. Similarly, molybdenum diselenide ($MoSe_2$) presents a bandgap ranging from 1.1 eV (1128 nm) in bulk to 1.57 eV (790 nm) in a monolayer; tungsten disulfide (WS_2) shows a bandgap ranging from 1.3 eV (954 nm) in bulk to 2.1 eV (590 nm) in a monolayer; tungsten diselenide (WSe_2) shows a bandgap ranging from 1.2 eV (1034 nm) in bulk to 1.65 eV (752 nm) in a monolayer.

As monolayer TMDs exhibit a direct bandgap corresponding to visible frequencies, prospects for their photonic and optoelectronic application in the visible spectral region are exciting. In addition, the introduction of edge state or the crystallographic defect in the material facilitates the absorption of photons with energies lower than the material bandgap, i.e., sub-bandgap absorption.[2] By exploiting the sub-bandgap absorption effect, TMD-based SA could thus support pulsed laser in the near-infrared and mid-infrared wavelengths where photon energies are lower than the expected material bandgap.

3.3.1 Mode-locked laser based on TMDs

Table 3.1 lists the performance of different TMD-SA mode-locked lasers when these were first reported at respective wavelength regions. As a representative, layered MoS_2 is the first member in TMDs that has been characterized with nonlinear optical properties, exhibiting more significant SA effect than graphene.[20] Thereafter, MoS_2 became the mostly investigated material in TMDs in the laser community. Fiber laser, with the assistance of well-developed fiber systems and the advantage of high integration, has always been a preferential platform for 2D material mode-locking exploration. The first demonstration of an MoS_2 mode locker was achieved in an ytterbium (Yb)-doped fiber laser,[3] delivering 800 ps dissipative soliton pulse

Table 3.1 Performance of mode-locked lasers based on TMD SAs when these were first reported at respective wavelength regions.

2D materials	Gain medium	Wavelength regime (center wavelength)	3-dB bandwidth	Pulse width	Integration platform	Time reported [Ref.]
MoS_2	Yb-doped fiber	1 μm (1054.3 nm)	2.7 nm	800 ps	Fiber facet	2014.3[3]
	Er-doped fiber	1.5 μm (1568.9 nm)	2.6 nm	1.28 ps	Fiber facet	2014.7[4]
	Tm-doped fiber	2 μm (1905 nm)	17.3 nm	843 ps	Mirror	2015.6[5]
	Nd:$GdVO_4$ crystal	1 μm	–	10.6 ps	Mirror	2015.9[6]
	Pr:$GdLiF_4$ crystal	Visible			Quartz glass	2017.2[7]
		522 nm	0.09 nm	46 ps		
		607 nm	0.16 nm	30 ps		
		639 nm	0.20 nm	55 ps		
WS_2	Yb-doped fiber	1 μm (1057.5 nm)	14.5 nm	21.1 ps	Side-polished fiber	2015.10[8]
	Er-doped fiber	1.5 μm (1557 nm)	2.3 nm	1.32 ps	Side-polished fiber	2015.1[9]
	Tm-doped fiber	2 μm (1941 nm)	5.6 nm	1.3 ps	Side-polished fiber	2015.7[10]
	Nd:YVO_4 crystal	1 μm (–)	–	8.6 ps	Mirror	2015.9[6]

$MoSe_2$	Yb-doped fiber	1 μm (1040 nm)	4.26 nm	471 ps	Fiber facet	2016.8[11]
	Er-doped fiber	1.5 μm (1558 nm)	1.76 nm	1.45 ps	Fiber facet	2015.6[12]
	Tm/Ho-doped fiber	2 μm (1912 nm)	4.62 nm	920 fs	Side-polished fiber	2017.8[13]
WSe_2	Er-doped fiber	1.5 μm (1556.7 nm)	2 nm	1.31 ps	Side-polished fiber	2016.3[14]
	Tm-doped fiber	2 μm (1886.22 nm)	3.29 nm	1.18 ps	Tapered fiber	2017.11[15]
$MoTe_2$	Er-doped fiber	1.5 μm (1561 nm)	2.4 nm	1.2 ps	Side-polished fiber	2016.3[16]
	Tm-doped fiber	2 μm (1930.22 nm)	4.45	952 fs	Tapered fiber	2018.5[17]
WTe_2	Er-doped fiber	1.5 μm (–)	–	–	Side-polished fiber	2016.3[16]
	Tm-doped fiber	2 μm (1915.5 nm)	3.13 nm	1.25 ps	Tapered fiber	2017.12[18]
TiS_2	Er-doped fiber	1.5 μm (1569.5 nm)	2.63 nm	1.48 ps	Fiber facet	2018.2[19]

at 1054 nm. Stimulated by the initial work at 1 μm, research interests of MoS_2 rapidly expanded to longer-wavelength from near-infrared to mid-infrared regions. Xia et al. and Liu et al. reported the generation of mode-locked pulse at the telecommunication waveband.[4, 21] By managing intracavity dispersion, an MoS_2 mode-locked erbium (Er)-doped fiber laser with a bandwidth of 20.5 nm and a pulse duration as short as 200 fs was demonstrated.[22] In addition, a wideband tunable (1535–1565 nm) ultrafast mode-locked Er-doped fiber laser cavity was demonstrated using a free-standing few-layer MoS_2–polymer composite,[23] as shown in Fig. 3.3. It was believed that the tunable wavelength could be further extended provided that a filter with larger tunable range is employed. The first presence of mode-locking with MoS_2 in 2 μm was realized in a compact, linear, thulium (Tm)-doped fiber laser,[5] and the output pulse energy of 15.5 nJ represents the highest among MoS_2 SA mode-locked fiber lasers to date. These applications of MoS_2 SA, either at discrete wavelengths or with wideband tunability, suggest that MoS_2 can be a broadband SA.

In addition to the broadband saturable absorption effect, large third-order nonlinear susceptibilities are another attractive merit of the few-layer MoS_2.[24] For mode-locked fiber lasers, the large third-order nonlinear susceptibilities combined with the nonlinear absorption effect render MoS_2-SA a promising device for producing high-repetition-rate pulses, which would have important applications in astronomical frequency combs and fiber communication systems. By taking advantage of the highly nonlinear properties of MoS_2, the 369th harmonic mode-locking of Er-doped fiber lasers produced pulses with a repetition rate as high as 2.5 GHz.[25] Wu et al. reported the fundamental mode-locking operation of a linear Er-doped fiber laser with a repetition rate of 463 MHz based on MoS_2-PVA SA, and the mode-locking provided an extinction ratio of up to 97 dB in the repetition frequency spectrum, much purer than that of harmonic mode-locking.[26]

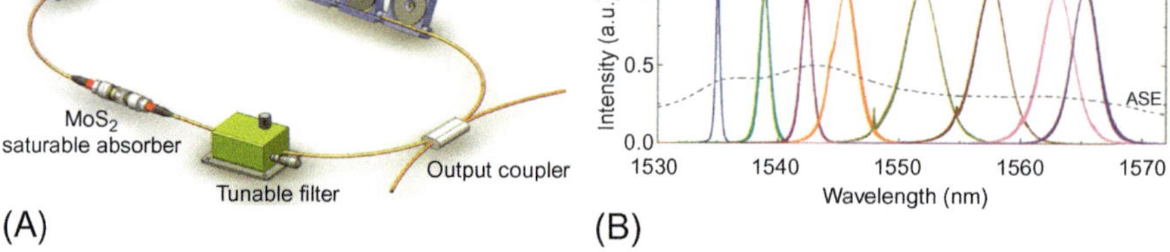

Fig. 3.3 (A) Schematic of the wideband tunable ultrafast laser. EDFA, Er-doped fiber amplifier. (B) Tunable output spectra at eight representative wavelengths. Reproduced under the terms of a Creative Commons Attribution License.[23]

Apart from conventional soliton, a growing number of researchers explored more various soliton patterns mode-locked by MoS_2 SA, such as bound soliton, soliton molecules, localized chaotic multipulses, and double-scale soliton clusters. Although the diverse soliton patterns in fiber lasers are independent of what type of mode locker is inserted, a research indicated that the large nonlinear refractive index offered by MoS_2 could reduce the pump threshold for versatile pattern generation.[27]

While fruitful achievements based on MoS_2 SA have been made, there remain issues, i.e., low optical damage threshold due to their poor thermal and mechanical property. A number of works have attempted to circumvent these shortcomings. The lateral interaction scheme, enabled by evanescent field interacting with materials deposited/transferred on side-polished fiber or tapered fiber, was employed to promote the optical damage threshold of MoS_2. Compared to perpendicular interaction where incident light is focused to a size of micrometer on the surface of MoS_2 and therefore result in poor heat dissipation, the lateral interaction scheme could bear with much larger power. A MoS_2-taper fiber device that can withstand a strong laser illumination up to 1 W without optical damage was reported by Du et al.[28] However, despite the high power tolerance, the output power was still limited, mainly caused by the high insertion loss of the tapered fiber. Research is ongoing, with the aim to improve the output power level of mode-locked pulses.

In the solid laser field, the MoS_2 SA mode-locking progress lags a little behind their fiber competitor; this is mainly ascribed to the large nonsaturable loss of MoS_2. The first demonstration is achieved in a $Nd:GdVO_4$ laser mode-locked by MoS_2-SAM, producing 10.6-ps pulses. It is interesting to note that by utilizing the direct bandgap of MoS_2, a mode-locked Pr:GLF visible laser was demonstrated,[7] as shown in Fig. 3.4. The work significantly boosts the prospect of TMD-based ultrafast photonics in the visible range, which have extensive applications and requirements in various topical ranges, including medicine, biomedical images, microscopy, scientific research, and our daily life.

Similar to MoS_2, WS_2 is another TMD material that has attracted tremendous attention for ultrafast applications. Broadband saturable absorption effect ranging from visible wavelength to infrared wavelength has been confirmed in fiber lasers and solid-state lasers, as shown in Table 3.1. Various mode-locking phenomena, such as harmonic mode-locking, dark soliton, dual-wavelength soliton, and dissipative soliton, are routinely reported. A hybrid mode-locked scheme based on nonlinear polarization evolution and WS_2 SA technique produced a mode-locked pulse with a pulse duration as short as 67 fs and 3 dB spectral width of 114 nm.[29] WS_2 has a higher thermal conductivity ($2.2\ Wm^{-1}\ K^{-1}$) compared with MoS_2 ($1.05\ Wm^{-1}\ K^{-1}$),

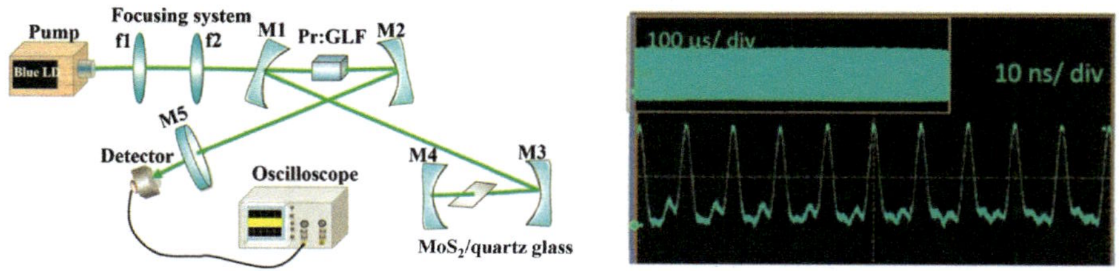

(A) (B)

Fig. 3.4 (A) Setup of the visible mode-locked laser. (B) Mode-locked pulse train recorded in 10 ns/div (inset: in 100 µs/div). *Reproduced with permission. Zhang, Y. X.; Yu, H. H.; Zhang, R.; Zhao, G.; Zhang, H. J.; Chen, Y. X.; Mei, L. M.; Tonelli, M.; Wang, J. Y. Broadband Atomic-Layer MoS2 Optical Modulators for Ultrafast Pulse Generations in the Visible Range. Opt. Lett. 2017, 42, 547-550.Y. X. Zhang, H. H. Yu, R. Zhang, G. Zhao, H. J. Zhang, Y. X. Chen, L. M. Mei, M. Tonelli, and J. Y. Wang, "Broadband atomic-layer MoS2 optical modulators for ultrafast pulse generations in the visible range," Opt. Lett. 42, 547-550 (2017).*

implying a better thermal stability; thus, this triggered a series of studies on scaling up of output power. A WS_2 SA based on a fluorine mica (FM) substrate with a high damage threshold (406 MW/cm^2) was proposed. By incorporating WS_2 FM-SA into an Yb-doped fiber laser cavity, mode-locking operation with 30 mW average output power was obtained.[30] In another study, a WS_2 film grown by CVD was directly transferred to the fiber facet to avoid being affected by organic host polymer materials. The WS_2 film was then embedded into an Er-doped fiber laser system and an average output power as high as 62.5 mW was obtained, which is the highest in fiber lasers based on TMD materials.[31]

Encouraged by the performance of MoS_2 and WS_2, other members in layered TMD families, including $MoSe_2$, WSe_2, $MoTe_2$, WTe_2, TiS_2, ReS_2, and $TiSe_2$, have also been successively exploited as mode lockers. Although the number of reports on these members is now lesser than those of MoS_2 and WS_2, significant outcomes are achieved. For instance, an $MoSe_2$ SA-based fiber laser was utilized for the production of femtosecond pulses at a repetition rate of 3.27 GHz,[32] which is the highest frequency that has ever been demonstrated from a passively mode-locked femtosecond fiber laser. An $MoTe_2$ SA mode-locked, Tm-doped fiber laser with 212 mW average power was also demonstrated.[33] This is the highest in Tm-doped fiber lasers based on 2D materials. Ongoing research on mode lockers, whether based on the well-developed MoS_2 and WS_2 or emerging TMDs, should yield more exciting results.

3.3.2 Q-switched lasers based on TMDs

Q-switched lasers produce low-repetition-rate (kHz) pulses with μs-ns pulse duration and normally larger pulse energy than those from mode-locked lasers. TMDs have been extensively exploited for both Q-switched fiber lasers and solid-state lasers. Table 3.2 lists the performance of TMD Q-switched lasers when first reported and at their respective wavelength regions. Initial work was done in the near-infrared and mid-infrared wavelengths by Wang et al..[37] The authors reported on a broadband MoS_2 modulator prepared by pulsed laser deposition, which Q-switched a Nd:$GdVO_4$ solid-state laser at 1.06 μm, a Nd:YGG solid-state laser at 1.42 μm, and a Tm:Ho:YGG solid-state laser at 2.1 μm, respectively. Subsequently, an MoS_2 modulator was incorporated into solid-state lasers based on a variety of bulk gain medium, including Nd:YAP, Tm:CLNGG, Yb:LGGG, Yb: GAB, Nd:YVO_4, Er:LuAG, Nd:YVO_4, and Nd:LuAG, covering the broad spectrum range from 1 to 2.8 μm.

Table 3.2 Performance of Q-switched lasers based on TMD SAs when they were first reported at each respective wavelength regions.

2D materials	Gain medium	Wavelength regime (center wavelength)	Pulse width	Max. output power	Integration platform	Time reported [Ref.]
MoS_2	Yb-doped fiber	1 μm (1066.5 nm)	5.8 μs	0.9 mW	Fiber facet	2014.12[34]
	Er-doped fiber	1.5 μm (1519.6–1567.7 nm)	3.3 μs	5.91 mW	Fiber facet	2014.10[35]
	Tm-doped fiber	2 μm (2032 nm)	1.76 μs	47.3 mW	Fiber facet	2014.12[34]
	Pr-doped fiber	Visible (635 nm)	227 ns	7.1 mW	Fiber facet	2015.11[36]
	Nd:GDdVO4 crystal	1 μm (1.06 μm)	970 ns	227 mW	Quartz glass	2014.6[37]
	Nd:YGG crystal	1.5 μm (1.42 μm)	729 ns	52 mW	Quartz glass	2014.6[37]
	Tm:Ho:YGG crystal	2 μm (2.1 μm)	410 ns	206 mW	Quartz glass	2014.6[37]
	Er:Lu_2O_3	3 μm (2.84 μm)	335 ns	1.03 W	YAG substrate	2016.02[38]
	Pr:GdLiF4 crystal	Visible			Quartz glass	2015.6[39]
		605 nm	278 ns	17 mW		
		639 nm	403 ns	9 mW		
		721 nm	382 ns	8 mW		
WS_2	Yb-doped fiber	1 μm (1207–1065 nm)	110 ns	–	Fiber facet	2015.11[40]
	Er-doped fiber	1.5 μm (1567.8 nm)	0.71 μs	2.5 mW	Side-polished fiber	2015.2[41]
	Ho/Pr co-doped fluoride fiber	3 μm (2865.7 nm)	1.73 μs	48.4 mW	Mirror	2016.10[42]
	Pr-doped fiber	Visible (635 nm)	207 ns	8.7 mW	Fiber facet	2015.11[36]
	Nd:GdVO4 crystal	1 μm (–)	110 ns	–	Crystal	2015.9[6]

	Er:YAG crystal	1.6 μm (1645 nm)	1.8 μs	420 mW	YAG substrate	2018.1[43]
	Tm:LuAG crystal	2 μm (2012.9 nm)	660 ns	1.08 W	Quartz substrate	2016.8[44]
	Pr: YLF crystal	Visible (640 nm)	630 ns	21.5 mW	BK7 glass plate	2016.6[45]
$MoSe_2$	Yb-doped fiber	1 μm (1060 nm)	2.8 μs	8.72 mW	Fiber facet	2015.7[46]
	Er-doped fiber	1.5 μm (1566 nm)	4.8 μs	29.2 mW	Fiber facet	2015.7[46]
	Tm-doped fiber	2 μm (1942 nm)	5.5 μs	0.13 mW	Fiber facet	2015.7[46]
	Pr-doped fiber	Visible (635 nm)	240 ns	6.2 mW	Fiber facet	2015.11[36]
	Yb:GMB crystal	1 μm (1062.8 nm)	446 ns	91.8 mW	Quartz substrate	2018.3[47]
	Tm:YAP crystal	2 μm (1936 nm)	360 ns	466.3 mW	Quartz substrate	2018.3[47]
WSe_2	Er-doped fiber	1.5 μm (~1560 nm)	1.82 μs	3.16 mW	Fiber facet	2015.10[48]
	Er:Lu_2O_3 crystal	3 μm (2.85 μm)	280 ns	776 mW	YAG substrate	2018.8[49]
	Yb:$LuPO_4$ crystal	1 μm (1005–1020 nm)	28 ns	3.32 W	Sapphire	2018.9[50]
$MoTe_2$	Er-doped fiber	1.5 μm (1559 nm)	677 ns	25 mW	Tapered fiber	2018.2[51]
	Ho,Pr:LLF	3 μm (2.95 μm)	670 ns	73 mW	YAG substrate	2018.3[52]
$TiSe_2$	Er-doped fiber	1.5 μm (1530 nm)	1.126 μs	11.54 mW	Mirror	2018.5[53]
	Ho,Pr:LLF	3 μm (2.95 μm)	160.5 ns	387 mW	Sapphire	2018.7[54]

For the fiber field, soon after Woodward's first report on an MoS_2-polymer composite SA to generate Q-switched pulses in a Yb-doped fiber laser at 1068 nm,[55] research interest rapidly expanded to Er-doped fiber lasers at 1.5 μm and Tm-doped fiber lasers at 2 μm. The wavelength tunable operation of MoS_2 Q-switched fiber lasers was also achieved by inserting a tunable filter in the cavity,[35] or by controlling the cavity loss with a strained microfiber coupler.[56]

In addition to the well-reported infrared spectrum region, MoS_2 SA has also found remarkable applications in the visible range. In general, visible lasers could be realized by the nonlinear optical process of infrared lasers. However, the pursuit of compactness, efficiency, and simplicity promotes the demand for visible optical modulators that enable direct generation at the visible range. While the theoretical prediction of TMD-based modulators for visible pulsed laser existing for years as TMDs has a direct bandgap at the visible frequency regions, recent works finally demonstrated MoS_2 visible modulators for pulse emission in solid-state Praseodymium (Pr) lasers.[39] The reported visible wavelength ranges from 522 nm (green) to 721 nm (deep red). Visible pulse generation was also achieved from a Pr-doped all-fiber laser using WS_2, Mos_2, and $MoSe_2$.[36] These initial advances of TMD visible optical modulators, associated with previously developed TMD SAs in the infrared range from 1.06 to 3 μm, provide a strong proof that TMDs function as a broadband optical switcher in the spectral range from green to mid-infrared.

While early studies primarily exploited MoS_2 as a Q-switcher, other TMD materials have also been developed later, as shown in Table 3.2. As a large number of TMD Q-switchers have been demonstrated, comparison between these different TMD materials could be very meaningful. Wu et al. systematically compared the nonlinear absorption effect of four TMDs (i.e., MoS_2, $MoSe_2$, WS_2, WSe_2) and their Q-switching performance in an identical fiber laser scheme.[48] Of the four TMD materials that are processed with the same procedures and parameters, $MoSe_2$ was found to have the highest modulation depth, while WS_2 exhibited the best thermal stability. Further research is required to understand the comparative advantages of different TMDs and provide a reference for special requirements.

3.4 BP in pulsed lasers

The broadband nonlinear optic response of few-layer BPs was first revealed by Lu et al..[57] BP also shows a strong layer count-dependent bandgap varying from 0.3 eV in bulk to 2 eV in monolayer, which covers the near-infrared and mid-infrared regions intrinsically rather than introduced by sub-bandgaps in TMDs. This bandgap bridges the gap between zero-gap graphene and large-gap TMDs and thus offers an ideal potential for applications in pulsed lasers at long wavelength, particularly those beyond 3 μm where SA effect in graphene and TMDs is not strong.

3.4.1 Mode-locked lasers based on BP

Table 3.3 lists the performance of mode-locking operation of BP SA-based lasers that were first reported at respective wavelength regions. Chen et al. first reported both mode-locking and Q-switching operation from a BP SA mode-locked all-fiber laser at the telecommunications band.[59] Soon after the first report, the wavelength tunable operation from 1532 to 1570 nm was demonstrated in an Er-doped fiber laser.[64] By employing BP SAM, Zhang et al. demonstrated the ultrafast pulse generation from a mode-locked bulk laser for the first time.[65] The initial works triggered a new wave of research on BP as a broadband SA from 1 to 3 μm. Particularly benefiting from its decreased bandgap in the bulk form down to ~0.3 eV, BP became a promising SA candidate for the mid-infrared wavelength region beyond 3 μm. Mode-locked pulses in 3.5 μm using BP SA were first achieved in an Er:ZBLAN fiber laser,[62] as shown in Fig. 3.5. Among BP SA mode-locked lasers, the reported pulse duration achieved can be as short as 102 fs in fiber lasers,[66] and 75 fs in solid-state lasers.[67] As for high repetition operation, 290-MHz harmonic mode-locking (up to 10th harmonic) was obtained from a fiber laser.[68] Some groups developed BP quantum dots (BPQDs), i.e., ultrasmall BP nanosheets, as a new form of SA for laser applications. As quantum confinement and edge effects play an important role in its electronic and optical properties, BPQDs show a "flat band" near the Fermi energy and are expected to be a promising candidate for wideband ultrafast photonics.[69]

While early interest mainly lay in extending the available wavelength region of BP SA, later research focus shifted to its drawbacks, i.e., lack of stability in air and ease of oxidation, which greatly hindered the practical application beyond laboratory. The protection methods could roughly be divided into two ways. One way is optimizing the interaction between the light and materials using a lateral interaction scheme such as side-polished fiber or tapered fiber. The lateral interaction scheme could boost optical damage threshold and reduce the thermal effect of BP SA, and therefore prevent it from oxidation. A research has found that by replacing the

Table 3.3 Performance of mode-locked lasers based on BP SAs when they were first reported at respective wavelength regions.

Gain medium	Wavelength regime (center wavelength)	3-dB bandwidth	Pulse width	Integration platform	Time reported [Ref.]
Yb-doped fiber	1 μm (1085.58 nm)	0.23 nm	7.54 ps	Fiber facet	2017.1 [58]
Er-doped fiber	1.5 μm (1571.45 nm)	2.9 nm	946 fs	Fiber facet	2015.5 [59]
Tm-doped fiber	2 μm (1910 nm)	5.8 nm	739 fs	Fiber facet	2015.8 [60]
Er:ZBLAN fiber	3 μm (2783 nm)	2.8 nm	42 ps	Mirror	2016.1 [61]
Er:ZBLAN fiber	3.5 μm (3489 nm)	4.7 nm	–	Mirror	2018.4 [62]
Nd:YVO_4 crystal	1 μm (1064 nm)	0.288 nm	6.1 ps	Mirror	2015.8 [63]

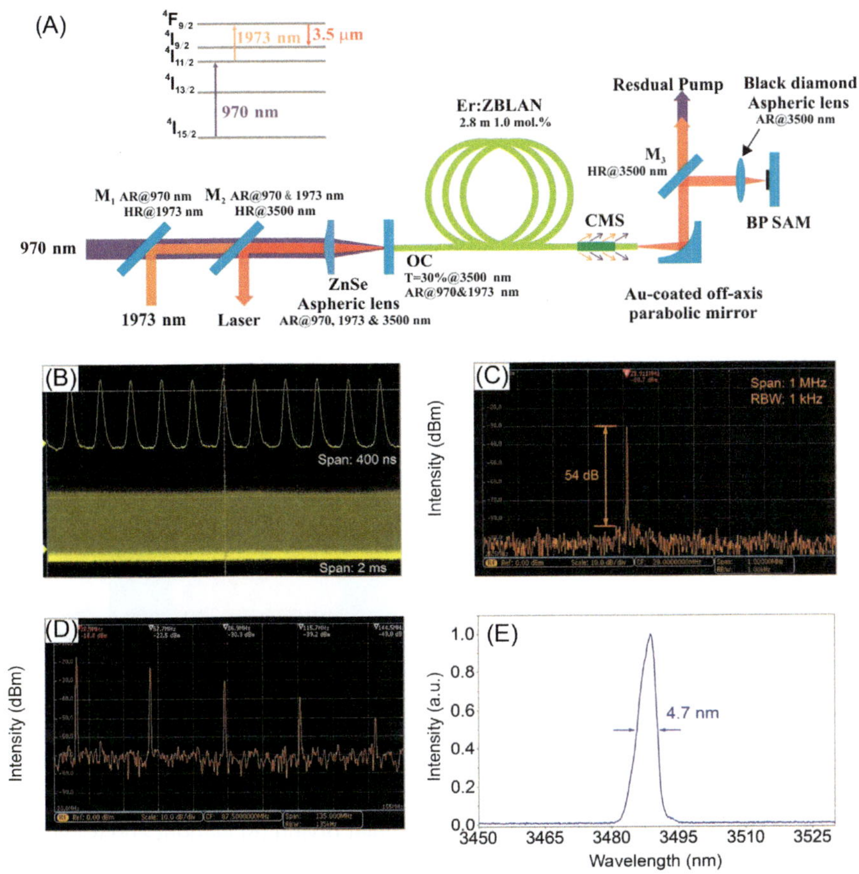

Fig. 3.5 (A) Schematic of BP mode-locked mid-infrared fiber lasers. *OC*, output coupler; *CMS*, cladding mode stripper; *BP SAM*, black phosphorus saturable absorber mirror. (B) Mode-locked pulse train in microsecond and millisecond time scales. (C) Fundamental radio frequency (RF) spectrum with a resolution bandwidth (RBW) of 1 kHz. (D) RF spectrum with broad span. (E) Mode-locked spectrum. *Reproduced with permission. Qin, Z. P.; Hai, T.; Xie, G. Q.; Ma, J. G.; Yuan, P.; Qian, L. J.; Li, L.; Zhao, L. M.; Shen, D. Y. Black Phosphorus Q-Switched and Mode Locked Mid-Infrared Er: ZBLAN Fiber Laser at 3.5 μm Wavelength. Opt. Express 2018, 26, 8224-8231.Z. P. Qin, T. Hai, G. Q. Xie, J. G. Ma, P. Yuan, L. J. Qian, L. Li, L. M. Zhao, and D. Y. Shen, "Black phosphorus Q-switched and mode locked mid-infrared Er:ZBLAN fiber laser at 3.5 μm wavelength," Optics Express 26, 8224-8231 (2018).*

fiber end scheme with a lateral interaction scheme, the thermal damage threshold of BP SA increased from 21.8 to 23.3 dBm, and stable mode-locking operation maintained for 168 h without any significant performance degradation.[70] Another way is enhancing its antioxidation stability in the BP SA fabrication process. For instance, by carefully optimizing the chemical

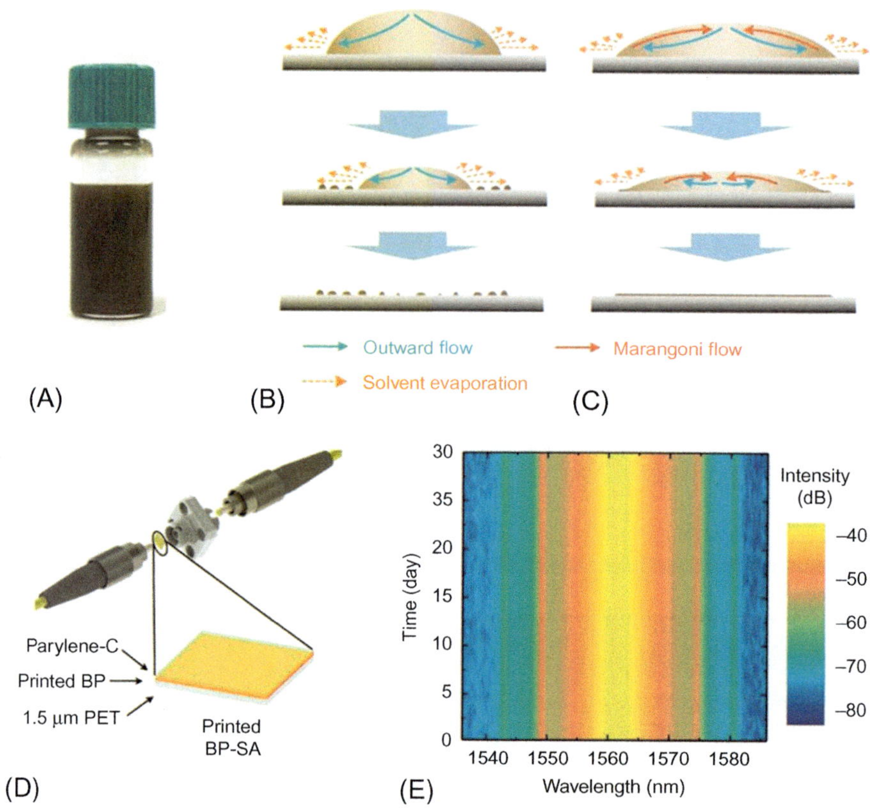

Fig. 3.6 (A) Photograph of produced BP ink. Droplet drying (B) without and (C) with a recirculating Marangoni flow induced. (D) Integration of a printed BP SA. (E) Spectra of long-term stable operation over 30 days. Reproduced under the terms of a Creative Commons Attribution 4.0 International License.[71]

composition through the balance of complex and competing fluidic effects, a stable BP ink compatible with conventional inkjet printing techniques was demonstrated,[71] as shown in Fig. 3.6A–C. The low boiling point of the ink promotes rapid drying, leading to minimal oxidation. The printed BP was then encapsulated by the pinhole free parylene-C and transferred onto a fiber tip to form BP SA (Fig. 3.6D), which enabled long-term (>30 days) mode-locking stability against degradation (Fig. 3.6E).

3.4.2 Q-switched lasers based on BP

In addition to mode-locking, BP has also found its wide-range applications as a reliable Q-switcher from visible to mid-infrared wavelength. Table 3.4

Table 3.4 Performance of Q-switched lasers based on BP SAs when they were first reported at respective wavelength regions.

Gain medium	Wavelength regime (center wavelength)	Min. pulse width	Max. output power	Integration platform	Time reported [Ref.]
Yb-doped fiber	1 μm (1056.6–1083.3 nm)	4 μs	0.3 mW	Fiber facet	2016.9[72]
Er-doped fiber	1.5 μm (1562.87 nm)	10.32 μs	~1.5 mW	Fiber facet	2015.5[59]
Tm-doped fiber	2 μm (1948 nm)	5.6 μs	3.8 mW	Microfiber	2016.6[73]
Ho-doped fluoride fiber	3 μm (2970 nm)	2.41 μs	308.7 mW	Mirror	2016.7[74]
Er:ZBLAN fiber	3.5 μm (~3465 nm)	2.05 μs	120 mW	Mirror	2018.4[62]
Pr-doped fiber	Visible (635 nm)	383 ns	165 mW	Fiber facet	2017.1[75]
Yb:CYA crystal	1 μm (1046 nm)	620 ns	37 mW	Mirror	2015.8[76]
Er:YAG crystal	1.6 μm (1645 nm)	2.8 μs	330 mW	YAG substrate	2018.3[77]
Tm:Ho:YGG crystal	2 μm (2.1 μm)	636 ns	18 mW	Quartz wafer	2015.12[78]
Cr:ZnSe crystal	2.4 μm (2411 nm)	189 ns	36 mW	Mirror	2016.6[79]
Er:Lu_2O_3 crystal	3 μm (2.84 μm)	359 ns	755 mW	YAG substrate	2015.10[80]
Pr:$GdLiF_4$ crystal	Visible (639 nm)	189 ns	27 mW	Quartz wafer	2015.12[78]

lists the BP SA Q-switching operations first reported at respective wavelength regions from both fiber lasers and solid-state lasers. The shortest Q-switched wavelength achieved is 635 nm from a Pr-doped fiber,[75] and the longest Q-switched wavelength achieved is 3465 nm from a Er:ZBLAN fiber.[62] The reported largest pulse energy and average output power are 39.5 μJ and 3.1 W, respectively,[81] and the largest repetition rate is 625 KHz.[82]

3.5 Conclusion

In this chapter, the highly active field of pulsed lasers based on TMDs and BP has been reviewed. The integration strategies of 2D SA into lasers are summarized. The wideband operations of TMDs and BP are verified by their demonstrations in mode-locked lasers and Q-switched lasers with operating wavelengths ranging from visible to mid-infrared regions. The best performance on some specifications of pulsed lasers is highlighted.

The van der Waals heterostructures of 2D materials are promising for engineering functional SA devices with remarkable optical properties for desirable applications. This area is still in its infancy.[83–85] For the commercial applications of 2D material SAs, investigations should be targeted to improve the output performance in terms of not only pulse duration and pulse energy but also the environmental stability and the possibility of mass device fabrication.

References

1. Bao, Q. L.; Zhang, H.; Wang, Y.; Ni, Z. H.; Yan, Y. L.; Shen, Z. X.; Loh, K. P.; Tang, D. Y. Atomic-Layer Graphene as a Saturable Absorber for Ultrafast Pulsed Lasers. *Adv. Funct. Mater.* **2009**, *19*, 3077–3083.
2. Trushin, M.; Kelleher, E. J. R.; Hasan, T. Theory of Edge-State Optical Absorption in Two-Dimensional Transition Metal Dichalcogenide Flakes. *Phys. Rev. B* **2016**, *94*, 9.
3. Zhang, H.; Lu, S. B.; Zheng, J.; Du, J.; Wen, S. C.; Tang, D. Y.; Loh, K. P. Molybdenum Disulfide (MoS_2) as a Broadband Saturable Absorber for Ultra-Fast Photonics. *Opt. Express* **2014**, *22*, 7249–7260.
4. Xia, H. D.; Li, H. P.; Lan, C. Y.; Li, C.; Zhang, X. X.; Zhang, S. J.; Liu, Y. Ultrafast Erbium-Doped Fiber Laser Mode Locked by a CVD-Grown Molybdenum Disulfide (MoS_2) Saturable Absorber. *Opt. Express* **2014**, *22*, 17341–17348.
5. Tian, Z.; Wu, K.; Kong, L. C.; Yang, N.; Wang, Y.; Chen, R.; Hu, W. S.; Xu, J. Q.; Tang, Y. L. Mode Locked Thulium Fiber Laser With MoS_2. *Laser Phys. Lett.* **2015**, *12*, 5.
6. Zhao, G.; Han, S.; Wang, A. Z.; Wu, Y. Z.; Zhao, M. W.; Wang, Z. P.; Hao, X. P. "Chemical Weathering" Exfoliation of Atom-Thick Transition Metal Dichalcogenides and Their Ultrafast Saturable Absorption Properties. *Adv. Funct. Mater.* **2015**, *25*, 5292–5299.
7. Zhang, Y. X.; Yu, H. H.; Zhang, R.; Zhao, G.; Zhang, H. J.; Chen, Y. X.; Mei, L. M.; Tonelli, M.; Wang, J. Y. Broadband Atomic-Layer MoS2 Optical Modulators for Ultrafast Pulse Generations in the Visible Range. *Opt. Lett.* **2017**, *42*, 547–550.

8. Mao, D.; Zhang, S. L.; Wang, Y. D.; Gan, X. T.; Zhang, W. D.; Mei, T.; Wang, Y. G.; Wang, Y. S.; Zeng, H. B.; Zhao, J. L. WS_2 Saturable Absorber for Dissipative Soliton Mode Locking at 1.06 and 1.55 μm. *Opt. Express* **2015**, *23*, 27509–27519.
9. Mao, D.; Wang, Y. D.; Ma, C. J.; Han, L.; Jiang, B. Q.; Gan, X. T.; Hua, S. J.; Zhang, W. D.; Mei, T.; Zhao, J. L. WS_2 Mode Locked Ultrafast Fiber Laser. *Sci. Rep.* **2015**, *5*, 7.
10. Jung, M.; Lee, J.; Park, J.; Koo, J.; Jhon, Y. M.; Lee, J. H. Mode Locked, 1.94-μm, All-Fiberized Laser Using WS_2-Based Evanescent Field Interaction. *Opt. Express* **2015**, *23*, 19996–20006.
11. Samikannu, S.; Sivaraj, S. Dissipative Soliton Generation in an All-Normal Dispersion Ytterbium-Doped Fiber Laser Using Few-Layer Molybdenum Diselenide as a Saturable Absorber. *Opt. Eng.* **2016**, *55*, 7.
12. Luo, Z. Q.; Li, Y. Y.; Zhong, M.; Huang, Y. Z.; Wan, X. J.; Peng, J.; Weng, J. Nonlinear Optical Absorption of Few-Layer Molybdenum Diselenide ($MoSe_2$) for Passively Mode Locked Soliton Fiber Laser Invited. *Photonics Res.* **2015**, *3*, A79–A86.
13. Lee, J.; Koo, J.; Lee, J.; Jhon, Y. M.; Lee, J. H. All-Fiberized, Femtosecond Laser at 1912 nm Using a Bulk-Like $MoSe_2$ Saturable Absorber. *Opt. Mater. Express* **2017**, *7*, 2968–2979.
14. Mao, D.; She, X. Y.; Du, B. B.; Yang, D. X.; Zhang, W. D.; Song, K.; Cui, X. Q.; Jiang, B. Q.; Peng, T.; Zhao, J. L. Erbium-Doped Fiber Laser Passively Mode Locked With Few-Layer $WSe_2/MoSe_2$ Nanosheets. *Sci. Rep.* **2016**, *6*, 9.
15. Yin, J. D.; Li, J. R.; Chen, H.; Wang, J. T.; Yan, P. G.; Liu, M. L.; Liu, W. J.; Lu, W.; Xu, Z. H.; Zhang, W. F.; Wang, J. Z.; Sun, Z. P.; Ruan, S. C. Large-Area Highly Crystalline WSe_2 Atomic Layers for Ultrafast Pulsed Lasers. *Opt. Express* **2017**, *25*, 30020–30031.
16. Mao, D.; Du, B. B.; Yang, D. X.; Zhang, S. L.; Wang, Y. D.; Zhang, W. D.; She, X. Y.; Cheng, H. C.; Zeng, H. B.; Zhao, J. L. Nonlinear Saturable Absorption of Liquid-Exfoliated Molybdenum/Tungsten Ditelluride Nanosheets. *Small* **2016**, *12*, 1489–1497.
17. Wang, J. T.; Chen, H.; Jiang, Z. K.; Yin, J. D.; Wang, J. Z.; Zhang, M.; He, T. C.; Li, J. Z.; Yan, P. G.; Ruan, S. C. Mode Locked Thulium-Doped Fiber Laser With Chemical Vapor Deposited Molybdenum Ditelluride. *Opt. Lett.* **2018**, *43*, 1998–2001.
18. Wang, J. T.; Jiang, Z. K.; Chen, H.; Li, J. R.; Yin, J. D.; Wang, J. Z.; He, T. C.; Yan, P. G.; Ruan, S. C. Magnetron-Sputtering Deposited WTe_2 for an Ultrafast Thulium-Doped Fiber Laser. *Opt. Lett.* **2017**, *42*, 5010–5013.
19. Ge, Y. Q.; Zhu, Z. F.; Xu, Y. H.; Chen, Y. X.; Chen, S.; Liang, Z. M.; Song, Y. F.; Zou, Y. S.; Zeng, H. B.; Xu, S. X.; Zhang, H.; Fan, D. Y. Broadband Nonlinear Photoresponse of 2D TiS_2 for Ultrashort Pulse Generation and All-Optical Thresholding Devices. *Adv. Opt. Mater.* **2018**, *6*, 10.
20. Wang, K. P.; Feng, Y. Y.; Chang, C. X.; Zhan, J. X.; Wang, C. W.; Zhao, Q. Z.; Coleman, J. N.; Zhang, L.; Blau, W. J.; Wang, J. Broadband Ultrafast Nonlinear Absorption and Nonlinear Refraction of Layered Molybdenum Dichalcogenide Semiconductors. *Nanoscale* **2014**, *6*, 10530–10535.
21. Liu, H.; Luo, A. P.; Wang, F. Z.; Tang, R.; Liu, M.; Luo, Z. C.; Xu, W. C.; Zhao, C. J.; Zhang, H. Femtosecond Pulse Erbium-Doped Fiber Laser by a Few-Layer MoS_2 Saturable Absorber. *Opt. Lett.* **2014**, *39*, 4591–4594.
22. Aiub, E. J.; Steinberg, D.; de Souza, E. A. T.; Saito, L. A. M. 200-fs Mode Locked Erbium-Doped Fiber Laser by Using Mechanically Exfoliated MoS_2 Saturable Absorber Onto D-Shaped Optical Fiber. *Opt. Express* **2017**, *25*, 10546–10552.
23. Zhang, M.; Howe, R. C. T.; Woodward, R. I.; Kelleher, E. J. R.; Torrisi, F.; Hu, G. H.; Popov, S. V.; Taylor, J. R.; Hasan, T. Solution Processed MoS_2-PVA Composite for Sub-Bandgap Mode Locking of a Wideband Tunable Ultrafast Er:fiber Laser. *Nano Res.* **2015**, *8*, 1522–1534.

24. Wang, R.; Chien, H. C.; Kumar, J.; Kumar, N.; Chiu, H. Y.; Zhao, H. Third-Harmonic Generation in Ultrathin Films of MoS_2. *ACS Appl. Mater. Interfaces* **2014**, *6*, 314–318.
25. Liu, M.; Zheng, X. W.; Qi, Y. L.; Liu, H.; Luo, A. P.; Luo, Z. C.; Xu, W. C.; Zhao, C. J.; Zhang, H. Microfiber-Based Few-Layer MoS_2 Saturable Absorber for 2.5 GHz Passively Harmonic Mode Locked Fiber Laser. *Opt. Express* **2014**, *22*, 22841–22846.
26. Wu, K.; Zhang, X. Y.; Wang, J.; Chen, J. P. 463-MHz Fundamental Mode Locked Fiber Laser Based on Few-Layer MoS_2 Saturable Absorber. *Opt. Lett.* **2015**, *40*, 1374–1377.
27. Luo, A. P.; Liu, M.; Wang, X. D.; Ning, Q. Y.; Xu, W. C.; Luo, Z. C. Few-Layer MoS_2-Deposited Microfiber as Highly Nonlinear Photonic Device for Pulse Shaping in a Fiber Laser Invited. *Photonics Res.* **2015**, *3*, A69–A78.
28. Du, J.; Wang, Q. K.; Jiang, G. B.; Xu, C. W.; Zhao, C. J.; Xiang, Y. J.; Chen, Y.; Wen, S. C.; Zhang, H. Ytterbium-Doped Fiber Laser Passively Mode Locked by Few-Layer Molybdenum Disulfide (MoS_2) Saturable Absorber Functioned With Evanescent Field Interaction. *Sci. Rep.* **2014**, *4*, 7.
29. Liu, W.; Pang, L.; Han, H.; Liu, M.; Lei, M.; Fang, S.; Teng, H.; Wei, Z. Tungsten Disulfide Saturable Absorbers for 67 fs Mode Locked Erbium-Doped Fiber Lasers. *Opt. Express* **2017**, *25*, 2950–2959.
30. Li, L.; Jiang, S. Z.; Wang, Y. G.; Wang, X.; Duan, L. N.; Mao, D.; Li, Z.; Man, B. Y.; Si, J. H. WS_2/Fluorine Mica (FM) Saturable Absorbers for All-Normal-Dispersion Mode Locked Fiber Laser. *Opt. Express* **2015**, *23*, 28698–28706.
31. Yan, P. G.; Chen, H.; Yin, J. D.; Xu, Z. H.; Li, J. R.; Jiang, Z. K.; Zhang, W. F.; Wang, J. Z.; Li, I. L.; Sun, Z. P.; Ruan, S. C. Large-Area Tungsten Disulfide for Ultrafast Photonics. *Nanoscale* **2017**, *9*, 1871–1877.
32. Koo, J.; Park, J.; Lee, J.; Jhon, Y. M.; Lee, J. H. Femtosecond Harmonic Mode Locking of a Fiber Laser at 3.27 GHz Using a Bulk-Like, $MoSe_2$-Based Saturable Absorber. *Opt. Express* **2016**, *24*, 15.
33. Wang, J. T.; Jiang, Z. K.; Chen, H.; Li, J. R.; Yin, J. D.; Wang, J. Z.; He, T. C.; Yan, P. G.; Ruan, S. C. High Energy Soliton Pulse Generation by a Magnetron-Sputtering-Deposition-Grown $MoTe_2$ Saturable Absorber. *Photonics Res.* **2018**, *6*, 535–541.
34. Luo, Z. Q.; Huang, Y. Z.; Zhong, M.; Li, Y. Y.; Wu, J. Y.; Xu, B.; Xu, H. Y.; Cai, Z. P.; Peng, J.; Weng, J. 1-, 1.5-, and 2-μm Fiber Lasers Q-Switched by a Broadband Few-Layer MoS_2 Saturable Absorber. *J. Lightwave Technol.* **2014**, *32*, 4077–4084.
35. Huang, Y. Z.; Luo, Z. Q.; Li, Y. Y.; Zhong, M.; Xu, B.; Che, K. J.; Xu, H. Y.; Cai, Z. P.; Peng, J.; Weng, J. Widely-Tunable, Passively Q-Switched Erbium-Doped Fiber Laser With Few-Layer MoS_2 Saturable Absorber. *Opt. Express* **2014**, *22*, 25258–25266.
36. Luo, Z. Q.; Wu, D. D.; Xu, B.; Xu, H. Y.; Cai, Z. P.; Peng, J.; Weng, J.; Xu, S.; Zhu, C. H.; Wang, F. Q.; Sun, Z. P.; Zhang, H. Two-Dimensional Material-Based Saturable Absorbers: Towards Compact Visible-Wavelength All-Fiber Pulsed Lasers. *Nanoscale* **2016**, *8*, 1066–1072.
37. Wang, S. X.; Yu, H. H.; Zhang, H. J.; Wang, A. Z.; Zhao, M. W.; Chen, Y. X.; Mei, L. M.; Wang, J. Y. Broadband Few-Layer MoS_2 Saturable Absorbers. *Adv. Mater.* **2014**, *26*, 3538–3544.
38. Fan, M. Q.; Li, T.; Zhao, S. Z.; Li, G. Q.; Ma, H. Y.; Gao, X. C.; Krankel, C.; Huber, G. Watt-Level Passively Q-Switched Er:Lu2O3 Laser at 2.84 μm Using MoS_2. *Opt. Lett.* **2016**, *41*, 540–543.
39. Zhang, Y. X.; Wang, S. X.; Yu, H. H.; Zhang, H. J.; Chen, Y. X.; Mei, L. M.; Di Lieto, A.; Tonelli, M.; Wang, J. Y. Atomic-Layer Molybdenum Sulfide Optical Modulator for Visible Coherent Light. *Sci. Rep.* **2015**, *5*, 7.

40. Lin, J.; Hu, Y. Y.; Chen, C. J.; Gu, C.; Xu, L. X. Wavelength-Tunable Yb-Doped Passively Q-Switching Fiber Laser Based on WS_2 Saturable Absorber. *Opt. Express* **2015**, *23*, 29059–29064.
41. Kassani, S. H.; Khazaeizhad, R.; Jeong, H.; Nazari, T.; Yeom, D. I.; Oh, K. All-Fiber Er-Doped Q-Switched Laser Based on Tungsten Disulfide Saturable Absorber. *Opt. Mater. Express* **2015**, *5*, 373–379.
42. Wei, C.; Luo, H. Y.; Zhang, H.; Li, C.; Xie, J. T.; Li, J. F.; Liu, Y. Passively Q-Switched Mid-Infrared Fluoride Fiber Laser Around 3 μm Using a Tungsten Disulfide (WS_2) Saturable Absorber. *Laser Phys. Lett.* **2016**, *13*, 7.
43. Guo, L.; Li, M.; Li, T.; Zhang, S. Y.; Yang, K. J.; Fan, M. Q.; Zhao, S. Z. Inband Pumped Passively Q-Switched Er: YAG Laser at 1645 nm Using WS_2. *Opt. Commun.* **2018**, *406*, 230–233.
44. Luan, C.; Yang, K. J.; Zhao, J.; Zhao, S. Z.; Song, L.; Li, T.; Chu, H. W.; Qiao, J. P.; Wang, C.; Li, Z.; Jiang, S. Z.; Man, B. Y.; Zheng, L. H. WS_2 as a Saturable Absorber for Q-Switched 2 Micron Lasers. *Opt. Lett.* **2016**, *41*, 3783–3786.
45. Cheng, Y. J.; Peng, J.; Xu, B.; Yang, H.; Luo, Z. Q.; Xu, H. Y.; Cai, Z. P.; Weng, J. Passive Q-Switching of a Diode-Pumped Pr:$LiYF_4$ Visible Laser Using WS_2 as Saturable Absorber. *IEEE Photonics J.* **2016**, *8*, 6.
46. Woodward, R. I.; Howe, R. C. T.; Runcorn, T. H.; Hu, G.; Torrisi, F.; Kelleher, E. J. R.; Hasan, T. Wideband Saturable Absorption in Few-Layer Molybdenum Diselenide ($MoSe_2$) for Q-Switching Yb-, Er- and Tm-Doped Fiber Lasers. *Opt. Express* **2015**, *23*, 20051–20061.
47. Wang, X. H.; Xu, J. L.; Xia, H. P.; You, Z. Y.; Huang, Y. S.; Lin, Z. B.; Tu, C. Y. $MoSe_2$ Nanoflower as the Saturable Absorber for 1 μm and 2 μm Passively Q-Switched Bulk State Laser. *Mater. Res. Bull.* **2018**, *99*, 168–173.
48. Chen, B. H.; Zhang, X. Y.; Wu, K.; Wang, H.; Wang, J.; Chen, J. P. Q-Switched Fiber Laser Based on Transition Metal Dichalcogenides MoS_2, $MoSe_2$, WS_2, and WSe_2. *Opt. Express* **2015**, *23*, 26723–26737.
49. Yan, Z. Y.; Li, T.; Zhao, J.; Yang, K. J.; Li, D. C.; Li, G. Q.; Fan, M. Q.; Zhang, S. Y.; Zhao, S. Z. Passively Q-Switched 2.85μm Er:Lu2O3 Laser With WSe2. *Laser Phys. Lett.* **2018**, *15*, 5.
50. Dou, X. D.; Ma, Y. J.; Zhu, M.; Yi, H. Y.; Xu, H. H.; Zhong, D. G.; Teng, B.; Yu, H. H.; Liu, J. H. Passively Q-Switched Yb:LuPO4 Laser Based on a Coupled Cavity With Few-Layer WSe_2 Acting as Saturable Absorber. *Laser Phys. Lett.* **2018**, *15*, 6.
51. Liu, M. L.; Liu, W. J.; Yan, P. G.; Fang, S. B.; Teng, H.; Wei, Z. Y. High-Power $MoTe_2$-Based Passively Q-Switched Erbium-Doped Fiber Laser. *Chin. Opt. Lett.* **2018**, *16*, 5.
52. Yan, Z. Y.; Li, T.; Zhao, S. Z.; Yang, K. J.; Li, D. C.; Li, G. Q.; Zhang, S. Y.; Gao, Z. J. $MoTe_2$ Saturable Absorber for Passively Q-SWITCHED Ho,Pr:$LiLuF_4$ Laser at Similar to 3 μm. *Opt. Laser Technol.* **2018**, *100*, 261–264.
53. Liu, W. J.; Liu, M. L.; Lei, M.; Fang, S. B.; Wei, Z. Y. Titanium Selenide Saturable Absorber Mirror for Passive Q-Switched Er-Doped Fiber Laser. *IEEE J. Sel. Top. Quantum Electron.* **2018**, *24*, 5.
54. Nie, H. K.; Sun, X. L.; Zhang, B. T.; Yan, B. Z.; Li, G. R.; Wang, Y. R.; Liu, J. T.; Shi, B. N.; Liu, S. D.; He, J. L. Few-Layer $TiSe_2$ as a Saturable Absorber for Nanosecond Pulse Generation in 2.95μm Bulk Laser. *Opt. Lett.* **2018**, *43*, 3349–3352.
55. Woodward, R. I.; Kelleher, E. J. R.; Runcorn, T. H.; Popov, S. V.; Torrisi, F.; Howe, R. C. T.; Hasan, T. Q-Switched Fiber Laser With MoS2 Saturable Absorber. *CLEO: Science and Innovations, CLEO_SI 2014, June 8, 2014–June 13, 2014, Optics InfoBase Conference Papers (Optical Society of America (OSA)*; 2014.
56. Chen, J. H.; Deng, G. Q.; Yan, S. C.; Li, C.; Xi, K.; Xu, F.; Lu, Y. Q. Microfiber-Coupler-Assisted Control of Wavelength Tuning for Q-Switched Fiber Laser With Few-Layer Molybdenum Disulfide Nanoplates. *Opt. Lett.* **2015**, *40*, 3576–3579.

57. Lu, S. B.; Miao, L. L.; Guo, Z. N.; Qi, X.; Zhao, C. J.; Zhang, H.; Wen, S. C.; Tang, D. Y.; Fan, D. Y. Broadband Nonlinear Optical Response in Multi-Layer Black Phosphorus: An Emerging Infrared and Mid-Infrared Optical Material. *Opt. Express* **2015**, *23*, 11183–11194.
58. Hisyam, M. B.; Rusdi, M. F. M.; Latiff, A. A.; Harun, S. W. Generation of Mode locked Ytterbium Doped Fiber Ring Laser Using Few-Layer Black Phosphorus as a Saturable Absorber. *IEEE J. Sel. Top. Quantum Electron.* **2017**, *23*.
59. Chen, Y.; Jiang, G. B.; Chen, S. Q.; Guo, Z. N.; Yu, X. F.; Zhao, C. J.; Zhang, H.; Bao, Q. L.; Wen, S. C.; Tang, D. Y.; Fan, D. Y. Mechanically Exfoliated Black Phosphorus as a New Saturable Absorber for Both Q-Switching and Mode Locking Laser Operation. *Opt. Express* **2015**, *23*, 12823–12833.
60. Sotor, J.; Sobon, G.; Kowalczyk, M.; Macherzynski, W.; Paletko, P.; Abramski, K. M. Ultrafast Thulium-Doped Fiber Laser Mode Locked With Black Phosphorus. *Opt. Lett.* **2015**, *40*, 3885–3888.
61. Qin, Z. P.; Xie, G. Q.; Zhao, C. J.; Wen, S. C.; Yuan, P.; Qian, L. J. Mid-Infrared Mode Locked Pulse Generation With Multilayer Black Phosphorus as Saturable Absorber. *Opt. Lett.* **2016**, *41*, 56–59.
62. Qin, Z. P.; Hai, T.; Xie, G. Q.; Ma, J. G.; Yuan, P.; Qian, L. J.; Li, L.; Zhao, L. M.; Shen, D. Y. Black Phosphorus Q-Switched and Mode Locked Mid-Infrared Er:ZBLAN Fiber Laser at 3.5 μm Wavelength. *Opt. Express* **2018**, *26*, 8224–8231.
63. Wang, S. X.; Yu, H. H.; Zhang, H. J. Band-Gap Modulation of Two-Dimensional Saturable Absorbers for Solid-State Lasers. *Photonics Res.* **2015**, *3*, A10–A20.
64. Luo, Z. C.; Liu, M.; Guo, Z. N.; Jiang, X. F.; Luo, A. P.; Zhao, C. J.; Yu, X. F.; Xu, W. C.; Zhang, H. Microfiber-Based Few-Layer Black Phosphorus Saturable Absorber for Ultra-Fast Fiber Laser. *Opt. Express* **2015**, *23*, 20030–20039.
65. Zhang, B. T.; Lou, F.; Zhao, R. W.; He, J. L.; Li, J.; Su, X. C.; Ning, J.; Yang, K. J. Exfoliated Layers of Black Phosphorus as Saturable Absorber for Ultrafast Solid-State Laser. *Opt. Lett.* **2015**, *40*, 3691–3694.
66. Jin, X. X.; Hu, G. H.; Zhang, M.; Hu, Y. W.; Albrow-Owen, T.; Howe, R. C. T.; Wu, T. C.; Wu, Q.; Zheng, Z.; Hasan, T. 102 fs Pulse Generation From a Long-Term Stable, Inkjet-Printed Black Phosphorus-Mode Locked Fiber Laser. *Opt. Express* **2018**, *26*, 12506–12513.
67. Lou, F.; Zhang, B. T.; Sun, S. J.; Hu, C.; Lin, Z. B.; Jiang, J. Y.; Zhang, S. Y.; Wang, X.; Teng, B.; He, J. L. Black Phosphorus Mode Locked Sub-100 fs Bulk Laser Based on Heterostructured Yb Composite Crystal. *Appl. Phys. Express* **2018**, *11*.
68. Pawliszewska, M.; Ge, Y. Q.; Li, Z. J.; Zhang, H.; Sotor, J. Fundamental and Harmonic Mode Locking at 2.1 μm With Black Phosphorus Saturable Absorber. *Opt. Express* **2017**, *25*, 16916–16921.
69. Du, J.; Zhang, M.; Guo, Z.; Chen, J.; Zhu, X.; Hu, G.; Peng, P.; Zheng, Z.; Zhang, H. Phosphorene Quantum Dot Saturable Absorbers for Ultrafast Fiber Lasers. *Sci. Rep.* **2017**, *7*.
70. Lee, D.; Park, K.; Debnath, P. C.; Kim, I.; Song, Y. W. Thermal Damage Suppression of a Black Phosphorus Saturable Absorber for High-Power Operation of Pulsed Fiber Lasers. *Nanotechnology* **2016**, *27*.
71. Hu, G.; Albrow-Owen, T.; Jin, X.; Ali, A.; Hu, Y.; Howe, R. C. T.; Shehzad, K.; Yang, Z.; Zhu, X.; Woodward, R. I.; Wu, T.-C.; Jussila, H.; Wu, J.-B.; Peng, P.; Tan, P.-H.; Sun, Z.; Kelleher, E. J. R.; Zhang, M.; Xu, Y.; Hasan, T. Black Phosphorus Ink Formulation for Inkjet Printing of Optoelectronics and Photonics. *Nat. Commun.* **2017**, *8*, 278.
72. Ahmad, H.; Salim, M. A. M.; Thambiratnam, K.; Norizan, S. F.; Harun, S. W. A Black Phosphorus-Based Tunable Q-Switched Ytterbium Fiber Laser. *Laser Phys. Lett.* **2016**, *13*.

73. Wang, Y. Z.; Li, J. F.; Han, L.; Lu, R. G.; Hu, Y. X.; Li, Z.; Liu, Y. Q-Switched Tm^{3+}-Doped Fiber Laser With a Micro-Fiber Based Black Phosphorus Saturable Absorber. *Laser Phys.* **2016**, *26*.
74. Li, J. F.; Luo, H. Y.; Zhai, B.; Lu, R. G.; Guo, Z. N.; Zhang, H.; Liu, Y. Black Phosphorus: A Two-Dimension Saturable Absorption Material for Mid-Infrared Q-Switched and Mode Locked Fiber Lasers. *Sci. Rep.* **2016**, *6*.
75. Wu, D. D.; Cai, Z. P.; Zhong, Y. L.; Peng, J.; Cheng, Y. J.; Weng, J.; Luo, Z. Q.; Xu, H. Y. Compact Passive Q-Switching Pr^{3+}-Doped ZBLAN Fiber Laser With Black Phosphorus-Based Saturable Absorber. *IEEE J. Sel. Top. Quantum Electron.* **2017**, *23*.
76. Ma, J.; Lu, S. B.; Guo, Z. N.; Xu, X. D.; Zhang, H.; Tang, D. Y.; Fan, D. Y. Few-Layer Black Phosphorus Based Saturable Absorber Mirror for Pulsed Solid-State Lasers. *Opt. Express* **2015**, *23*, 22643–22648.
77. Guo, L.; Li, T.; Zhang, S. Y.; Wang, M. J.; Yang, K. J.; Fan, M. Q.; Zhao, S. Z.; Li, M. Black Phosphorus Saturable Absorber for Q-Switched Er:YAG Laser at 1645 nm. *Opt. Laser Technol.* **2018**, *100*, 225–229.
78. Zhang, R.; Zhang, Y. X.; Yu, H. H.; Zhang, H. J.; Yang, R. L.; Yang, B. C.; Liu, Z. Y.; Wang, J. Y. Broadband Black Phosphorus Optical Modulator in the Spectral Range From Visible to Mid-Infrared. *Adv. Opt. Mater.* **2015**, *3*, 1787–1792.
79. Wang, Z. W.; Zhao, R. W.; He, J. L.; Zhang, B. T.; Ning, J.; Wang, Y. R.; Su, X. C.; Hou, J.; Lou, F.; Yang, K. J.; Fan, Y. S.; Bian, J. T.; Nie, J. S. Multi-Layered Black Phosphorus as Saturable Absorber for Pulsed Cr:ZnSe Laser at 2.4 μm. *Opt. Express* **2016**, *24*, 1598–1603.
80. Fan, M. Q.; Li, T.; Zhao, S. Z.; Li, G. Q.; Gao, X. C.; Yang, K. J.; Li, D. C.; Krankel, C. Multilayer Black Phosphorus as Saturable Absorber for an Er:Lu_2O_3 Laser at Similar to 3μm. *Photonics Res.* **2016**, *4*, 181–186.
81. Zhang, H. K.; He, J. L.; Wang, Z. W.; Hou, J.; Zhang, B. T.; Zhao, R. W.; Han, K. Z.; Yang, K. J.; Nie, H. K.; Sun, X. L. Dual-Wavelength, Passively Q-Switched Tm:YAP Laser With Black Phosphorus Saturable Absorber. *Opt. Mater. Express* **2016**, *6*, 2328–2335.
82. Sun, X. L.; Nie, H. K.; He, J. L.; Zhao, R. W.; Su, X. C.; Wang, Y. R.; Zhang, B. T.; Wang, R. H.; Yang, K. J. Passively Q-Switched Nd:GdVO4 1.3 mu m Laser with Few-Layered Black Phosphorus Saturable Absorber. *IEEE J. Sel. Top. Quantum Electron.* **2018**, *24*.
83. Chen, H.; Yin, J. D.; Yang, J. W.; Zhang, X. J.; Liu, M. L.; Jiang, Z. K.; Wang, J. Z.; Sun, Z. P.; Guo, T.; Liu, W. J.; Yan, P. G. Transition-metal Dichalcogenides Heterostructure Saturable Absorbers for Ultrafast Photonics. *Opt. Lett.* **2017**, *42*, 4279–4282.
84. Zhao, G.; Hou, J.; Wu, Y. Z.; He, J. L.; Hao, X. P. Preparation of 2D MoS_2/Graphene Heterostructure Through a Monolayer Intercalation Method and its Application as an Optical Modulator in Pulsed Laser Generation. *Adv. Opt. Mater.* **2015**, *3*, 937–942.
85. Liu, W. J.; Zhu, Y. N.; Liu, M. L.; Wen, B.; Fang, S. B.; Teng, H.; Lei, M.; Liu, L. M.; Wei, Z. Y. Optical Properties and Applications for MoS_2-Sb_2Te_3-MoS_2 Heterostructure Materials. *Photonics Res.* **2018**, *6*, 220–227.

Further reading

86. Khazaeizhad, R.; Kassani, S. H.; Jeong, H.; Yeom, D. I.; Oh, K. Mode Locking of Er-Doped Fiber Laser Using a Multilayer MoS_2 Thin Film as a Saturable Absorber in Both Anomalous and Normal Dispersion Regimes. *Opt. Express* **2014**, *22*, 23732–23742.

CHAPTER 4

Infrared photodetectors

Peng Wang[a], Qiaoliang Bao[b], Weida Hu[a]
[a]State Key Laboratory of Infrared Physics, Shanghai Institute of Technical Physics, Chinese Academy of Sciences, Shanghai, China
[b]Department of Materials Science and Engineering, ARC Centre of Excellence in Future Low-Energy Electronics Technologies (FLEET), Monash University, Clayton, VIC, Australia

4.1 Introduction

Infrared (IR) light with wavelengths longer than 0.78 μm is beyond the visual range of the eyes of a human being. IR photodetectors are, therefore, important devices that allow us to discover the wider electromagnetic spectrum and explore information carried by IR light. To detect IR light, the photoelectric effect or photothermal effect is often employed. The difference is that the former originated from an optical transition that can be a very fast process, while the latter is attributed to the heating effect of IR light and can be a much slower process. Traditional high-performance thin-film IR detectors are based on semiconductors with narrow bandgaps, such as InGaAs, mercury cadmium telluride, or materials fashioned into artificial structures such as super lattice and quantum well, resulting in narrow subbands. To achieve a desired sensitivity, these types of IR detectors require a low operating temperature. Recently, a number of 2D material-based room-temperature IR detectors have emerged, with broad detection waveband, fast response speed, and high responsivity. To achieve a decentz signal-to-noise ratio, the dark current of IR detectors should be limited to a low level, while the quantum efficiency should be considerable. Because of the ultrasmall volume, the dark current of 2D materials can theoretically be much lower than that of thin-film materials. However, due to the ultrathin thickness (one or few atomic layers), induced low light absorption can result in unsatisfactory quantum efficiency. In this chapter, we discuss how 2D materials can be utilized in high-performance IR detectors and whether these detectors can have practical applications.

2D Materials for Photonic and Optoelectronic Applications
https://doi.org/10.1016/B978-0-08-102637-3.00004-8

4.2 Recent progress

Typical thin-film semiconductors for IR detection have narrow bandgaps <1 eV. Basically, the excess carriers, excited by absorbed photons, contribute to the photoconductance. Photoconductors and photovoltaic devices work in this regime. Table 4.1 lists the layered TMDCs with small bandgaps that could be potentially harnessed to fabricate IR photodetectors. For the most commonly explored MX_2 (M=Mo, W; X=S, Se, Te), sulfides and selenides are semiconductors with bandgaps of ~1 eV in bulk or larger in monolayer, while tellurides can be semimetallic. Consequently, photonics and optoelectronics based on these materials work in the visible and near-IR waveband. Generally, MX_2 presented in blue in Table 4.1 have desired bandgaps. However, to date, few of these have been configured into devices or further studied for optoelectronic applications except noble metals Pd and Pt. Apart from TMDCs, black phosphorous (b-P), the rediscovered layered material with a narrow bandgap, as well as its arsenic alloy, bAsP, has been attempted to realize mid-IR detection. Specifically, the first discovered layered material, graphene, a semimetal, exhibits a wide optical absorption covering a spectrum from visible to IR and even to terahertz. Hence, graphene can also be harnessed to detect IR light. Moreover, plasmons in graphene that could be tuned by gate voltage will further enhance MIR absorption, demonstrating the potency of graphene for IR optoelectronics.

In the following sections, we focus on the aforementioned materials and discuss their recent progress on IR detectors, especially mid-IR detectors. Note that in this primary research stage, most devices are photoconductors as a unit photosensing pixel without a complex structure design and optimization. In addition to the photoelectric effect, graphene can also achieve fast and sensitive IR detection through photothermal effect, which will be briefly introduced. In the near-IR range beyond the detection waveband of silicon or particularly in the two optical communication bands, some

Table 4.1 Semiconductors among the TMDCs with bandgaps <1 eV.[1–4]

	Mo	W	Ti	Zr	Hf	V	Nb	Ta	Tc	Re	Ni	Pd	Pt
S													
Se													
Te													

2D materials have emerged for practical applications; we select some cases for interpretation.

4.2.1 Black phosphorous

The rediscovered 2D b-P draws attention from 2014. Experimentally, it shows a strong in-plane anisotropy with a high hole mobility of 1000 $cm^2 V^{-1} s^{-1}$ along the light effective mass direction and 600 $cm^2 V^{-1} s^{-1}$ along the heavy effective mass direction at 120 K, and exhibits considerable optical conductivity in samples with thickness from 2 to 5 μm.[5] Almost at the same time and theoretically, it is expected that the bandgap of b-P with a thicknesses of ~10–60 layers resides in the mid- and near-IR spectrum, and the optical conductivity of multilayer b-P is sensitive to the thickness, doping, and light polarization.[6] Besides, the arsenic alloy of b-P, bAsP, possesses similar properties and an even narrower bandgap (Fig. 4.1).[9]

Early in 2015, Yuan et al. used ionic gel to tune the interfacial band bending of b-P and realized a vertical p-n junction with polarization-sensitive photoresponse over a bandwidth from 400 to 3750 nm.[10] However, this broadband photodetector did not show sufficient quantum efficiency probably owing to the low light absorption and short carrier diffusion length compared to the channel length. Soon, Guo et al. achieved a b-P photoconductor with room temperature responsivity of 82 A/W at a wavelength of 3.39 μm. This high responsivity is attributed to the shallow trap level-induced long carrier lifetime, named the photogating effect, especially under weak light illumination. Generally, high gain means limited bandwidth. The internal gain of this photodetector is not reduced in a

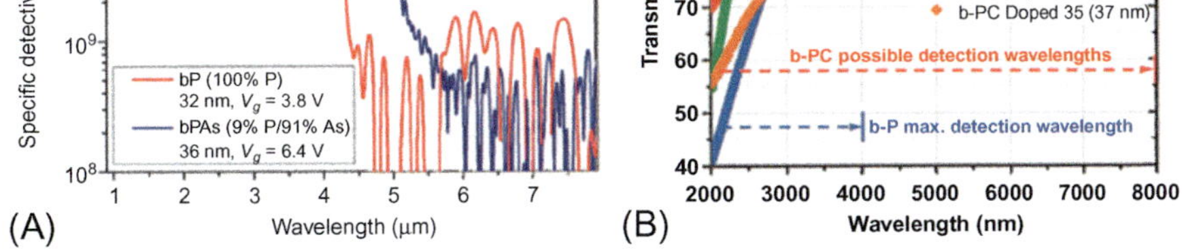

Fig. 4.1 (A) Detectivity of b-P and b-PA photoconductors measured by FTIR at 297 K.[7] (B) Transmission spectrum of b-PC.[8]

bandwidth of 200 Hz. Even for a chopping frequency over 10 kHz, the responsivity reaches 60 mA/W, which is originated from the intrinsic photovoltaic effect. This work is of great significance and recovers the potency of b-P for practical mid-IR applications. By applying bAsP instead of pure b-P, Long et al. achieved a room temperature van der Waals heterojunction with detectivity $>10^9$ Jones and extended detection wavelength up to 8.2 μm.[11] Taking advantage of an external vertical electric filed can also tune the optical absorption edge of 5-nm b-P film and extend the detection waveband beyond 7.7 μm,[12] while the quantum efficiency could be undesired. In 2017, Amani et al. used the Fourier transform IR spectrometer (FTIR) microscope and traditional black body emission to check the response spectrum and responsivity of b-P and b-PAs alloys.[7] It was found that the cut-off wavelength can be tuned from 3.9 to 4.6 μm over a composition range 0%–91% of As, with a peak room temperature detectivity of $\sim 10^{10}$ Jones. This is a record achievement because the detectivity is even one order of magnitude larger than that of most commercial mid-IR photodetectors. Moreover, this detector is based on the photovoltaic effect implying a higher response speed than thermal detectors. The main performance of reported b-P IR photodetectors is listed in Table 4.2.

From the current point of view, b-P is a very suitable layered candidate that could be employed for practical mid-IR optoelectronics. First, b-P and its alloys have decent bandgaps. The reported results have demonstrated a widely tunable detection spectrum based on As composition. Additionally, another b-P derivative, b-P carbide,[8] has been proved to show a wide optical absorption spectrum up to 8 μm. Second, b-P field effect transistors (FETs) have a high mobility of hundreds of $cm^2\ V^{-1}\ s^{-1}$ or even higher. The theoretical mobility of b-P carbide FET is even as high as $10^5\ cm^2\ V^{-1}\ s^{-1}$.[15] Large mobility indicates short carrier transit time and great potency of high optical gain. Thin 2D materials suffer from low light absorption; thus, optical gain is an efficient approach to extracting the weak light signal. The recently reported graphene detector enhanced by the photoconductive effect even achieved a considerable responsivity covering the spectrum from visible to far-IR.[16] Third, the dark current of a b-P channel can be tuned to a relatively low level.[13] Dark current directly suggests the noise level of a photodetector. Though the mobility of graphene is even higher, the on/off ratio of a graphene FET (including monolayer) is very low, leading to an uncontrolled dark current. Fortunately, b-P FET is able to work at an off-state simultaneously with a high mobility. Fourth, the intrinsic anisotropy of b-P reveals the potency for IR polarization-sensitive detection. Currently,

Table 4.2 Performance comparison between various IR b-P detectors.

Composition	Thickness (nm)	Detection wavelength (μm)	Responsivity (A/W)	Time constant/ bandwidth	Test condition	Ref.
b-P	11.5	1.55	0.135	3 GHz	RT, laser	13
b-P	30∼50	3.75	∼0.001	–	RT, laser	10
b-P	∼10	3.39	82/0.062	1.1 kHz/>10 kHz	RT, 1.6 nW incident Power	14
b-P	∼5	3.4/5/7.7	0.518/0.03/0.0022	1.3 GHz (predicted)	77 K, laser	12
b-P/b-$P_{0.09}As_{0.91}$	∼28	Up to 3.9/4.6	–	117 kHz	RT, FTIR	7
b-$P_{0.17}As_{0.83}$	5∼20	Up to 8.2	–	0.5 ms	RT, laser	11
b-P carbide	7.5	Up to 8	2163 A/W at 2004 nm	–	RT, Laser	8

high-performance thin-film IR detectors develop toward a tendency, wherein more information carried by light such as polarization and phase can be explored. In the case of polarization sensitivity, polaroid polarizer or metal nanostructures are required for traditional thin-film detectors. The inner anisotropy of b-P depicts a more efficient approach.

4.2.2 Graphene

Several mechanisms can be harnessed to fabricate a graphene-based IR photodetector. Collecting the photoexcited free carriers is the most common operation. Owing to the zero bandgap, graphene shows broadband light absorption from visible to even terahertz. However, most graphene detectors have a responsivity <10 mA/W because of low light absorption and short carrier lifetime. In 2013, Zhang et al. processed graphene nanosheet into a graphene quantum dot-like structure, thus introducing electron trapping centers.[17] These trapping centers lead to a long carrier lifetime and help to achieve high responsivity from visible to IR (0.4 A/W in the mid-IR range). In 2014, Liu et al. demonstrated a graphene double-layer photodetector, in which a thin tunnel layer was inserted between the two monolayer graphene, and photoexcited electron-hole pairs could be spatially separated into the two layers.[18] This also leads to a considerable mid-IR responsivity higher than 1 A/W. The drawback of these two typical graphene mid-IR detectors is that the photoresponse speed is quite slow, which is undesired for practical applications. To ensure both responsivity and bandwidth are satisfactory, the collection efficiency of photoexcited carriers should be improved, while the carrier lifetime cannot be very long. Structural design such as using metallic antenna-like nanoarrays has been demonstrated to be an effective way.[16, 19] By fabricating metallic antenna structures on graphene, Yao et al. realized a room temperature mid-IR detector with 200% enhancement of photoresponsivity (0.4 V/W at a wavelength of 4.45 μm).[19] These nanostructures not only help couple light into graphene but also efficiently separate the photoexcited electron-hole pairs as small p-n-p units. The temporal response of this detector reaches ~60 ns, implying the potency for high-speed detection. Very recently, Cakmakyapan et al. designed gold nanostripes as contact electrodes on graphene wherein the distance between the stripe tips was only 50 nm. This structure enables ultrashort carrier transit time and high photoconductive gain as well as good optical absorption.[16] Consequently, a high responsivity of 11.5 A/W was even achieved at a far-IR wavelength of 20 μm. Meanwhile,

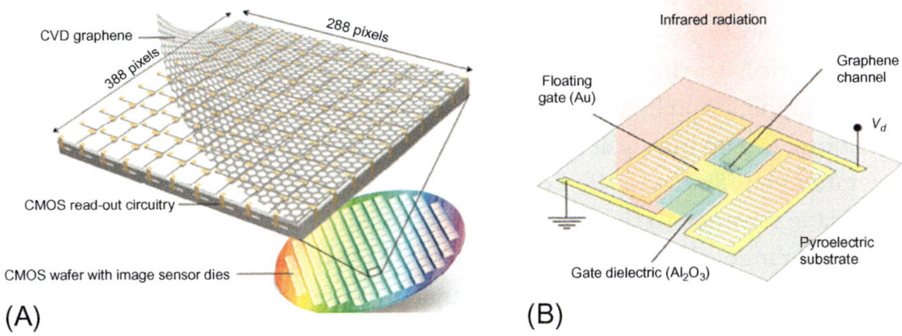

Fig. 4.2 (A) Integrated graphene/CMOS focal array.[20] (B) Schematic of pyroelectric graphene detectors.[21]

this ultra-broadband detector shows a high bandwidth exceeding 50 GHz (Fig. 4.2).

Utilizing the photothermoelectric effect instead of the photovoltaic effect can also achieve room temperature mid-IR graphene detectors. Nonuniform heating or Seebeck coefficient difference in the photosensitive region of devices will cause the difference of carrier diffusion velocity, leading to a photovoltage. Nonlocal hot carrier transport in graphene has been applied to elucidate the photocurrent mechanism.[22, 23] By integrating graphene-based photothermoelectric detectors with IR absorber silicon nitride membranes as a whole thermal isolating structure, Hsu et al. achieved a room temperature responsivity of 7–9 V/W ($\lambda = 10.6$ μm, tested by CO_2 laser) and a fast response time of ~23 ms.[24] Moreover, this detector is sensitive to black body and human targets with a detectivity on the order of 10^8 Jones.

Generally, graphene shows minimal light absorption in the mid-IR range, because the interband transition in graphene locates in the visible to near-IR waveband and the intraband transition is corresponding to the far-IR to terahertz, while in the mid-IR range, Pauli blocking works.[25] However, it has been demonstrated that the plasmons in graphene can be tuned by gate voltage, thus could be at technologically relevant mid-IR frequencies.[26, 27] The observed graphene plasmon wavelength is on the order of 200 nm,[26, 27] which is ~40 times shorter than the incident wavelength (a 20 times shorter plasmon wavelength also has ever been reported[28]), indicating a highly confined optical field in a small volume. Enhanced mid-IR absorption[29] and photocurrent[30] based on this regime have been experimentally characterized, paving the way toward novel manipulation of mid-IR light.

In addition to the several graphene-based mid-IR detectors discussed earlier, graphene can be indirectly used to detect IR photons. High mobility and favorable gate tuning are the two advantages of graphene. Goossens et al. integrated graphene with PbS quantum dots and achieved a 388×288 focal array plane sensitive to the ultraviolet, visible, and IR light (up to 2.0 μm).[20] In the IR range, the light absorber is the quantum dots, while graphene is used to help one of the photoexcited carriers to speed up transit, thus leading an ultrahigh gain as much as 10^7. Though several difficulties such as transfer and encapsulation of graphene and uniformity of pixels remain to be improved, this achievement tells a bright future of 2D material-based room temperature IR detectors. By integrating graphene with a traditional pyroelectric material, $LiNbO_3$, a room temperature mid-IR bolometer has been demonstrated.[21] In this design, graphene is tuned by the temperature change induced and further amplified floating gate voltage, and thus is sensitive to thermal irradiation, including black body and human hands. Through this design, an ultralow noise equivalent temperature difference of 15 μK was estimated, while the detectivity was on the order of 10^4 Jones at 1 Hz bandwidth, which is much lower than the commercial thermal detectors ($\sim 10^9$ Jones).

4.2.3 Noble metal dichalcogenides

The search for layered semiconductors with narrow bandgaps and high mobility among the TMDCs is of great significance toward the realization of high-speed logic electronics and high-performance IR detectors. Noble metal dichalcogenides provide such an opportunity. The theoretical mobility of PtX_2 (X=Se, S) at room temperature is over 1000 $cm^2\ V^{-1}\ s^{-1}$,[31] and the bandgaps of their bilayers and bulks could be very small.[4, 32–34] Even for the monolayer $PdSe_2$, its bandgap could be as low as 0.39 eV.[4] Experimental results based on PtS_2 FETs show that the bandgap, mobility, and transistor on/off ratio are very sensitive to the number of layers.[34] Larger film thickness means a narrower bandgap, higher mobility, but harder gate tunability. For a 9-nm $PdSe_2$ film, its bandgap is about 0.16 eV, and the corresponding FET shows a high mobility of 216 $cm^2\ V^{-1}\ s^{-1}$ and a large on/off ratio of 10^3 (Fig. 4.3).[33]

Currently, few optoelectronic properties of noble metal dichalcogenides have been explored. In 2018, Yu et al. reported an Ar-plasmon-treated

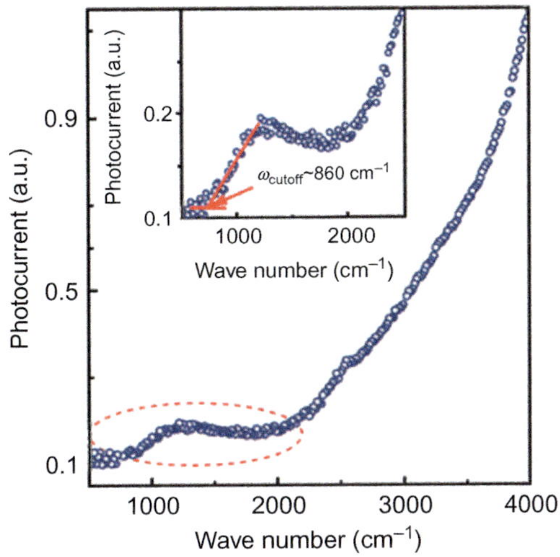

Fig. 4.3 Photocurrent response spectrum of $PtSe_2$ photoconductor measured by FTIR.[32]

bilayer $PtSe_2$ mid-IR photodetector with a responsivity and response time of 4.5 A/W and 1 ms at a wavelength of 10.0 μm, respectively.[32] Defect engineering plays an important role in determining the material bandgap. Further photocurrent measurement by FTIR confirmed that the bilayer $PtSe_2$ showed a wide response from visible to IR with a cut-off frequency of 860 cm^{-1}. This work is of great significance for revealing that narrow-bandgap 2D noble metal dichalcogenides could be novel platforms for room temperature IR optoelectronics.

4.3 Challenges

Despite the achievements mentioned earlier, the biggest challenge that limits the practical application of 2D semiconductors lies in material synthesis and processing. The large area and uniform growth of the desired 2D materials is an issue that remains to be tackled. Nevertheless, 2D IR detectors have a long way to go, and we look forward to future developments.

References

1. Chhowalla, M.; Shin, H. S.; Eda, G.; Li, L. J.; Loh, K. P.; Zhang, H. *Nat. Chem.* **2013**, *5*, 263–275.
2. Wilson, J. A.; Yoffe, A. D. *Adv. Phys.* **1969**, *18*, 193–335.
3. Duan, X.; Wang, C.; Pan, A.; Yu, R.; Duan, X. *Chem. Soc. Rev.* **2015**, *44*, 8859–8876.

4. Miro, P.; Ghorbani-Asl, M.; Heine, T. *Angew. Chem. Int. Ed. Eng.* **2014**, *53*, 3015–3018.
5. Xia, F.; Wang, H.; Jia, Y. *Nat. Commun.* **2014**, *5*, 4458.
6. Low, T.; Rodin, A. S.; Carvalho, A.; Jiang, Y.; Wang, H.; Xia, F.; Castro Neto, A. H. *Phys. Rev. B* **2014**, 90.
7. Amani, M.; Regan, E.; Bullock, J.; Ahn, G. H.; Javey, A. *ACS Nano* **2017**, *11*, 11724–11731.
8. Tan, W. C.; Huang, L.; Ng, R. J.; Wang, L.; Hasan, D. M. N.; Duffin, T. J.; Kumar, K. S.; Nijhuis, C. A.; Lee, C.; Ang, K. W. *Adv. Mater.* **2018**, 30.
9. Liu, B.; Kopf, M.; Abbas, A. N.; Wang, X.; Guo, Q.; Jia, Y.; Xia, F.; Weihrich, R.; Bachhuber, F.; Pielnhofer, F.; Wang, H.; Dhall, R.; Cronin, S. B.; Ge, M.; Fang, X.; Nilges, T.; Zhou, C. *Adv. Mater.* **2015**.
10. Yuan, H.; Liu, X.; Afshinmanesh, F.; Li, W.; Xu, G.; Sun, J.; Lian, B.; Curto, A. G.; Ye, G.; Hikita, Y.; Shen, Z.; Zhang, S. C.; Chen, X.; Brongersma, M.; Hwang, H. Y.; Cui, Y. *Nat. Nanotechnol.* **2015**, *10*, 707–713.
11. Long, M.; Gao, A.; Wang, P.; Xia, H.; Ott, C.; Pan, C.; Fu, Y.; Erfu Liu, X. C.; Lu, W.; Nilges, T.; Xu, J.; Wang, X.; Hu, W.; Miao, F. *Sci. Adv.* **2017**, *3,* e1700589.
12. Chen, X.; Lu, X.; Deng, B.; Sinai, O.; Shao, Y.; Li, C.; Yuan, S.; Tran, V.; Watanabe, K.; Taniguchi, T.; Naveh, D.; Yang, L.; Xia, F. *Nat. Commun.* **2017**, *8*, 1672.
13. Youngblood, N.; Chen, C.; Koester, S. J.; Li, M. *Nat. Photonics* **2015**, *9*, 247–252.
14. Guo, Q.; Pospischil, A.; Bhuiyan, M.; Jiang, H.; Tian, H.; Farmer, D.; Deng, B.; Li, C.; Han, S. J.; Wang, H.; Xia, Q.; Ma, T. P.; Mueller, T.; Xia, F. *Nano Lett.* **2016**, *16*, 4648–4655.
15. Tan, W. C.; Cai, Y. Q.; Ng, R. J.; Huang, L.; Feng, X. W.; Zhang, G.; Zhang, Y. W.; Nijhuis, C. A.; Liu, X. K.; Ang, K. W. *Adv. Mater.* **2017**, 29.
16. Cakmakyapan, S.; Lu, P. K.; Navabi, A.; Jarrahi, M. *Light Sci. Appl.* **2018**, 7.
17. Zhang, B. Y.; Liu, T.; Meng, B.; Li, X.; Liang, G.; Hu, X.; Wang, Q. J. *Nat. Commun.* **2013**, *4*, 1811.
18. Liu, C. H.; Chang, Y. C.; Norris, T. B.; Zhong, Z. *Nat. Nanotechnol.* **2014**, *9*, 273–278.
19. Yao, Y.; Shankar, R.; Rauter, P.; Song, Y.; Kong, J.; Loncar, M.; Capasso, F. *Nano Lett.* **2014**, *14*, 3749–3754.
20. Goossens, S.; Navickaite, G.; Monasterio, C.; Gupta, S.; Piqueras, J. J.; Pérez, R.; Burwell, G.; Nikitskiy, I.; Lasanta, T.; Galán, T.; Puma, E.; Centeno, A.; Pesquera, A.; Zurutuza, A.; Konstantatos, G.; Koppens, F. *Nat. Photonics* **2017**, *11*, 366–371.
21. Sassi, U.; Parret, R.; Nanot, S.; Bruna, M.; Borini, S.; De Fazio, D.; Zhao, Z.; Lidorikis, E.; Koppens, F. H.; Ferrari, A. C.; Colli, A. *Nat. Commun.* **2017**, *8*, 14311.
22. Song, J. C. W.; Rudner, M. S.; Marcus, C. M.; Levitov, L. S. *Nano Lett.* **2011**, *11*, 4688–4692.
23. Gabor, N. M.; Song, J. C. W.; Ma, Q.; Nair, N. L.; Taychatanapat, T.; Watanabe, K.; Taniguchi, T.; Levitov, L. S.; Jarillo-Herrero, P. *Science* **2011**, *334*, 648–652.
24. Hsu, A. L.; Herring, P. K.; Gabor, N. M.; Ha, S.; Shin, Y. C.; Song, Y.; Chin, M.; Dubey, M.; Chandrakasan, A. P.; Kong, J.; Jarillo-Herrero, P.; Palacios, T. *Nano Lett.* **2015**, *15*, 7211–7216.
25. Low, T.; Avouris, P. *ACS Nano* **2014**, *8*, 1086–1101.
26. Fei, Z.; Rodin, A. S.; Andreev, G. O.; Bao, W.; McLeod, A. S.; Wagner, M.; Zhang, L. M.; Zhao, Z.; Thiemens, M.; Dominguez, G.; Fogler, M. M.; Castro Neto, A. H.; Lau, C. N.; Keilmann, F.; Basov, D. N. *Nature* **2012**, *487*, 82–85.
27. Chen, J.; Badioli, M.; Alonso-Gonzalez, P.; Thongrattanasiri, S.; Huth, F.; Osmond, J.; Spasenovic, M.; Centeno, A.; Pesquera, A.; Godignon, P.; Elorza, A. Z.; Camara, N.; Garcia de Abajo, F. J.; Hillenbrand, R.; Koppens, F. H. *Nature* **2012**, *487*, 77–81.

28. Brar, V. W.; Jang, M. S.; Sherrott, M.; Lopez, J. J.; Atwater, H. A. *Nano Lett.* **2013**, *13*, 2541–2547.
29. Emani, N. K.; Chung, T. F.; Ni, X. J.; Kildishev, A. V.; Chen, Y. P.; Boltasseva, A. *Nano Lett.* **2012**, *12*, 5202–5206.
30. Freitag, M.; Low, T.; Zhu, W.; Yan, H.; Xia, F.; Avouris, P. *Nat. Commun.* **2013**, *4*, 1951.
31. Zhang, W.; Huang, Z.; Zhang, W.; Li, Y. *Nano Res.* **2014**, *7*, 1731–1737.
32. Yu, X.; Yu, P.; Wu, D.; Singh, B.; Zeng, Q.; Lin, H.; Zhou, W.; Lin, J.; Suenaga, K.; Liu, Z.; Wang, Q. J. *Nat. Commun.* **2018**, *9*, 1545.
33. Chow, W. L.; Yu, P.; Liu, F.; Hong, J.; Wang, X.; Zeng, Q.; Hsu, C. H.; Zhu, C.; Zhou, J.; Wang, X.; Xia, J.; Yan, J.; Chen, Y.; Wu, D.; Yu, T.; Shen, Z.; Lin, H.; Jin, C.; Tay, B. K.; Liu, Z. *Adv. Mater.* **2017**, *29*.
34. Zhao, Y.; Qiao, J.; Yu, P.; Hu, Z.; Lin, Z.; Lau, S. P.; Liu, Z.; Ji, W.; Chai, Y. *Adv. Mater.* **2016**, *28*, 2399–2407.

CHAPTER 5

Two-dimensional materials toward future photovoltaic devices

Lin Wang, Kah-Wee Ang
Department of Electrical and Computer Engineering, National University of Singapore, Singapore, Singapore
Centre for Advanced 2D Materials, National University of Singapore, Singapore, Singapore

5.1 Introduction

Currently, growing use of renewable energy is strongly demanded to meet the rapidly increasing global energy requirement constrained by limited nonrenewable energy sources and environmental crisis. Photovoltaics, which convert solar energy into electric energy, represent one of the most practical and sustainable solutions to overcome such challenge.[1] To realize economically feasible solar energy conversion at global scale, it is important to develop inexpensive semiconducting materials that enable facile, scalable, and economical technical process.[2] In the meantime, the society and technology developments have envisioned next-generation electronic devices with particular properties such as flexibility and wearability,[3] for a wide range of applications. However, realizing these proposed applications with conventional rigid semiconducting materials is difficult.[4, 5] Following the discovery of single-atom-thick graphene in 2004, a vast number of materials in two-dimensional (2D) layered forms have been (re)discovered. Their bandgaps span from zero (graphene) to as large as 6.0 eV (hexagonal boron nitride [hBN]) and their electronic properties vary from (semi)metallic to semiconducting and to insulating (Fig. 5.1).[6–10] Generally, in these layered materials, the atoms are held together by strong in-plane covalent bonds, whereas the atomically thin layers are stacked perpendicularly to the 2D plane with weak van der Waals (vdW) interactions to form a three-dimensional (3D) crystal. Compared with traditional 3D semiconducting materials, 2D materials present many exceptional properties that are appealing for future optoelectronic applications, including photovoltaics.[11, 12] First, the atomic thickness renders these materials inherently flexible and almost transparent, which is of high interest for new generation of

2D Materials for Photonic and Optoelectronic Applications
https://doi.org/10.1016/B978-0-08-102637-3.00005-X

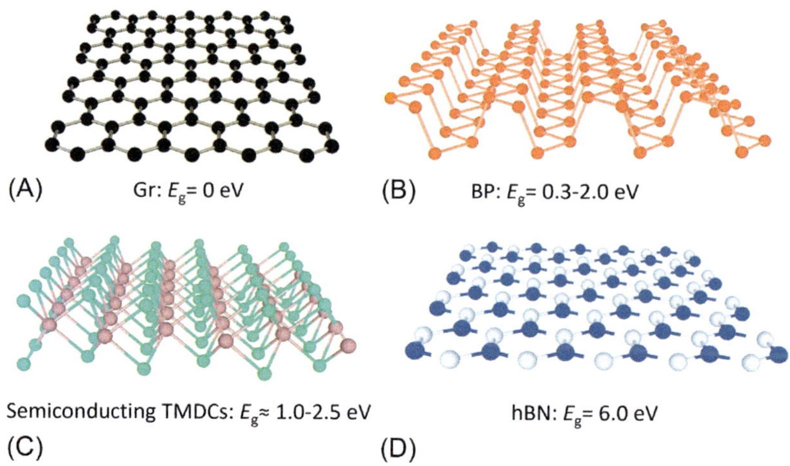

Fig. 5.1 Atomic structures of representative 2D materials covering an array of electrical properties and their bandgaps. (A) Graphene (Gr), which is a semimetal having a zero-gap energy band and ultrahigh carrier mobility. (B) Black phosphorus (BP), known for its relatively large carrier mobility and layer-dependent bandgap ranging from 0.3 eV (bulk) to 2.0 eV (monolayer), corresponding to a spectral range from visible to mid-infrared. (C) Semiconducting TMDCs, which have thickness-dependent bandgaps from ~1.0 to 2.5 eV, covering a wide electromagnetic spectrum from visible to near-infrared regime. (D) hBN, whose large bandgap of ~6.0 eV makes it suitable for gate dielectric in 2D materials system. *Reproduced with permission from Wang, L.; Huang, L.; Tan, W. C.; Feng, X.; Chen, L.; Huang, X.; Ang, K.-W., 2D Photovoltaic Devices: Progress and Prospects. Small Methods* ***2018****, 2(3), 1700294. Copyright 2018, Wiley-VCH.*

optoelectronics where lightweight, flexibility, bendability, and transparency are desired. As for the rigid Si and GaAs wafers, they are non-transparent and non-flexible; thus, specific mechanical structures have to be designed to obtain desired flexibility. Second, quantum confinement effects in the out-of-plane direction give rise to novel electronic and optical properties distinctively different from their bulk counterparts. For instance, the reduced thickness results in strongly bound excitons in semiconducting transition metal dichalcogenides (TMDCs), largely increasing their absorption efficiency. Third, the bandgap of 2D materials exhibits a layer number-dependent character, making it possible for optical absorption engineering. Finally and importantly, the naturally passivated surfaces of 2D materials without any dangling bonds open up abundant possibilities to use different layered materials to construct artificial materials for diverse applications, in the absence of "lattice mismatch" trouble encountered in conventional heterostructures.[13] Particularly, in terms of photovoltaic application, 2D layered materials hold some unique advantages that are not offered by present mature technology. For example, although Si-based photovoltaic technology is very mature, it suffers from high material cost per output power due to its low spectral absorption of the solar spectrum. Another concern of present solar cell industry is its dependence on certain materials such as Pt and In (for transparent indium tin oxide [ITO] electrode), which suffer scarcity, high cost, and even toxicity. In contrast, the new class of 2D materials are ultrathin, highly absorbent per thickness, eco-friendly, abundant, mechanically strong, as well as semi-transparent, rendering them well suitable for future-generation photovoltaic cells that can be readily integrated to various surfaces including even human bodies. Furthermore, the large variety of 2D materials with diverse bandgaps offer good possibility of bandgap engineering for high-efficiency photovoltaic process.

As a matter of fact, owing to their large optical absorption ($>10^7\ m^{-1}$) across the visible spectral range, the parental materials of 2D crystals have been investigated for solar energy conversion decades ago.[14] For example, WSe_2 and $MoSe_2$ crystals were used as photoanodes, and solar-to-electrical power conversion efficiencies up to 10.2% and 9.4% have been achieved.[15] Photovoltaic effect arising from metal/TMDC Schottky barriers was investigated, and efficient solar energy harvesting has been demonstrated.[16, 17] Nevertheless, the successful isolation of graphene has led to renewed interest in these layered crystals in their single- or few-layer form, which shows distinct optical properties. For instance, unlike their bulk form possessing an indirect bandgap, TMDC monolayers show a direct bandgap that is very attractive for optoelectronics.[18] Moreover, as a result of strong light-matter interactions from dipole transitions with large joint density of states as well as oscillator strengths between localized d-states with strong spatial overlap, these monolayer or few-layer crystals exhibit superior light-harvesting characteristics.[19, 20]

Because of the exceptional potential in optoelectronics promised by 2D materials, considerable effort has been made to demonstrate and realize their applications in light energy conversion. Until now, a large number of semiconducting crystals in the monolayer or few-layer form, currently featured by graphene, TMDCs, and BP, have been exploited for next-generation ultrathin photovoltaic devices.[11, 21, 22] Moreover, in virtue of their difference from conventional bulk materials, various architectures are proposed and utilized for investigating their photovoltaic characteristics.

5.2 Photovoltaics in 2D materials

5.2.1 2D Schottky junction-based photovoltaics

Typically, the photovoltaic process happens within a junction (p-n junction or Schottky junction) possessing internal built-in electric field. Upon illumination, electron-hole pairs are generated (known as photogenerated carriers) by incident photons equal to or more energetic than the semiconductor bandgap. They are separated as a result of the junction internal electric field and then collected by electrodes, generating electric power. Table 5.1 summarizes important figures of merit used to characterize a photovoltaic process (or device).

For metal/semiconductor contact, a Schottky junction can be formed with internal electric field as a result of the difference between metal work function and semiconductor Fermi level, allowing photocurrent generation (Fig. 5.2A). While a symmetric metal contact structure gives rise to no net photocurrent, asymmetric Schottky barriers can be realized by using metals with different work functions, thus yielding a nonzero internal electrical field and allowing net photocurrent generation (Fig. 5.2B). Fontana et al. fabricated an MoS_2-based (50 nm thick) phototransistor with gold (Au) and palladium (Pd) for source and drain contact, respectively.[23] As expected,

Table 5.1 Important figures of merit for a photovoltaic device.

Figures of merit	Definition
Power conversion efficiency (PCE)	The ratio of electrical power generated by the device to the incident light power on it
Short-circuit current (I_{sc})	Current flowing through the device under illumination and at zero external bias with its contacts shorted
External quantum efficiency (EQE)	The ratio of the number of charge carriers flowing through the device under short-circuit current conditions to the total number of impinging photons on it
Internal quantum efficiency (IQE)	The ratio of the number of charge carriers flowing through the device under short-circuit current conditions to the total number of absorbed photons
Open-circuit voltage (V_{oc})	The voltage generated by the device under illumination and with no current flowing
Fill factor (FF)	The ratio of the maximum electric power generated by the device to the product of its short-circuit current and its open-circuit voltage

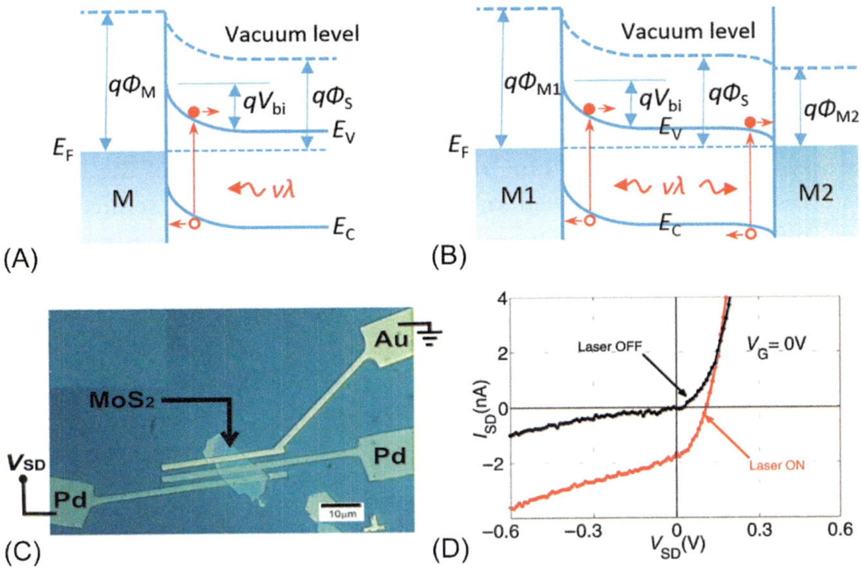

Fig. 5.2 (A) Schematic of Schottky junction-based photovoltaic process. (B) Schematic of net photocurrent generation process in an asymmetric Schottky barrier structure caused by contact metals with different work functions. (C) Optical image of an example device with Pd-Au asymmetric contact configuration. (D) I_d-V_d curve at 0 V V_g of the device in the dark and under illumination showing diode-like behavior and evident photovoltaic effect. *Panels (C, D) reproduced with permission from Fontana, M.; Deppe, T.; Boyd, A. K.; Rinzan, M.; Liu, A. Y.; Paranjape, M.; Barbara, P., Electron-Hole Transport and Photovoltaic Effect in Gated MoS2 Schottky Junctions. Sci. Rep.* ***2013****, 3, 1634. Copyright 2013, Nature Publishing Group.*

the device exhibited photovoltaic effect under global illumination (Fig. 5.2).[23, 24] However, the photovoltaic efficiency of the device was extremely poor, as reflected by the low open-circuit voltage (V_{oc}) (only 0.1 V for 100 mW/cm^2 at 532 nm wavelength), which is much smaller than the semiconductor's bandgap (1.2 eV for 50 nm MoS_2). Similarly, a vertical-structured Au/MoS_2 Schottky barrier-based solar cell was fabricated with chemical vapor deposition (CVD)-grown MoS_2 nanomembranes as a photoactive layer and ITO as transparent electrode.[25] The device can efficiently absorb light in the 350–950 nm range to generate electrical power, and the power conversion efficiency reached 0.7% for ~110 nm MoS_2 and 1.8% for ~220 nm MoS_2. Furthermore, the optimization of band alignment within the phototransistor can be realized to attain higher photovoltaic performance, as demonstrated in solar cells based on $Mo_{0.5}W_{0.5}Se_2$, which have shown enhanced photovoltaic efficiency than $MoSe_2$ and WSe_2 devices.[26]

5.2.2 2D p-n homojunction-based photovoltaics

Photovoltaic effect within a Schottky junction is relatively limited and results in low efficiency. Therefore, more efforts are devoted to utilize semiconducting p-n junction in layered materials for improved photovoltaic process. This requires obtaining different conducting types of the layered materials of interest. As is well known, in conventional bulk semiconductors, different conduction types (namely electron-majority n-type and hole-majority p-type) for creating p-n junctions are realized by selectively incorporating particular substitutional impurity atoms. However, such techniques are relatively difficult to be implemented in nanostructured materials due to their ultrasmall size and fragility. Nevertheless, the ultrathin nature of 2D layered materials brings novel possibilities for their conduction engineering.

In 2014, Pospischil et al., Baugher et al., and Ross et al. independently demonstrated p-n junctions within monolayer WSe_2 by using a split-gate configuration (Fig. 5.3).[27, 28] Principally, WSe_2 monolayer flake was prepared by mechanical exfoliation and then transferred onto a pair of split gates covered with previously grown gate dielectric materials such as HfO_2 or SiN. Since the conduction type and carrier density within the monolayer-thick channel can be modulated through electrostatic control by applying different voltages on the two local gates, an atomically thin p-n junction was created, which exhibited considerable current-rectifying diode behavior and was capable of photovoltaic power generation. Particularly in the work of Pospischil et al., under illumination of 140 mW/cm^2 (halogen lamp light source), a V_{oc} of 0.64 V and a power conversion of 0.5% were recorded by assuming the gap between the two gates as photoactive area. Considering a 95% transparency of the single-layer WSe_2, the result indicates highly efficient photovoltaic energy conversion, which makes monolayer TMDC attractive for semitransparent solar cells. Furthermore, by using other layered materials, e.g., few-layer hBN as gate dielectric and graphene as gate electrode, an all-2D system can be established for optoelectronic energy devices.[29, 30] In order to obtain TMDC p-n junctions with superior photovoltaic performance, relatively thick TMDC sheets can be employed. For example, electrostatically generated p-n junctions based on multilayer $MoSe_2$ have given rise to V_{oc} in the range of 0.7–0.85 V, already comparable to the typical value for high-quality monocrystalline Si solar cells. Moreover, the photovoltaic efficiencies exceed 14% (~24 times larger than the maximum value observed for monolayers) under standard AM 1.5 illumination

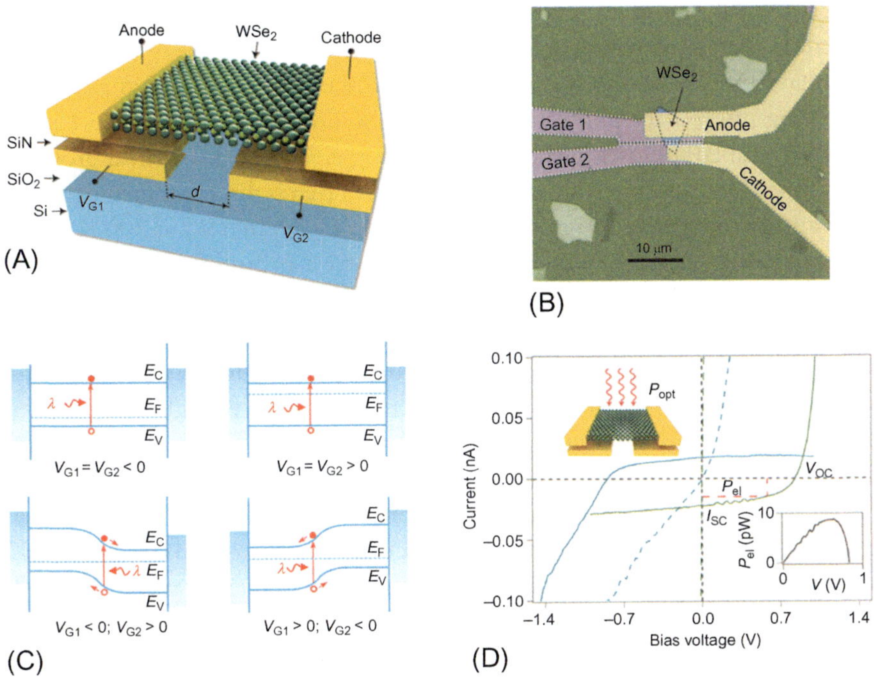

Fig. 5.3 Photovoltaic effect in a WSe_2 monolayer diode device defined by split-gate electrodes. (A) Schematic of device structure. (B) Colored microscope image of the device. The gap between the two gates is 460 nm wide. (C) Four configurations of WSe_2 channel modulated by different biasing conditions of the two gates. (D) Device operation as solar cell and photodiode. *I-V* characteristics of the device under optical illumination with 140 mW/cm^2. The biasing conditions: p-n (solid *green* line; $V_{G1}=-40$ V, $V_{G2}=40$ V), n-p (solid *blue* line; $V_{G1}=40$ V, $V_{G2}=-40$ V), n-n (dashed *green* line; $V_{G1}=V_{G2}=40$ V), p-p (dashed *blue* line; $V_{G1}=V_{G2}=-40$ V). When operated as a diode (solid lines), electrical power (P_{el}) can be extracted. Top inset: Schematic of experiment. Lower inset: P_{el} vs. voltage under incident illumination of 140 mW/cm^2. Maximum PCE is obtained for $V=0.64$ V and $I=14$ pA. The *red* dashed rectangle in the main panel shows the corresponding power area. *Reproduced with permission from Pospischil, A.; Furchi, M. M.; Mueller, T., Solar-Energy Conversion and Light Emission in an Atomic Monolayer p-n Diode. Nat. Nanotechnol.* ***2014****, 9(4), 257-61. Copyright 2014, Nature Publishing Group.*

(100 mW/cm^2) with ~0.7 fill factor (FF), implying the remarkable potential of multilayer TMDC junction devices.[31] Here the higher power conversion efficiencies (PCEs) were attributed to the increased light absorption and consequent larger-amount electron-hole pair generation owing to the larger flake thickness, suggesting a potential of trade-off engineering between

device optical transparency and efficiency by choosing appropriate film thickness to meet various application requirements.

In addition to TMDCs, the recently rediscovered 2D BP has attracted increasing interest for photovoltaics as well. Compared with semiconducting TMDCs, BP holds a number of appealing electrical and optical properties, especially its sizable bandgap (0.3 ~ 2.0 eV) that is directly dependent on the number of layers, its high carrier mobility (1000 cm^2/V s at 300 K), and its in-plane anisotropy; thus, BP is considered a promising candidate for broadband optoelectronic applications.[8, 32, 33] Lateral BP p-n junction has been demonstrated by means of electrostatically local gating, showing potential for photovoltaic application.[34] Moreover, due to the relatively narrow bandgap of few-layer BP, its p-n junction can convert light energy into electrical power up to 940 nm, making it attractive for energy harvesting in the near-infrared regime.

Undoubtedly, the pioneering works were of significant importance in the demonstration of atomically thin p-n junctions as well as photovoltaic process. However, it is easy to learn that the realization of electrostatically defined 2D p-n junctions requires complicated split-gate structure followed by a transfer process and suffers from non-abrupt junction interface due to the divergence of static electric field. To address this issue, one alternative way is to create p-n junction in layered material via chemical doping or surface functionalization. As a matter of fact, various chemicals have been studied for doping purpose in semiconductor 2D materials to obtain improved electrical and optoelectronic properties. For example, intrinsic MoS_2 is normally n-type. To realize its p-n junction, Choi et al. utilized $AuCl_3$ solution as an effective electron acceptor.[35] Owing to the large positive reduction potential of $AuCl_3$, the reduction of $AuCl_4^-$ ions receives electrons from the MoS_2 layer, thereby leading to p-doping of the material through surface charge transfer. By selectively doping MoS_2 in this way, MoS_2 lateral p-n junction was obtained, which showed largely enhanced photovoltaic behavior, with 0.2 V V_{oc} and 31.2 nA I_{sc} at $V_G = 40$ V and 0.5 V V_{oc} and 46.7 nA I_{sc} at $V_G = -40$ V at 600 nm wavelength. In regard to BP, which intrinsically shows hole-dominating conduction, multiple chemicals have been proved helpful to achieve its n-type, with examples of Cu adatoms doping, Al atoms doping, potassium modification, benzyl viologen (BV) doping, Cs_2CO_3, and cross-linked poly(methyl methacrylate) surface modification.[36–41] Consequently, its p-n junction can be realized. In Ref. 37, Liu et al. realized lateral BP p-n junction by partially doping BP into n-type via Al doping through the ALD process (TMA was used as precursor). The resulting BP diode exhibits photovoltaic at infrared 1550 nm, and a V_{oc} of 0.14 V and I_{sc} of 1.75 nA were observed at 451 nW incident light power (Fig. 5.4). PCE was estimated to be 0.66% under the assumption that the power conversion took place in the 100-nm adjacent region near the junction interface.

It is to be noted that the pioneering photovoltaic demonstrations are based on lateral p-n junction architecture. These works, although of great importance toward ultimate 2D material photovoltaic application, usually

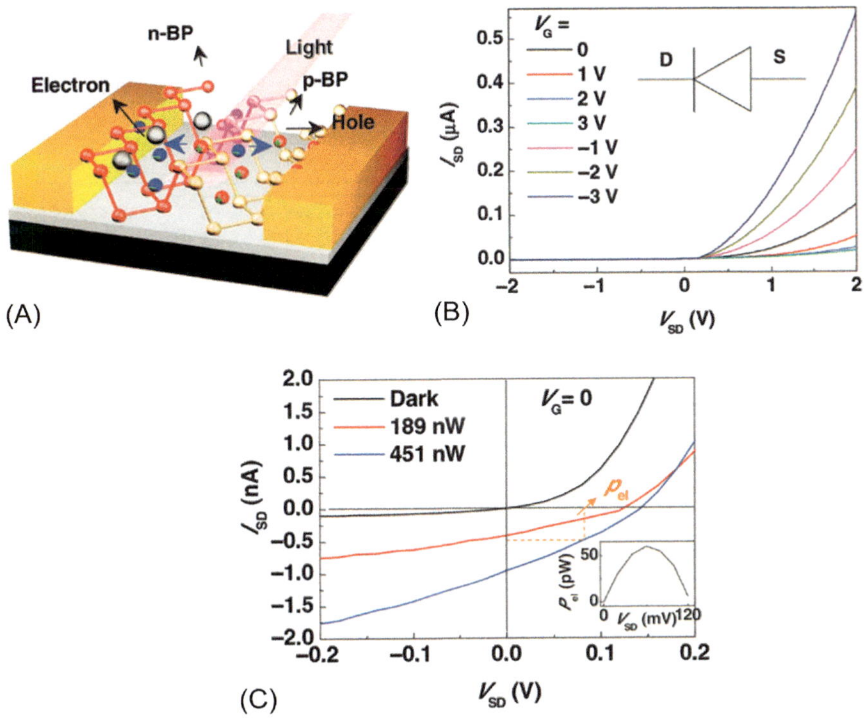

Fig. 5.4 BP p-n homojunction diode operation under photovoltaic effect with Al doping for n-type BP realization. (A) Schematic showing the separation of photogenerated electron-hole pairs due to the built-in electric field at the homojunction interface. The *blue* arrows show the electron and hole drift directions. (B) Gate-tunable rectifying characteristics of the device, indicating the presence of a p-n homojunction. (C) I_{SD}-V_{SD} curves measured under dark and light illumination without gate modulation at $V_g = 0$ V. The inset shows electrical power as a function of V_{SD} under light illumination of 451 nW. Maximum conversion efficiency of ~0.66% is obtained at $V_{SD} = 0.08$ V, $I_{SD} = 67.9$ nA. The yellow, dashed-line rectangle in the main panel shows the corresponding power area. *Reproduced with permission from Liu, Y.; Cai, Y.; Zhang, G.; Zhang, Y.-W.; Ang, K.-W., Al-Doped Black Phosphorus p-n Homojunction Diode for High Performance Photovoltaic. Adv. Funct. Mater.* ***2017****, 27(7), 1604638. Copyright 2017, Wiley-VCH.*

exhibit low PCEs because of their limited photoactive area, hence are unlikely to be adopted in practical photovoltaic devices. By comparison, the vertical junction structure is more reasonable for real application. In this regard, Wi et al. reported photovoltaic devices with a vertically stacked ITO/multilayer MoS_2/metal electrode structure.[42] In their work, p-n junctions of MoS_2 were created by utilizing a plasma-induced p-doping approach. The devices have exhibited an enhanced photovoltaic response with V_{oc}, J_{sc}, FF, PCE values of 0.28 V, 20.9 mA/cm^2, 0.47, and 2.8% obtained under AM 1.5G illumination, respectively. Li et al. fabricated an MoS_2 vertical p-n homojunction based on a single flake in another way (Fig. 5.5).[43] They used BV to n-dope the bottom surface of MoS_2 flake

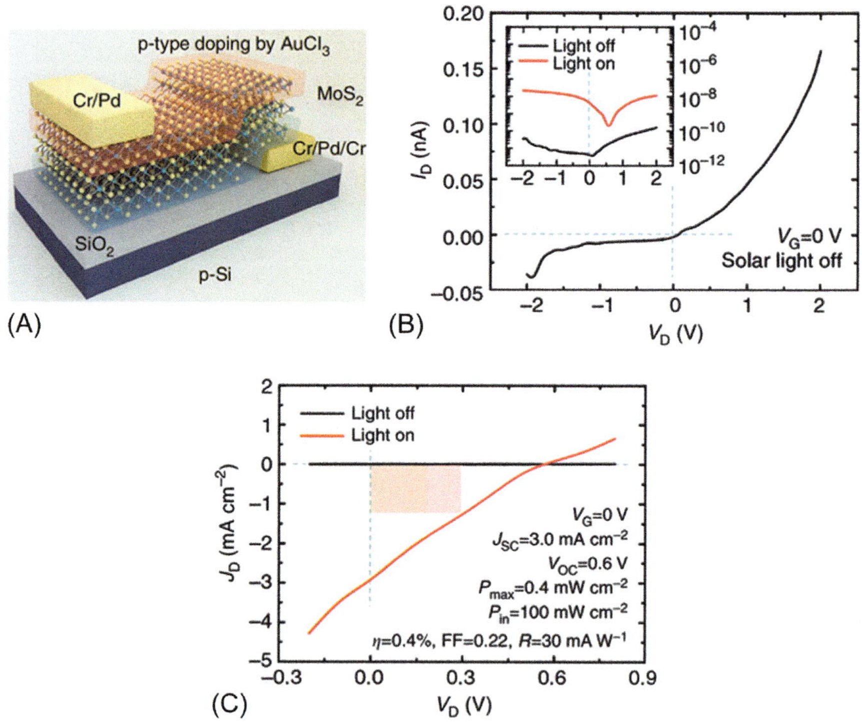

Fig. 5.5 (A) Creation of vertical p-n junction in the MoS_2 flake via BV doping (for n-type) and $AuCl_3$ doping (for p-type). (B) The MoS_2 p-n junction operates as a solar cell for light harvesting at 0 V V_G. (C) Current density as a function of V_D illustrating the energy conversion properties. *Reproduced with permission from Li, H. M.; Lee, D.; Qu, D.; Liu, X.; Ryu, J.; Seabaugh, A.; Yoo, W. J., Ultimate Thin Vertical p-n Junction Composed of Two-Dimensional Layered Molybdenum Disulfide. Nat. Commun.* ***2015****, 6, 6564. Copyright 2015, Nature Publishing Group.*

and $AuCl_3$ to dope its top surface into p-type. Cr/Pd metal contacts were made on each surface for device fabrication. The device shows potential for photovoltaic application with 0.6 V V_{oc} and 0.4% PCE demonstrated on a 11-nm-thick MoS_2 vertical p-n junction diode. By investigating the thickness-dependent vertical p-n junction formation, they found that the chemical doping depth was ~1.5 nm, and ultimate thickness limits for MoS_2 p-n homogeneous vertical junctions were about 3 nm. In another work, Jin et al. succeeded in synthesizing p-type $MoSe_2$ (intrinsically n-type) flakes via Nb elemental doping and fabricated ideal $MoSe_2$ vertical p-n junctions by vertically stacking the p-type and n-type $MoSe_2$ together.[44] The p-n junction exhibited ideal diode characteristics and can be well used for photovoltaics where a V_{oc} of 0.35 V and I_{sc} of 20 mA/cm^2 have been observed under 89 mW/cm^2 light intensity at 405 nm. In fact, the vertical stacking process well reflected the flexibility and diversity of combining two (or more) different 2D materials together, implying a brand-new platform of materials engineering to create new material systems with desired properties. This will be emphatically discussed later.

5.2.3 2D Heterojunction-based photovoltaics

5.2.3.1 Thickness-modulated 2D heterojunction

A well-known nature of 2D layered materials is their thickness-dependent (or layer number-dependent) bandgaps.[45–47] Such dependence implies energy band discontinuities at the interface of different layer-number films, forming a natural heterojunction distinct from conventional ones.[48–50] The study on this kind of structure is likely to expand the realms of 2D materials for diverse applications, including photovoltaics. In an early study, Xu et al. experimentally investigated a graphene monolayer-bilayer interface junction and observed photocurrent arising from the graphene heterostructure, which was attributed to photothermoelectric effect.[51] For TMDCs, Howell et al. investigated monolayer-multilayer MoS_2 heterojunctions exhibiting current-rectifying characteristics.[52] In scanning photocurrent microscopy measurement, a clear photocurrent was observed correlating to the MoS_2 thickness junction. Based on the comparison of experimental results with finite element device simulations, a type II band alignment between monolayer and multilayer MoS_2 was suggested, and the junction-induced current at zero bias was ascribed to hot electron contribution. However, in another work on MoS_2 lateral-thickness heterojunction, Tosun et al. observed a photocurrent dominated by the MoS_2 monolayer-multilayer interface rather than the MoS_2-metal Schottky contacts.[48] Furthermore, they attributed this photocurrent to the formation of a type I band alignment at the MoS_2 heterointerface. Using scanning tunneling microscopy and spectroscopy, Zhang et al. experimentally examined the lateral heterostructures between TMDC monolayer and bilayer ($MoSe_2$ and WSe_2).[49] It was shown that a type I band alignment is indeed formed at the atomically sharp heterointerfaces. The discrepancy of the results from these works indicates the complexity of naturally formed lateral-thickness heterojunctions in terms of their physical and optoelectrical properties, which still need to be further understood.[53] In addition, a lateral 1L-2L WSe_2 heterojunction transistor was created by selectively removing one layer of WSe_2 using photolithography and Ar plasma-etching process.[54] The device demonstrated enhanced photoresponse performance compared to those based on 1L or 2L uniform WSe_2 nanosheets, which was attributed to a more efficient electron-hole pair separation process in the heterojunction structure.

In comparison with semiconducting TMDCs, BP exhibits a more sizable thickness-dependent bandgap. Therefore, a more pronounced band discontinuity is readily expected at the interface of BP flakes with different layer numbers (in the few-layer range). Additionally, considering its direct bandgap property, this feature is likely to unleash the potential of BP lateral heterostructures for optoelectronic applications. Theoretical analysis revealed that both type I and type II band alignment can possibly form at the BP monolayer-bilayer interfaces, depending on its orientation or hydrogen passivation or the width of the heterostructures.[50] Recently, Wang et al. have experimentally demonstrated a gate-tunable all-BP lateral heterojunction diode for enhanced-performance photovoltaics (Fig. 5.6).[55] Thickness modulation gives rise to bandgap difference and energy band misalignment between BP flakes, which eliminates the need for double-gate structure or selective chemical doping or a transfer process. By adjusting the back-gate voltage V_g, the heterojunction device exhibits diode characteristics at certain V_g range and can work for energy harvesting both in the visible and near-infrared spectral range. A V_{oc} of 0.21 V and external quantum efficiency (EQE) of 7.4% have been obtained at 3.6 W/cm^2 light power (660 nm wavelength), which are higher than both locally defined BP homojunctions and that formed by intentional chemical doping, indicating the potential of such lateral heterostructure for photovoltaic devices with improved performance. Furthermore, as can be expected, the device can also be used for infrared photovoltaics owing to its narrow bandgap.

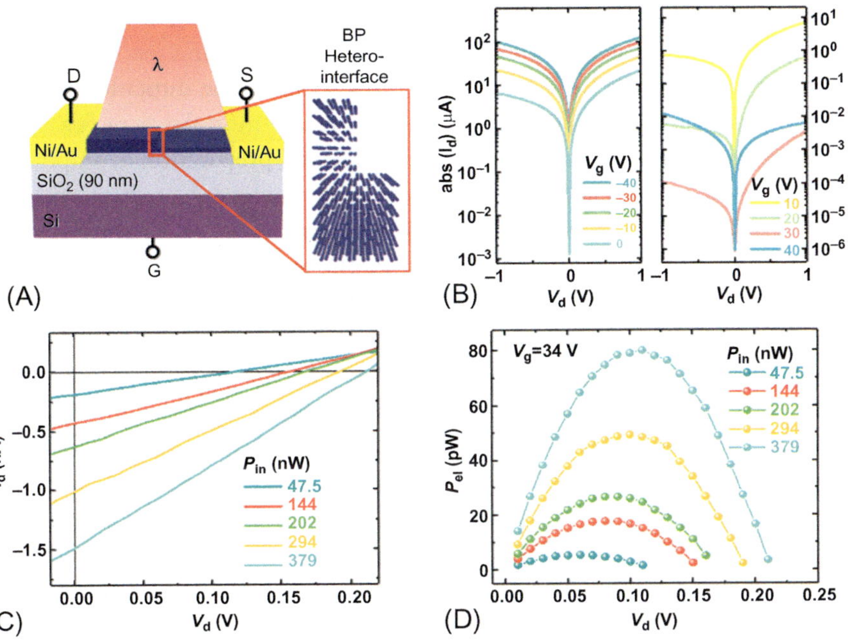

Fig. 5.6 Photovoltaics in all-BP lateral heterojunction diodes. (A) Schematic illustration of the thin-thick BP device structure. (B) I_d-V_d characteristics of the device at various V_g values. The device displays diode behavior at certain range of V_g. (C) I_d-V_d curves under varying incident laser power (660 nm) at $V_g = 34$ V, showing photovoltaic effect. (D) Corresponding generated electrical power vs. V_d for different incident laser power. *Reproduced with permission from Wang, L.; Huang, L.; Tan, W. C.; Feng, X.; Chen, L.; Ang, K.-W., Pronounced Photovoltaic Effect in Electrically Tunable Lateral Black-Phosphorus Heterojunction Diode. Adv. Electron. Mater.* ***2018****, 4(1), 1700442. Copyright 2018, Wiley-VCH.*

5.2.3.2 Lateral heterojunction involving different 2D materials

Heterogeneous p-n junction serves as a fundamental unit cell for many optoelectronic applications. In parallel with the advances in fabricating homogeneous p-n diodes in 2D materials, efforts have been made in realizing relevant heterojunction diode devices. To this end, one instinct approach is to successively grow another type of material along the edges of a previously grown 2D film. However, this process can be rather challenging due to its substantial difference from the case in traditional 3D materials.[56] First, the edge growth front of the first-grown 2D films can be easily passivated after the growth process and exposure to ambient conditions, and thus cannot function as an effective nucleation site for subsequent lateral epitaxy after being transferred into another growth chamber. Second, the atomically thin nanoflakes are in most cases too delicate to withstand the significant changes in growth temperature or conditions that are required for a second material growth to form the desired heterostructures. Despite the difficulty, following the growth of a large variety of TMDCs in the monolayer or few-layer form on common SiO_2 substrate and the successful preparation of graphene-hBN in-plane heterostructures,[57–59] analogical structures have been realized in the important semiconducting TMDC materials.[56, 60–66] Through in situ modulation of vapor phase reactants during the growth of 2D crystals, Duan et al. successfully grew compositionally modulated MoS_2-$MoSe_2$ and WS_2-WSe_2 lateral heterostructures in a thermal CVD process.[56] The well-defined p-type WSe_2 and n-type WS_2 forms a natural heterojunction p-n diode exhibiting obvious current rectification behavior. The ultrathin nature of the heterojunction allows gate tunability of diode characteristics. From the measurements of such heterojunction p-n diode type in the dark and under laser illumination (514 nm, 30 nW), evident photovoltaic effect with a V_{oc} of ~0.47 V and I_{sc} of ~1.2 nA was demonstrated (Fig. 5.7). EQE and internal quantum efficiency (IQE) were determined to be 9.9% and 43%, respectively, and the active region was identified to be the lightly doped WS_2 and the WS_2-WSe_2 interface region by photocurrent mapping, indicating that a large fraction of the depletion layer is localized to the lightly doped WS_2 of the diode.[56] Similarly, the realization of lateral monolayer $MoSe_2/WSe_2$ and WS_2/MoS_2 heterostructures by in-plane heteroepitaxy using vapor phase growth process has been demonstrated by different groups.[60, 61, 67] Moreover, atomically sharp in-plane WS_2/MoS_2 interface was revealed by high-magnification STEM, and photovoltaic effect of the intrinsic monolayer p-n heterojunction was evidenced.[61, 67] These works have contributed a significant step in employing layered semiconductor heterostructures toward the development of functional optoelectronics. Following the creation of lateral heterostructures by maintaining either the same metal or chalcogen, p-n heterojunction has been successfully synthesized where neither metal nor chalcogen is the same. Taking monolayer WSe_2-MoS_2 lateral junction for example,[62] typically, sapphire was used as substrate, and the growth of the two TMDCs was carried out in separate furnaces at different growth temperatures (925°C for WSe_2 and 755°C for MoS_2). By using scanning Kelvin probe microscopy, depletion region width near the junction interface was determined to be ~320 nm. After transferring the flake onto a Si/SiO_2 substrate, electrical and optoelectrical properties of the heterostructure were characterized. It turned out that the intrinsic monolayer p-n junction displayed good rectification capability, and

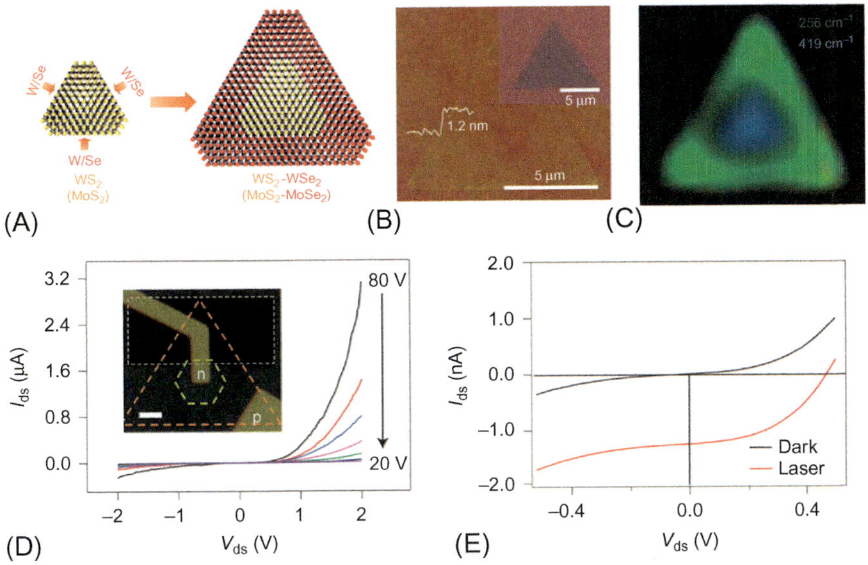

Fig. 5.7 (A) Schematic of lateral epitaxial growth of WS_2-WSe_2 and MoS_2-$MoSe_2$ heterostructures. A triangular domain of WS_2 (MoS_2) is first grown using a CVD process. With an in situ switch of the chemical vapor source for WSe_2 ($MoSe_2$) in the middle of growth, lateral heteroepitaxial growth occurs at the peripheral active growth front to form WS_2-WSe_2 (MoS_2-$MoSe_2$) lateral heterostructures. (B) AFM image of a triangular domain with a thickness of 1.2 nm. Inset: optical image of a triangular domain. (C) Composite image consisting of Raman mapping at 256 and 419 cm^{-1}, showing no apparent overlap or gap between WS_2 and WSe_2 signals, demonstrating that the WS_2 inner triangle and WSe_2 peripheral areas are laterally connected. (D) Gate-tunable output characteristics of a lateral WSe_2-WS_2 heterojunction p-n diode. The gate voltage varies from 80 to 20 V in steps of 10 V. Inset: optical image of a heterojunction p-n diode device. The orange dashed line outlines the triangular heterostructure domain and the white dashed rectangle outlines the 50-nm Al_2O_3 deposited on WSe_2 to insulate the WS_2 contact electrodes. Scale bar, 2 μm. (E) Experimental output (I_{ds}-V_{ds}) characteristics of the lateral WSe_2-WS_2 heterojunction p-n diode in the dark (black line) and under illumination (*red* line; wavelength 514 nm; power 30 nW). Inset: temporal photocurrent response under periodic on/off laser illumination. *Reproduced with permission from Duan, X.; Wang, C.; Shaw, J. C.; Cheng, R.; Chen, Y.; Li, H.; Wu, X.; Tang, Y.; Zhang, Q.; Pan, A.; Jiang, J.; Yu, R.; Huang, Y.; Duan, X., Lateral Epitaxial Growth of Two-Dimensional Layered Semiconductor Heterojunctions. Nat. Nanotechnol.* ***2014****, 9(12), 1024-30. Copyright 2014, Nature Publishing Group.*

photovoltaic effect was observed with a V_{oc} of 0.22 V and I_{sc} of 7.7 pA (white light illumination at 1 mW/cm^2 power density), corresponding to a PCE >0.2%. In a follow-up of this work, Tsai et al. carefully explored the potential of monolayer WSe_2-MoS_2 planar p-n heterojunction for highly efficient photovoltaics, and a PCE as high as 2.56% under AM 1.5G condition was achieved together with excellent omnidirectional light-harvesting characteristics.[63] Additionally in their work, by means of electrode-spacing optimization, gate-tuning controllability and environment-independent PCE became attainable.[40] Recently, a simple approach to produce large-scale optoelectronic devices with graded composition was reported by Li et al.[68] Using a single-step chemical substitution method in ambient pressure CVD, graded doped lateral WSe_2/WS_2 heterostructures were grown, which exhibited increased photoresponse together with much larger active area, implying the potential of optoelectronic device fabrication by graded band structure engineering.

5.2.3.3 2D/2D vdW materials

The in-plane stability of 2D layered crystals is guaranteed by strong covalent bonds, whereas relatively weak, vdW-like forces are present to hold the planes together, which enables the possibility to isolate them by simple scotch-tape cleaving technique.[6, 69] This approach has allowed us to prepare a large variety of 2D layered materials in monolayer and/or few-layer forms, study their unique physical properties, and explore their potential for versatile applications.[9, 11, 70, 71] However, in addition to such "top-down" approach with a focus on individual 2D materials after isolation from their parental material, there is another "bottom-up" strategy. Simply, it is possible to reassemble isolated atomic planes layer by layer in a precisely chosen sequence. This allows us to design artificial vdW heterostructures without the "lattice mismatch" concern.[72, 73] Especially, considering the large quantity of 2D family members covering a wide range of electrical and optical properties, a greater number of their combinations become potentially achievable.[74] Their construction with unprecedented complexity opens up enormous opportunities in materials engineering to fabricate devices with desired properties (Fig. 5.8).

In a typical configuration, gapless graphene can serve as electrode owing to its outstanding carrier mobility, whereas hBN being a perfect insulator with ~6 eV bandgap can serve for gate dielectric or encapsulation purpose.[75–77] Thus, together with the rich family of semiconducting 2D materials, all-2D structures are within reach for future devices. As an important example, Yu et al. demonstrated efficient photon harvesting in vertically stacked graphene-MoS_2-graphene and graphene-MoS_2-metal junctions.[78] In such structures, an external gate bias can readily be used to modulate both

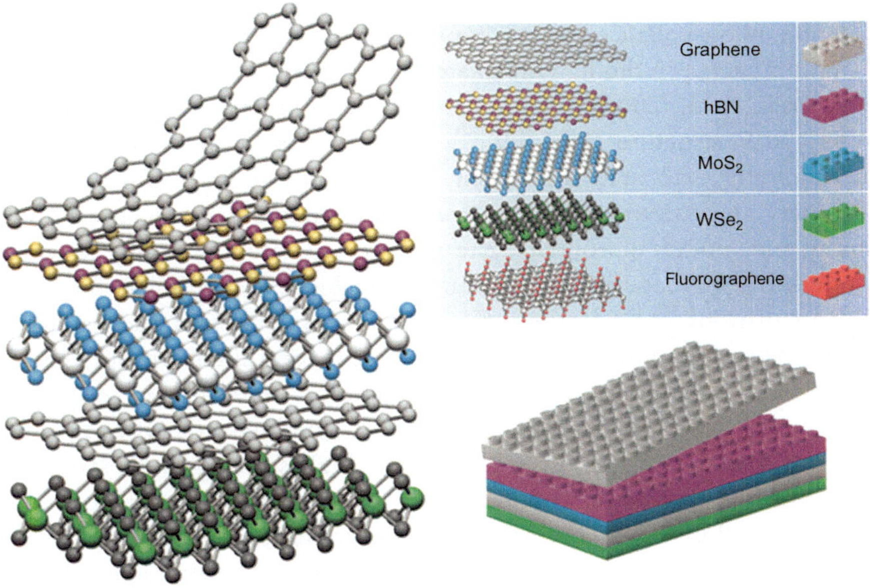

Fig. 5.8 The concept of building vdW heterostructures by vertically stacking various layered crystals together. *Reproduced with permission from Geim, A. K.; Grigorieva, I. V. Van der Waals heterostructures.* Nature ***2013****, 499(7459), 419–425. Copyright 2013, Nature Publishing Group.*

the amplitude and polarity of the photocurrent, and EQE and IQE can reach 55% and 85%, respectively.[78]

It is well known that, among 2D semiconductor crystals, some show inherently n-type (e.g., MoS_2, WS_2, and $MoSe_2$) while others exhibit intrinsically p-type (e.g., WSe_2, $MoTe_2$, and BP). While it is usually challenging to engineer their carrier type (e.g., requiring complicated split-gate structure or elemental doping, as previously discussed), a natural and vertical p-n junction can be readily created by stacking together two flakes of opposite conduction type. In such a simple way, p-n heterojunction as a building block for a wide range of optoelectronic applications can be constructed.[13, 79–81]

An investigation on sunlight absorption of TMDC monolayers (MoS_2, $MoSe_2$, and WS_2) has shown that they can absorb up to 5%–10% incident sunlight in a thickness of <1 nm (one order of magnitude higher than that for GaAs or Si with the same thickness), and generate electrical currents as high as 4.5 mA/cm^2.[19] Moreover, photovoltaic application of stacked monolayer TMDCs and graphene was demonstrated by theoretical calculation showing that the PCE for MoS_2/graphene bilayer and MoS_2/WS_2

bilayer can achieve 1% and 1.5%, respectively, in merely 1 nm thickness. Later on, photovoltaic effect was experimentally observed in TMDC vdW heterojunction p-n diode with an atomically sharp interface.[82–84] In Ref. 82, Furchi et al. reported light energy conversion in monolayer MoS_2/WSe_2 vdW structure (Fig. 5.9).[82] MoS_2 and WSe_2 monolayers were mechanically exfoliated from their bulk crystals and vertically stacked together. The constructed heterostructure exhibited good electrical tunability and, under appropriate gate voltage, operated as atomically thin diode with apparent current rectification capability. According to theoretical calculations,[46, 85, 86] the difference in work function and bandgap between the two monolayers results in a type II band alignment at the heterointerface. Upon optical illumination, electron-hole pairs are generated as a result of incident photon absorption. As the lowest-energy electron states are spatially located in MoS_2 and the highest-energy hole states lie in WSe_2, the opposite-type photogenerated carriers are then relaxed and separated, followed by their diffusion to the contacts, giving rise to a photocurrent. Such a process of electrical power generation from light energy harvesting was evidenced by the device I-V curves that entered the forth quadrant under illumination, where a V_{oc}, I_{sc}, and FF value of 0.58 V, 0.5 nA, and 0.55, respectively, were obtained at 640 mW/cm^2 light intensity (white light from a halogen lamp). PCE was calculated to be 0.2%, which is comparable to the reported values for WSe_2 and MoS_2 bulk samples and lateral WSe_2 monolayer p-n junctions.[82] In another atomically thin WSe_2/MoS_2 vertical heterojunction, Cheng et al. transferred few-layer MoS_2 flake on top of monolayer/bilayer WSe_2 that was previously synthesized on an Si/SiO_2 substrate. The resulting p-n diode exhibited excellent current rectification behavior with an ideality factor of 1.2 at zero gate bias. Significant photovoltaic effect of the diode with a maximum EQE of 12% (under an excitation power of 500 nW) was obtained.[83] This is much higher than those achieved in electrostatically defined WSe_2 p-n homojunctions, which can partly be attributed to the more efficient charge separation process occurring near the atomically sharp junction interface.[27, 28] Furthermore, from scanning photocurrent microscopy characterization, the photocurrent response was observed to arise from the entire overlapping region of the two TMDC materials, suggesting it as photoactive region for photocurrent generation. This is unlike lateral p-n junctions where active region is limited in the depletion part of the material. In Ref. 84, Lee et al. also studied MoS_2/WSe_2 vdW heterojunction. Besides the structure showing gate-tunable diode-like current

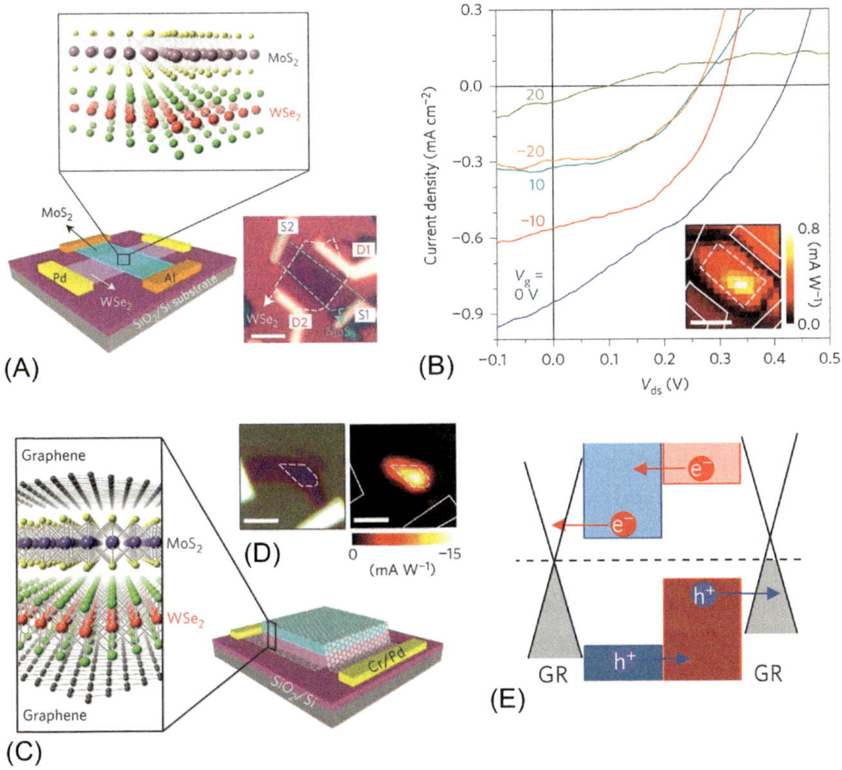

Fig. 5.9 (A) *Bottom left*: Schematic of a vdW-stacked MoS_2/WSe_2 heterojunction device with lateral metal contacts. Top: enlarged crystal structure, with *purple*, *red*, *yellow*, and *green* spheres representing Mo, S, W, and Se atoms, respectively. *Bottom right*: Optical image of the fabricated device, where D1 and D2 (S1 and S2) indicate the metal contacts for WSe_2 (MoS_2). Scale bar, 3 μm. (B) Photovoltaic response at various gate voltages under white light illumination. Inset: Photocurrent map of the device for $V_{ds}=0$ V and 532 nm laser excitation. (C) Schematics of the MoS_2/WSe_2 junction sandwiched between top and bottom graphene electrodes. (D) Optical image and corresponding photocurrent map (for $V_{ds}=0$ V) of the graphene-sandwiched monolayer p-n junction device (1L-1L). Photocurrent is uniformly observed only in the junction area indicated by the dashed lines, separated from the metal electrodes (indicated by solid lines). Scale bar, 3 μm. (E) Schematic of the band structure, with exciton dissociation and charge collection processes indicated by horizontal arrows. The dashed line represents the Fermi level. *Reproduced with permission from Lee, C. H.; Lee, G. H.; van der Zande, A. M.; Chen, W.; Li, Y.; Han, M.; Cui, X.; Arefe, G.; Nuckolls, C.; Heinz, T. F.; Guo, J.; Hone, J.; Kim, P. Atomically thin p-n junctions with van der Waals heterointerfaces.* Nat. Nanotechnol. ***2014**, 9(9), 676–681. Copyright 2014, Nature Publishing Group.*

rectification, the strongest photoresponse and quenching of MoS_2 and WSe_2 photoluminescence peaks happened in the junction area, which is an evidence of spontaneous charge separation occurring at the junction. After exciton dissociation, photocarriers are collected through lateral diffusion by the contacts to generate photocurrent. However, in such an atomically thin p-n heterojunction, interlayer tunneling-related recombination process still occurs, which can severely affect the carrier collection efficiency and thus the photovoltaic performance. By employing graphene as top and bottom transparent electrodes for the vertical p-n junction, the photogenerated carriers can be collected more efficiently by direct vertical charge transfer (recombination in lateral diffusion process is minimized simultaneously), thus offering an appealing option for boosting photovoltaic device performance. A very recent study by Wong et al. suggests that a careful engineering of the top/bottom contact and flake thickness in a vdW heterostructure can significantly improve its absorbance as well as photocarrier collection to enable high photovoltaic efficiency, with the possibility for EQE to exceed even 90%.[87] Furthermore, by sandwiching graphene in an atomically thin MoS_2-WSe_2 p-n junction, the spectral range can be broadened to infrared regime due to the gapless nature of graphene.[88] In addition to the most studied MoS_2/WSe_2 vdW heterojunction, vertical heterostructures based on other 2D crystals, such as BP/MoS_2,[89] $WSe_2/MoSe_2$,[90] α-$MoTe_2/MoS_2$,[91] SnS_2/MoS_2,[92] GaTe-MoS_2,[93] GaSe/$MoSe_2$,[66] $ReS_2/ReSe_2$,[94] MoS_2/Bi_2Te_3,[95] $MoS_2/WS_2/WSe_2$,[96] were also explored for photovoltaics. Particularly, TMDC can be assembled with organic materials for efficient 2D-based solar cells. Shastry et al. experimentally fabricated large-area single-layer MoS_2-polymer photovoltaic heterojunction.[97] At <20 nm total thickness, the MoS_2-polymer (PTB7) heterojunction solar cell provides a peak EQE exceeding 20% and a peak IQE exceeding 40%, higher than present photovoltaic technologies in terms of current density per thickness, demonstrating the prospect of this strategy to realize 2D-based large-area, ultrathin, and high-efficiency solar cells. These various heterostructures clearly manifest a large number of 2D crystals and their combinations as candidates for future photovoltaic devices, and imply the potential of further enhancement of photovoltaic performance by material choice and engineering.

5.2.3.4 Hybrid vdW heterostructures

In the previous section, we discussed all-2D vdW heterostructures. In fact, self-passivated, dangling bond-free surface of 2D crystals interacts with other

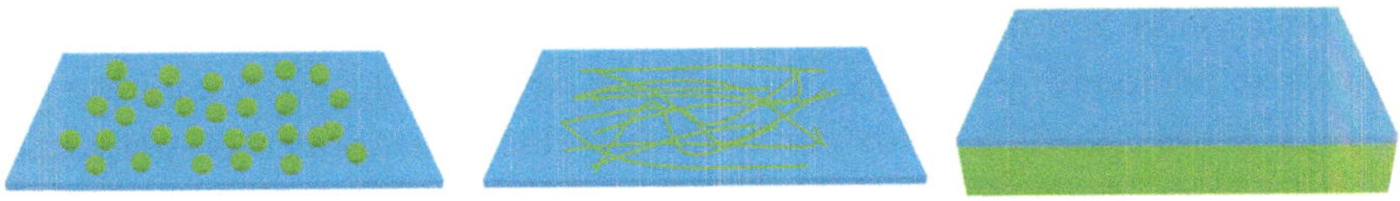

Fig. 5.10 Schematic illustrations of 2D materials-related hybrid vdW heterojunctions in (A) 2D/0D, (B) 2D/1D, and (C) 2D/3D systems.

materials of different dimensionality (0D, 1D, or 3D) by vdW forces as well. Thus, it is also possible to integrate layered 2D materials with them to construct hybrid (mix-dimensional) vdW heterostructures (Fig. 5.10).[98] Such kinds of combinations further broaden the class of vdW heterostructures that can be employed for photovoltaic purpose.

Among the 2D materials, graphene has been widely investigated to form hybrid vdW heterostructures with conventional semiconductors for photovoltaics.[99] Compared to ITO-based Schottky junctions, graphene has certain important advantages such as its superior electrical and optical properties, its cost-effectiveness, and its tunable work function.[100] Li et al. first reported solar cells based on graphene-on-silicon structure.[101] They etched a window in thermally grown SiO_2 on Si and transferred graphene sheet onto it to form a Schottky junction. Under AM 1.5 illumination, the devices were able to convert light energy into electricity with a V_{oc} of 0.42–0.48 V and I_{sc} of 4–6.5 mA/cm^2 and PCE near 2%. Later on, various techniques, including antireflective coatings,[102, 103] silicon passivation,[104] and pillar array patterning of silicon,[105, 106] have been used to improve graphene/Si solar cell performance, and PCE >7% was achieved.[106] Especially, Song et al. carefully studied the role of interfacial oxide in graphene-silicon Schottky barrier solar cells and found that an increase in efficiency is attainable by optimizing silicon oxide thickness.[103] Furthermore, by taking advantage of the Fermi level tunability of graphene, it is possible to desirably increase the Schottky barrier height of the graphene/semiconductor junction, thus optimizing photovoltaic efficiency. For example, by doping graphene with bis(trifluoromethanesulfonyl)amide, Miao et al. demonstrated graphene/Si Schottky junction solar cells exhibiting a PCE of 8.6%, which is 4.5 times higher compared with undoped devices.[100] Adjustment of graphene work function can also be achieved in an electrostatic way by introducing a gate electrode. Using this approach, photovoltaic performance enhancement has been demonstrated in graphene/Zn_3P_2,[107] graphene/InP,[108] and graphene/GaAs[109] heterojunction solar cells. In particular, PCE and V_{oc} of the graphene/GaAs Schottky barrier solar cells achieved as high as 18.5% and 0.96 V, respectively, promising for practical application.[109] In addition to forming heterojunctions with conventional 3D semiconductors, graphene was used to create 2D/0D and 2D/1D heterojunctions as well, exemplified by graphene/perovskite,[110] graphene/CdTe quantum dots,[111] graphene/CdSe nanobelt,[112, 113] and graphene/CdS nanowire[114] solar cells.

In addition to hybrid Schottky barrier heterostructures, mixed-dimensional p-n vdW heterostructures have also been constructed by combining 2D semiconducting materials with semiconductors of a different dimension. MoS_2/p-Si heterojunction was first reported by stacking exfoliated MoS_2 onto passivated p-type Si substrate (Fig. 5.11).[115] A type II p-n heterojunction formed between the two materials, which facilitates the photogenerated carrier separation process, and the fabricated device

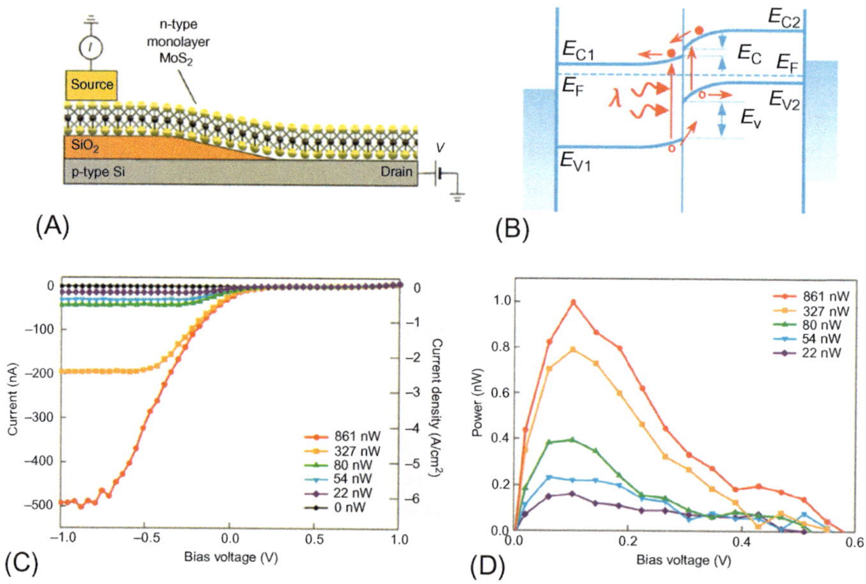

Fig. 5.11 MoS_2/Si heterojunction operates as a solar cell. (A) Schematic cross-sectional view of the structure of the MoS_2/Si heterojunction device. (B) Band diagram of the heterostructure illustrating photocurrent generation at zero bias voltage. (C) Current as a function of bias voltage under different illumination powers from a 541-nm laser. (D) Electrical power generated by the device as a function of bias voltage, recorded for different illumination powers. *Reproduced with permission from Lopez-Sanchez, O.; Alarcon Llado, E.; Koman, V.; Fontcuberta i Morral, A.; Radenovic, A.; Kis, A., Light Generation and Harvesting in a van der Waals Heterostructure. ACS Nano* ***2014****, 8(3), 3042-8. Copyright 2014, American Chemical Society. (doi/abs/10.1021/nn500480u; further permissions related to the material excerpted should be directed to the ACS.)*

exhibited a photovoltaic effect with ECE exceeding 4%. By employing large-area CVD-grown MoS_2, monolayer MoS_2/Si heterojunction solar cells yielding a PCE of 5.23% were reported by Tsai et al., with V_{oc} and J_{sc} reaching 0.41 V and 22.36 mA/cm^2, respectively.[116] Additionally, high-performance MoS_2/Si heterojunction solar cells can be obtained by incorporating a nanoscale SiO_2 buffer into the heterointerface. Using this strategy, Hao et al. fabricated n-MoS_2/i-SiO_2/p-Si heterojunction photovoltaic devices exhibiting 4.5% PCE. Further improvement in photovoltaic performance has been sought by integrating 2D semiconducting crystals with direct-bandgap materials such as GaAs,[117, 118] InP,[119, 120] and GaN.[121] Gehring et al. investigated the photovoltaic behavior of p-BP/n-GaAs 2D/3D heterostructure where GaAs was heavily doped with Te. The heterojunction device exhibited excellent diode behavior and photovoltaic effect with 0.6 V V_{oc} and 9.7% EQE.[117] In another work,[118] through interface engineering, i.e., inserting hBN monolayer into MoS_2/GaAs heterostructure, the resulting MoS_2/hBN/GaAs solar cell displayed an improved solar energy conversion efficiency of 5.42%. Furthermore, the solar cell efficiency can be further increased to 9.03% (under AM 1.5G illumination) by means of MoS_2 Fermi level tuning via chemical doping and electrical gating.[118]

5.3 Challenges and prospects toward practical 2D photovoltaic devices

Two-dimensional layered materials hold promise for future generation of photovoltaic energy devices. As previously discussed, over the past five years, their great potential has been widely demonstrated in a large variety of structures using diverse materials. However, alike 2D layered materials themselves in electronics and optoelectronics, the exploration of their use for light energy conversion is still in its infancy. Table 5.2 summarizes important 2D-based photovoltaic devices and their performances in the literature. Generally, it is noticeable that the reported 2D-related photovoltaic devices still suffer from a rather inferior performance in comparison to state-of-the-art photovoltaic technology. In particular, considering one of the most critical figures of merit for a photovoltaic device, i.e., PCE, its value is under 1% or even smaller for most 2D-based systems, which is much lower than other photovoltaic technologies (Si, GaAs, CdTe, CIGS, perovskite solar cells with >20% efficiency).[1] Although higher PCEs are achieved for the devices with 2D/3D hybrid structures, the integration with

Table 5.2 Reported 2D-based photovoltaic devices and their performances in literature.

Materials	Device architecture	V_{oc} (V)	I_{sc} or J_{sc}	EQE (IQE)	FF	PCE (%)	Ref.
50 nm MoS_2	Lateral Schottky	0.1	1.8 nA	NA	NA	1.25	23
ITO/MoS_2/Au	Vertical Schottky	0.597	5.37 mA/cm^2	52%	NA	1.8	25
$Mo_{0.5}W_{0.5}Se_2$	–	0.44	92 mA/cm^2	NA	0.32	5–16	26
ML WSe_2	Local-gated p-n	0.65	1 nA	0.2%		0.005	28
ML WSe_2	–	0.64	0.0014 nA	1.5%	0.5	0.5	27
ML WSe_2	–	0.7	0.41 nA	0.1%	0.29	0.14	29
~6 nm BP	–	0.05	1 nA	0.1%	0.3	<0.001	34
10L $MoSe_2$	–	0.36	0.6 nA	NA	0.7	14	31
Multilayer MoS_2	Doped lateral p-n	0.5	46.7 nA	NA	NA	NA	35
BP (8.5 nm)	–	0.14	3 nA	NA	0.38	0.66	37
BP (2.8 nm)	–	0.44	180 nA	NA	0.75	0.75	40
WS_2-WSe_2	Lateral hetero- p-n	0.47	1.2 nA	9.9% (43%)	NA	0.9	56
WSe_2-MoS_2	–	0.22	0.008 nA	NA	0.39	0.2	62
BP	–	0.21	1.5 nA	7.4%	0.29	0.25	55
6 nm $MoSe_2$	Doped vertical p-n	0.35	10 mA/cm^2	50%	NA	NA	44
MoS_2	–	0.28	20.9 mA/cm^2	37%–78%	0.47	2.8	42
MoS_2	–	0.6	5.1 nA	NA	0.22	0.4	43
Gr/MoS_2/Gr	Vertical vdw	0.28	1700 nA	27%	NA	NA	78
Gr/MoS_2/M				55% (85%)			
MoS_2/WSe_2	–	0.5	70 nA	34%			84
FL BP/ML MoS_2	–	0.3	20 nA	30%	0.5	0.57	89
ML-WSe_2/FL-MoS_2	–	0.27	200 nA	12%	NA	~0.4	83
ML WSe_2/MoS_2	Vertical 2D vdw	0.53	0.005 nA	1.5%	NA	0.2	82
α-$MoTe_2$/MoS_2	–	0.3	150 nA	85%	NA	NA	91

$WSe_2/MoSe_2$	–	0.46	0.004 nA	0.1%	NA	NA	90
$ReS_2/ReSe_2$	–	0.175	0.29 mA/cm^2	NA	NA	0.5	94
MoS_2/WSe_2	–	0.57	0.005 nA	NA	NA	0.2	96
ML MoS_2/Si	Vertical hybrid vdw	0.6	~10 nA	4.4%	NA	0.1	115
MoS_2/Si	–	0.41	22.36 mA/cm^2		0.57	5.23	116
MoS_2/h-BN/GaAs	–	0.76	21.1 mA/cm^2	56.3%	NA	9.03	118
MoS_2/InP	–	0.47	27.4 A/cm^2	NA	0.55	7.1	119
Gr/p-InP	–	0.63	21.9 mA/cm^2	45%–70%		5.6	108
$MoSe_2$/GaN	–	0.62	6.05 mA/cm^2	NA	0.35	1.29	121
GQDs/MoS_2/InP	–	0.27	24.2 mA/cm^2	NA	0.53	4.1	120
Gr/GaAs	–	0.81	24.9 mA/cm^2	NA	0.69	16.2	122
MoS_2/SiO_2/p-Si	–	0.3	5.5 mA/cm^2	NA	0.42	4.5	123
BP/GaAs	–	0.6	1250 nA	31%	0.3	0.24	117

conventional semiconductors renders the devices to lose the advantages promised by layered materials. Therefore, to realize practical photovoltaic applications with 2D materials requires some important challenging issues to be addressed.

5.3.1 Absorption enhancement for 2D materials

Although many 2D crystals exhibit stronger light-matter interaction than conventional semiconductor materials, the demonstrated solar cell devices based on them have shown relatively low PCEs. The primary factor responsible for such inferior photovoltaic performance of 2D material devices is their atomic-scale thickness that results in inefficient light absorption, highlighting the need of light-trapping techniques to overcome this issue. In conventional Si solar cells, light trapping is improved by typically using a pyramidal surface texture that causes scattering of light over a large angular range, thereby increasing the effective path length in the cell.[124] However, this strategy is hardly applicable to thin film cells and 2D-based ultrathin devices due to geometrical reasons and increased minority carrier recombination.[125] In the past few years, plasmonic has emerged as a rapidly expanding field for materials and device research.[126] Subsequently, plasmonic nanostructures are adopted in photovoltaic devices to enhance light trapping for their improved efficiency. Plasmon-enhanced light absorption can possibly be realized in three ways.[127] First, metallic nanoparticles can function as subwavelength scattering elements to couple and trap propagating sunlight into an absorbing semiconductor film by folding the light into a thin absorber layer. Second, metallic nanoparticles can be used as subwavelength antennas in which the plasmonic near-field is coupled to the semiconductor to increase its effective absorption cross-section. Third, a corrugated metallic film on the back surface of a thin active layer can couple sunlight into surface plasmon polaritons modes supported at the metal/semiconductor interface as well as guided modes in the semiconductor slab, whereupon light is converted to photocarriers in the semiconductor. In 2D materials research, plasmonic nanostructures have become a compelling solution to overcoming the inherently weak absorption bottleneck for practical device performances. In Ref. 20, after introducing Au nanoparticles (5–10 nm size) by depositing a thin Au layer on the upper graphene electrode followed by an annealing process, the authors observed a 10-fold enhancement in the photocurrent of hBN/Gr/MoS2/Gr heterostructure device under visible light illumination. This large increase can be well related to more light absorption as a result of surface plasmon resonances. Significant plasmonic enhancement of photocurrent was also observed in MoS_2 field effect transistor.[128, 129] In Ref. 128, the decoration with 15-nm-diameter Au nanoparticles to MoS_2 sheet gave rise to a 10-time larger photocurrent as a result of enhanced light trapping near the plasmonic nanospheres. In Ref. 129, Miao et al. demonstrated that in comparison to random nanoparticles, periodic Au nanoarrays are able to further boost the light-matter interaction on few-layer MoS_2 phototransistor, which can be attributed to a better light-trapping outcome. Recently, Butun et al. numerically and experimentally investigated the effect of periodic Ag plasmonic nanodisk arrays on light absorption in WS_2 monolayer.[130] As result, a fourfold enhancement of light absorption near WS_2 band edge was obtained, and the Ag/WS_2 heterostructure exhibited a short-circuit current 2.5 times higher than bare WS_2 film, suggesting the potential application of such hybrid plasmonic nanostructures for practical 2D photovoltaic devices. Besides integrating metallic nanoparticles or nanoarrays on top of 2D materials system, enhanced light absorption can also be achieved by fabricating 2D-based devices onto plasmonic metasurfaces. For example, Petoukhoff et al. prepared hybrid MoS_2-organic heterojunction for solar cell application (Fig. 5.12). While the system suffers poor light absorption constrained by its thin physical thickness, its efficiency can be

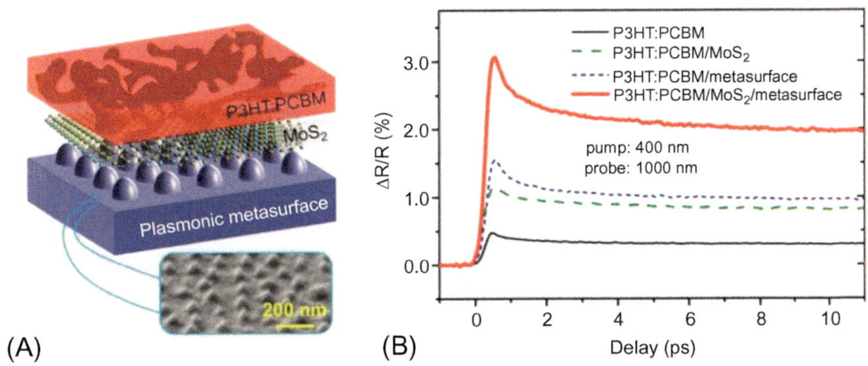

Fig. 5.12 (A) Schematic of the hybrid P3HT:PCBM/MoS_2 heterojunction on a plasmonic metasurface with a scanning electron microscope image of the metasurface shown at the bottom. (B) Picosecond-scale transient pump-probe reflection measurements from P3HT:PCBM and the hybrid P3HT:PCBM/MoS_2 active layers, demonstrating enhanced light absorption with the plasmonic metasurfaces. *Reproduced with permission from Petoukhoff, C. E.; Krishna, M. B.; Voiry, D.; Bozkurt, I.; Deckoff-Jones, S.; Chhowalla, M.; O'Carroll, D. M.; Dani, K. M., Ultrafast Charge Transfer and Enhanced Absorption in MoS2-Organic van der Waals Heterojunctions Using Plasmonic Metasurfaces. ACS Nano* ***2016****, 10(11), 9899-9908. Copyright 2016, American Chemical Society.*

improved sixfold when it is fabricated on Ag plasmonic metasurfaces, demonstrating a new approach toward efficient 2D-based hybrid solar cells.[131]

In addition to plasmonic nanostructure-based light trapping, some other strategies are also under exploration for absorption enhancement in 2D materials. Jariwala et al. prepared ultrathin semiconducting TMDCs (<20 nm) on flat Ag and Au substrates.[132] It was found that by coupling to strongly damped optical modes of the TMDC/metal heterostructures, near-unity light absorption is achievable in a wide spectrum range, rendering the system suitable for light-harvesting application. Furthermore, the fabricated photovoltaic devices based on such highly absorbing heterostructures exhibited an EQE of ~13%. Additionally, the employment of a photonic crystal slab or microcavity with a spectrum-splitting structure for improved light absorption has been justified by some simulation and theoretical work.[133, 134] Particularly, keeping the ultrathin and ultra-strong nature of 2D materials in mind, it is worth pointing out the great potential of strain engineering for bandgap tuning as well as broad-range light absorbance, which have already been demonstrated both theoretically and experimentally.[135–138]

5.3.2 Toward efficient charge separation and collection

After the absorption of photons with sufficient energy, electrons in valence band of the semiconductor are excited into the conduction band, forming electron-hole pairs (excitons) where the two oppositely charged carriers are still bounded together through Coulomb attraction. Then a solar cell performance will be determined by the following processes, i.e., excitons dissociation into free photocarriers, photocarrier separation, and their collection by electrodes. In the meantime, unwanted charge recombination process also exists in a photovoltaic device. Therefore, it is important for a solar cell to maximize the former processes and suppress the latter one. Besides the thickness-dependent bandgap of 2D semiconducting nanosheets as a result of strong quantum confinement, another important feature of them is the exciton binding energy, which increases with reduced layer number. Taking MoS_2 for example, while it has a 0.1 eV exciton binding energy in the bulk form, its value dramatically increases to as high as 1.1 eV in the monolayer.[139] Such large binding energy will make it less favorable for excitons to dissociate into free charge carriers, consequently affecting the attainable V_{oc} and ultimate efficiency in a negative way.[140] It is worth noting that this issue has also been encountered in the development of high-efficiency organic solar cells, where V_{oc} is limited by the difference between HOMO (highest occupied molecular orbital) of the donor and LUMO (lowest unoccupied molecular orbital) of the acceptor minus at least the large exciton binding energy (0.3∼0.8 eV).[140, 141] This fact can at least partly explain the obtained V_{oc} of 2D photovoltaic devices, which is smaller than expected considering their large bandgap in monolayer limit.[27–29] One possible solution to overcoming this barrier is the construction of a stronger built-in electric field in the 2D homojunction region. As an example, in Ref. 43, by doping with $AuCl_3$ for p-type and BV for n-type, Li et al. fabricated ultrathin (<11 nm) vertical MoS_2 homogeneous p-n junction that gave rise to a 0.6-V V_{oc}. However, this approach is unlikely to make an advisable option because of the complicated doping process. Speaking of this, doping techniques for 2D crystals should be further developed in terms of their feasibility and reliability. As mentioned before, doped p-n junction has been realized via various approaches, including substitutional doping, electrostatic doping, and surface charge transfer, among which the last method looks very attractive due to its simplicity. In contrast, electrostatic doping through complicated double-gate configuration is unlikely to be adopted for practical application, although it is an important strategy for device demonstration in initial photovoltaic studies.

Instead, as previously mentioned, vertical vdW heterostructures should be considered as a promising choice by offering a huge amount of bandgap engineering possibilities. First, it can provide a much larger active junction area that allows more photons to be absorbed, thus more excitons are generated. Second, the large band offset for electrons and holes across the atomically sharp interface is able to drive spontaneous dissociation of the photogenerated exciton into free carriers, without the need of exciton diffusion process. Such a rapid charge separation process (that might be on a timescale of ∼10 ps[142]) has been evidenced by photoluminescence characterization, where a considerable quenching of emission from direct-gap transition of monolayer MoS_2 and WSe_2 is observed at the MoS_2/WSe_2 overlapping junction area.[84] Furthermore, unlike conventional 3D heterojunction with a high density of interfacial traps, 2D-based vdW heterojunctions have the advantage of high-quality interfaces with extremely low density of dangling bonds and charge traps, readily alleviating the problem of interfacial recombination centers that impact photovoltaic efficiency. Finally, considering the fact that a huge amount of 2D semiconductors exist with tunable bandgaps, it will become completely possible to engineer the detailed band structure of vdW heterostructure, by carefully selecting the material system as well as their thickness, for obtaining higher V_{oc} and higher conversion efficiency.[82, 84, 90, 96]

After exciton dissociation, free photocarriers are separated and need to be collected by the electrodes for effective photocurrent generation. While Ohmic contact is important in p-n junction solar cells for efficient charge collection, metal electrode usually forms Schottky contact with 2D semiconducting materials. Contact engineering problem has started gaining much attention to improve the performance of layered material-related devices.[143] In practice, appropriate metals have been carefully chosen considering their work functions to minimize the barrier height and facilitate high-quality electrical contact.[144] Other approaches such as metalized or degenerately doped 2D materials for contact have also been investigated.[145] In any case, improved contact properties between electrode and active 2D semiconductors will benefit further performance enhancement of 2D photovoltaic devices. In particular, for future vertical photovoltaic structures, the rate of collection of photogenerated carriers can be largely improved by sandwiching it with graphene electrodes to take advantage of its ultrahigh carrier mobility, with a five-times more efficient charge collection being demonstrated.[84]

One special figure of merit for a photovoltaic cell is the FF, defined as $FF = P_{max}/V_{oc}I_{sc}$. While a large FF value is desirable for a photovoltaic device, at the present stage, its values for most 2D material devices are still relatively low in comparison to mature Si photovoltaic technology (~0.8) (see Table 5.2). This is reflected by the device *I*–*V* characteristics under optical excitation. Normally a more square-like *I*–*V* sweep in the fourth quadrant is expected, corresponding to a large FF. In contrast, near-straight curves are observed in a large part of 2D photovoltaic studies, a typical signature of the existence of a small shunt resistance (R_{sh}) and/or a large series resistance (R_s) of the device. In principle, a resistance in parallel to the junction will decrease V_{oc} by leakage of photogenerated carriers across it.[34] Pospischil et al. found that for a monolayer WSe_2 lateral p-n diode, R_{sh}-related carrier loss is dominated by monomolecular recombination, which is likely to be correlated to disorder-induced trap states and can cause a reduction of FF by 37%.[27] In another work of MoS_2 vertical junction, Wi et al. ascribed the limited V_{oc} to pinholes and edge leakage defects in MoS_2 flakes, which resulted in undesirable leakage between electrodes. In Ref. 84, Lee et al. pointed out that although Langevin recombination may be a dominant mechanism in an MoS_2/WSe_2 vdW structure due to the enhanced Coulomb interaction between electrons and holes confined in the 2D layers,

Shockley-Read-Hall recombination can still exist as a result of imperfections of the junction interface or defects in the materials. All these works underline the importance of material defect control and interface engineering in the design and fabrication process of photovoltaic devices.[143, 146] As a matter of fact, the atomic-scale thickness and pristine surface of 2D crystals makes them extremely sensitive to the surrounding environment.[143] Consequently, the related device performance is significantly affected by the interface of 2D crystals with another material. For photovoltaic devices, the defects and unwanted interface states not only lead to additional scattering, limiting carrier mobility, but also serve as recombination centers, severely affecting the device IQE and ultimate performance. Therefore, defect and interface engineering are required for improvement in 2D-based photovoltaic devices with desired performances. In this regard, it is believed that with the advances of TMDC material quality and near-unity luminescence quantum yield,[147] TMDC-based photovoltaics with meaningful efficiencies can be well expected. It is worth pointing out that a more controllable engineering of contact, doping, defect, and interface will certainly benefit from further advances of synthesis techniques and more profound understanding of 2D materials physics.

5.3.3 Some other issues

Motivated by the need for bendable/foldable, transparent, and functional electronic devices, flexible electronics has been an emerging applied research field. Due to their outstanding properties, including atomic thickness, high crystalline quality, transparency (and yet strong light-matter interaction), and large breaking strength, 2D layered materials will greatly benefit the nascent flexible nanotechnology, with many 2D material-based flexible devices fabricated and demonstrated.[12, 148–151] Regarding the PCE for ultrathin solar cells based on 2D nanomaterials (especially all-2D devices) constrained by their thickness-limited absorption, we expect that their commercialization potential is likely to be firstly and fully explored for photovoltaic devices in circumstances requiring light weight, flexibility, and/or transparency, all of which are readily promised by monolayer and few-layer 2D semiconducting crystals. Therefore, it is expected that 2D flexible photovoltaic devices will be fabricated and demonstrated in the very near future, although some new challenges may arise on achieving the desirable performance.

Another concern in realizing practical 2D materials-based photovoltaic devices in the future is their cost-effective, highly-controllable, large-area synthesis with desired structures and quality, to allow for scalable and reproducible device fabrication. Indeed, this is essential not only for photovoltaic devices, but for all potential electronic and optoelectronic devices that are expected from 2D layered materials. While the most used micromechanical cleavage technique gives the highest quality of 2D crystals, which is extremely important for materials study and device demonstration in pioneering work, it suffers a low yield and low production rate, and thus does not afford scalability at industry level. In this regard, the past few years have seen significant advances in preparing 2D materials at a large scale. In particular, graphene, hBN, and a number of TMDCs have been successfully synthesized via various growth techniques such as CVD, physical vapor transport, and layer-by-layer conversion (see Ref. 152, 153 and therein). Graphene and various TMDC nanosheets in single-layer and few-layer form, including those for potential photovoltaic applications such as MoS_2, WS_2, WSe_2, and $MoSe_2$, have been prepared by CVD process with high crystal quality, scalable sizes, controllable thicknesses, and excellent electronic properties. Moreover, photovoltaic devices have also been extensively fabricated and demonstrated on these 2D nanomaterials grown via a bottom-up synthesis process, as discussed earlier. The drawbacks of the CVD technique are its high cost, high growth temperature (>400°C) that is incompatible with certain device fabrication process (e.g., flexible substrate), and the need to transfer the synthesized 2D layers to another substrate for photovoltaic device application.[13] In contrast, liquid-based wet chemical approaches, owing to their extremely low cost and powerful controllability, are widely developed and considered as convenient strategies for the preparation of ultrathin 2D nanomaterials, except that substantial improvement in terms of crystal quality and film uniformity is demanded before they can be used for high-performance photovoltaic applications.[153–155]

Novel, perhaps more complicated structures should lead to a further development of 2D materials-based solar energy conversion. However, their construction will certainly rely on more controllable synthesis process or more advanced transfer technique. Very recently, the Duan group has succeeded in preparing more sophisticated TMDC multi-heterostructures with precisely controlled spatial modulation, which may contribute to the fabrication of photovoltaic devices with improved efficiency.[156] Moreover, since self-passivated 2D atomic crystals enable the construction of vdW heterostructures, the advancement in their growth and transfer technique will open up a new approach in designing tandem heterostructure solar cells by using a variety of semiconducting 2D crystals with different bandgaps, which is able to make full use of the solar spectrum and lead to a higher photovoltaic efficiency. With this in mind, besides currently studied 2D-formed candidates for solar energy harvesting, some other layered materials (such as $NiSe_2$ and $ZrSe_2$) with a suitable bandgap also have potential in photovoltaics and are worthy of investigation in the future.[157]

5.4 Conclusions and outlook

The fascinating properties of single-layer and few-layer 2D materials have made them promising candidates for future photovoltaic applications. Over the past years, significant progress has been made in demonstrating their potential through the fabrication of photovoltaic devices with different structures. For lateral configuration, homojunctions can be created by means of electrostatic doping, chemical doping, and surface functionalization, while heterojunctions have been fabricated through straightforward vapor phase synthesis and thickness modulation. In contrast, vertical junction structures, especially vdW heterostructures, can be constructed without lattice mismatch concern. Such configuration is more appealing for photovoltaic device fabrication due to its abundant materials option, larger amount of light absorption, as well as its much more efficient photocarrier separation process occurring at the heterointerface within atomic scale. Furthermore, graphene can be used as a perfect electrode for high-efficiency carrier collection. However, the photovoltaic efficiency of 2D crystal-based devices is, at present, much lower than the state-of-the-art solar cell technology primarily due to their thickness-limited light absorption. While 2D materials are extremely suitable for fabricating next-generation photovoltaic devices with transparency, flexibility, and light weight, absorption improvement techniques such as plasmon-enhanced absorption and strain engineering are needed to further enhance light-matter interaction for these ultrathin materials before they become competitive with conventional bulk semiconductors. Particularly, with PCE comparable to current solar cells, 2D/3D hybrid heterostructures coupled with 2D materials are highly promising for photovoltaic technology. However, although significant advances have been achieved, the use of 2D layered materials for solar energy conversion is still in its infancy. There are still many challenges that need to be overcome for their practical implementation in future photovoltaic devices. Especially, access to controllable high-quality materials growth, efficient transfer technology, more appropriate vdW structure construction, as well as scalable fabrication process with reliable doping technology and proper contact/

interface engineering is indispensable. In addition, integration with current mature microelectronic technology will, no doubt, be highly appreciable.

References

1. Polman, A.; Knight, M.; Garnett, E. C.; Ehrler, B.; Sinke, W. C. Photovoltaic Materials: Present Efficiencies and Future Challenges. *Science* **2016**, *352*(6283), aad4424.
2. Jean, J.; Brown, P. R.; Jaffe, R. L.; Buonassisi, T.; Bulović, V. Pathways for Solar Photovoltaics. *Energy Environ. Sci.* **2015**, *8*(4), 1200–1219.
3. Rogers, J. A.; Someya, T.; Huang, Y. Materials and Mechanics for Stretchable Electronics. *Science* **2010**, *327*(5973), 1603–1607.
4. Nathan, A.; Ahnood, A.; Cole, M. T.; Sungsik, L.; Suzuki, Y.; Hiralal, P.; Bonaccorso, F.; Hasan, T.; Garcia-Gancedo, L.; Dyadyusha, A.; Haque, S.; Andrew, P.; Hofmann, S.; Moultrie, J.; Daping, C.; Flewitt, A. J.; Ferrari, A. C.; Kelly, M. J.; Robertson, J.; Amaratunga, G. A. J.; Milne, W. I. Flexible Electronics: The Next Ubiquitous Platform. *Proc. IEEE* **2012**, *100*(Special Centennial Issue), 1486–1517.
5. Wong, W. S.; Salleo, A. *Flexible Electronics: Materials and Applications;* vol. 11; Springer Science & Business Media, 2009.
6. Novoselov, K. S.; Geim, A. K.; Morozov, S. V.; Jiang, D.; Zhang, Y.; Dubonos, S. V.; Grigorieva, I. V.; Firsov, A. A. Electric Field Effect in Atomically Thin Carbon Films. *Science* **2004**, *306*(5696), 666–669.
7. Radisavljevic, B.; Radenovic, A.; Brivio, J.; Giacometti, V.; Kis, A. Single-Layer MoS_2 Transistors. *Nat. Nanotechnol.* **2011**, *6*(3), 147–150.
8. Li, L.; Yu, Y.; Ye, G. J.; Ge, Q.; Ou, X.; Wu, H.; Feng, D.; Chen, X. H.; Zhang, Y. Black Phosphorus Field-Effect Transistors. *Nat. Nanotechnol.* **2014**, *9*(5), 372–377.
9. Xu, M.; Liang, T.; Shi, M.; Chen, H. Graphene-Like Two-Dimensional Materials. *Chem. Rev.* **2013**, *113*(5), 3766–3798.
10. Dean, C. R.; Young, A. F.; Meric, I.; Lee, C.; Wang, L.; Sorgenfrei, S.; Watanabe, K.; Taniguchi, T.; Kim, P.; Shepard, K. L.; Hone, J. Boron Nitride Substrates for High-Quality Graphene Electronics. *Nat. Nanotechnol.* **2010**, *5*(10), 722–726.
11. Xia, F.; Wang, H.; Xiao, D.; Dubey, M.; Ramasubramaniam, A. Two-Dimensional Material Nanophotonics. *Nat. Photonics* **2014**, *8*(12), 899–907.
12. Wang, L.; Huang, L.; Tan, W. C.; Feng, X.; Chen, L.; Huang, X.; Ang, K.-W. 2D Photovoltaic Devices: Progress and Prospects. *Small Methods* **2018**, *2*(3), 1700294.
13. Geim, A. K.; Grigorieva, I. V. Van der Waals Heterostructures. *Nature* **2013**, *499*(7459), 419–425.
14. Bucher, E. Photovoltaic Properties of Solid State Junctions of Layered Semiconductors. In *Photoelectrochemistry and Photovoltaics of Layered Semiconductors,* Springer: Netherlands, 1992, pp 1–81.
15. Kline, G.; Kam, K.; Canfield, D.; Parkinson, B. A. Efficient and Stable Photoelectrochemical Cells Constructed With WSe2 and MoSe2 Photoanodes. *Sol. Energy Mater.* **1981**, *4*(3), 301–308.
16. Clemen, C.; Saldaña, X. I.; Munz, P.; Bucher, E. Photovoltaic Properties of Some Semiconducting Layer Structures. *Phys. Status Solidi A* **1978**, *49*(2), 437–443.
17. Fortin, E.; Sears, W. M. Photovoltaic Effect and Optical Absorption in MoS_2. *J. Phys. Chem. Solids* **1982**, *43*(9), 881–884.
18. Splendiani, A.; Sun, L.; Zhang, Y.; Li, T.; Kim, J.; Chim, C. Y.; Galli, G.; Wang, F. Emerging Photoluminescence in Monolayer MoS_2. *Nano Lett.* **2010**, *10*(4), 1271–1275.

19. Bernardi, M.; Palummo, M.; Grossman, J. C. Extraordinary Sunlight Absorption and One Nanometer Thick Photovoltaics Using Two-Dimensional Monolayer Materials. *Nano Lett.* **2013**, *13*(8), 3664–3670.
20. Britnell, L.; Ribeiro, R. M.; Eckmann, A.; Jalil, R.; Belle, B. D.; Mishchenko, A.; Kim, Y. J.; Gorbachev, R. V.; Georgiou, T.; Morozov, S. V.; Grigorenko, A. N.; Geim, A. K.; Casiraghi, C.; Castro Neto, A. H.; Novoselov, K. S. Strong Light-Matter Interactions in Heterostructures of Atomically Thin Films. *Science* **2013**, *340*(6138), 1311–1314.
21. Choi, K.; Lee, Y. T.; Im, S. Two-Dimensional van der Waals Nanosheet Devices for Future Electronics and Photonics. *Nano Today* **2016**, *11*(5), 626–643.
22. Yu, X.; Sivula, K. Toward Large-Area Solar Energy Conversion with Semiconducting 2D Transition Metal Dichalcogenides. *ACS Energy Lett.* **2016**, *1*(1), 315–322.
23. Fontana, M.; Deppe, T.; Boyd, A. K.; Rinzan, M.; Liu, A. Y.; Paranjape, M.; Barbara, P. Electron-Hole Transport and Photovoltaic Effect in Gated MoS_2 Schottky Junctions. *Sci. Rep.* **2013**, *3*, 1634.
24. Kaushik, N.; Nipane, A.; Basheer, F.; Dubey, S.; Grover, S.; Deshmukh, M. M.; Lodha, S. Schottky Barrier Heights for Au and Pd Contacts to MoS_2. *Appl. Phys. Lett.* **2014**, *105*(11), 113505.
25. Shanmugam, M.; Durcan, C. A.; Yu, B. Layered Semiconductor Molybdenum Disulfide Nanomembrane Based Schottky-Barrier Solar Cells. *Nanoscale* **2012**, *4*(23), 7399–7405.
26. Yi, S. G.; Kim, S. H.; Park, S.; Oh, D.; Choi, H. Y.; Lee, N.; Choi, Y. J.; Yoo, K. H. $Mo_{1-x}W_xSe_2$-Based Schottky Junction Photovoltaic Cells. *ACS Appl. Mater. Interfaces* **2016**, *8*(49), 33811–33820.
27. Pospischil, A.; Furchi, M. M.; Mueller, T. Solar-Energy Conversion and Light Emission in an Atomic Monolayer p-n Diode. *Nat. Nanotechnol.* **2014**, *9*(4), 257–261.
28. Baugher, B. W.; Churchill, H. O.; Yang, Y.; Jarillo-Herrero, P. Optoelectronic Devices Based on Electrically Tunable p-n Diodes in a Monolayer Dichalcogenide. *Nat. Nanotechnol.* **2014**, *9*(4), 262–267.
29. Groenendijk, D. J.; Buscema, M.; Steele, G. A.; Michaelis de Vasconcellos, S.; Bratschitsch, R.; van der Zant, H. S.; Castellanos-Gomez, A. Photovoltaic and Photothermoelectric Effect in a Double-Gated WSe_2 Device. *Nano Lett.* **2014**, *14*(10), 5846–5852.
30. Ross, J. S.; Klement, P.; Jones, A. M.; Ghimire, N. J.; Yan, J.; Mandrus, D. G.; Taniguchi, T.; Watanabe, K.; Kitamura, K.; Yao, W.; Cobden, D. H.; Xu, X. Electrically Tunable Excitonic Light-Emitting Diodes Based on Monolayer WSe_2 p-n Junctions. *Nat. Nanotechnol.* **2014**, *9*(4), 268–272.
31. Memaran, S.; Pradhan, N. R.; Lu, Z.; Rhodes, D.; Ludwig, J.; Zhou, Q.; Ogunsolu, O.; Ajayan, P. M.; Smirnov, D.; Fernandez-Dominguez, A. I.; Garcia-Vidal, F. J.; Balicas, L. Pronounced Photovoltaic Response from Multilayered Transition-Metal Dichalcogenides PN-Junctions. *Nano Lett.* **2015**, *15*(11), 7532–7538.
32. Liu, H.; Neal, A. T.; Zhu, Z.; Luo, Z.; Xu, X.; Tomanek, D.; Ye, P. D. Phosphorene: An Unexplored 2D Semiconductor With a High Hole Mobility. *ACS Nano* **2014**, *8*(4), 4033–4041.
33. Xia, F.; Wang, H.; Jia, Y. Rediscovering Black Phosphorus as an Anisotropic Layered Material for Optoelectronics and Electronics. *Nat. Commun.* **2014**, *5*, 4458.
34. Buscema, M.; Groenendijk, D. J.; Steele, G. A.; van der Zant, H. S.; Castellanos-Gomez, A. Photovoltaic Effect in Few-Layer Black Phosphorus PN Junctions Defined by Local Electrostatic Gating. *Nat. Commun.* **2014**, *5*, 4651.
35. Choi, M. S.; Qu, D.; Lee, D.; Liu, X.; Watanabe, K.; Taniguchi, T.; Yoo, W. J. Lateral MoS_2 p-n Junction Formed by Chemical Doping for Use in High-Performance Optoelectronics. *ACS Nano* **2014**, *8*(9), 9332–9340.

36. Koenig, S. P.; Doganov, R. A.; Seixas, L.; Carvalho, A.; Tan, J. Y.; Watanabe, K.; Taniguchi, T.; Yakovlev, N.; Castro Neto, A. H.; Ozyilmaz, B. Electron Doping of Ultrathin Black Phosphorus With Cu Adatoms. *Nano Lett.* **2016**, *16*(4), 2145–2151.
37. Liu, Y.; Cai, Y.; Zhang, G.; Zhang, Y.-W.; Ang, K.-W. Al-Doped Black Phosphorus p-n Homojunction Diode for High Performance Photovoltaic. *Adv. Funct. Mater.* **2017**, *27*(7), 1604638.
38. Xiang, D.; Han, C.; Wu, J.; Zhong, S.; Liu, Y.; Lin, J.; Zhang, X. A.; Ping Hu, W.; Ozyilmaz, B.; Neto, A. H.; Wee, A. T.; Chen, W. Surface Transfer Doping Induced Effective Modulation on Ambipolar Characteristics of Few-Layer Black Phosphorus. *Nat. Commun.* **2015**, *6*, 6485.
39. Wang, G.; Bao, L.; Pei, T.; Ma, R.; Zhang, Y. Y.; Sun, L.; Zhang, G.; Yang, H.; Li, J.; Gu, C.; Du, S.; Pantelides, S. T.; Schrimpf, R. D.; Gao, H. J. Introduction of Interfacial Charges to Black Phosphorus for a Family of Planar Devices. *Nano Lett.* **2016**, *16*(11), 6870–6878.
40. Yu, X.; Zhang, S.; Zeng, H.; Wang, Q. J. Lateral Black Phosphorene P–N Junctions Formed via Chemical Doping for High Performance Near-Infrared Photodetector. *Nano Energy* **2016**, *25*, 34–41.
41. Han, C.; Hu, Z.; Gomes, L. C.; Bao, Y.; Carvalho, A.; Tan, S. J. R.; Lei, B.; Xiang, D.; Wu, J.; Qi, D.; Wang, L.; Huo, F.; Huang, W.; Loh, K. P.; Chen, W. Surface Functionalization of Black Phosphorus via Potassium toward High-Performance Complementary Devices. *Nano Lett.* **2017**, *17*(7), 4122–4129.
42. Wi, S.; Kim, H.; Chen, M.; Nam, H.; Guo, L. J.; Meyhofer, E.; Liang, X. Enhancement of Photovoltaic Response in Multilayer MoS_2 Induced by Plasma Doping. *ACS Nano* **2014**, *8*(5), 5270–5281.
43. Li, H. M.; Lee, D.; Qu, D.; Liu, X.; Ryu, J.; Seabaugh, A.; Yoo, W. J. Ultimate Thin Vertical p-n Junction Composed of Two-Dimensional Layered Molybdenum Disulfide. *Nat. Commun.* **2015**, *6*, 6564.
44. Jin, Y.; Keum, D. H.; An, S. J.; Kim, J.; Lee, H. S.; Lee, Y. H. A Van Der Waals Homojunction: Ideal p-n Diode Behavior in $MoSe_2$. *Adv. Mater.* **2015**, *27*(37), 5534–5540.
45. Cai, Y.; Zhang, G.; Zhang, Y. W. Layer-Dependent Band Alignment and Work Function of Few-Layer Phosphorene. *Sci. Rep.* **2014**, *4*, 6677.
46. Kang, J.; Tongay, S.; Zhou, J.; Li, J.; Wu, J. Band Offsets and Heterostructures of Two-Dimensional Semiconductors. *Appl. Phys. Lett.* **2013**, *102*(1), 012111.
47. Tran, V.; Soklaski, R.; Liang, Y.; Yang, L. Layer-Controlled Band Gap and Anisotropic Excitons in Few-Layer Black Phosphorus. *Phys. Rev. B* **2014**, *89*(23), 235319.
48. Tosun, M.; Fu, D.; Desai, S. B.; Ko, C.; Kang, J. S.; Lien, D. H.; Najmzadeh, M.; Tongay, S.; Wu, J.; Javey, A. MoS_2 Heterojunctions by Thickness Modulation. *Sci. Rep.* **2015**, *5*, 10990.
49. Zhang, C.; Chen, Y.; Huang, J. K.; Wu, X.; Li, L. J.; Yao, W.; Tersoff, J.; Shih, C. K. Visualizing Band Offsets and Edge States in Bilayer-Monolayer Transition Metal Dichalcogenides Lateral Heterojunction. *Nat. Commun.* **2016**, *6*, 10349.
50. Lin, H.-F.; Liu, L.-M.; Zhao, J. 2D Lateral Heterostructures of Monolayer and Bilayer Phosphorene. *J. Mater. Chem. C* **2017**, *5*(9), 2291–2300.
51. Xu, X.; Gabor, N. M.; Alden, J. S.; van der Zande, A. M.; McEuen, P. L. Photo-Thermoelectric Effect at a Graphene Interface Junction. *Nano Lett.* **2010**, *10*(2), 562–566.
52. Howell, S. L.; Jariwala, D.; Wu, C. C.; Chen, K. S.; Sangwan, V. K.; Kang, J.; Marks, T. J.; Hersam, M. C.; Lauhon, L. J. Investigation of Band-Offsets at Monolayer-Multilayer MoS_2 Junctions by Scanning Photocurrent Microscopy. *Nano Lett.* **2015**, *15*(4), 2278–2284.
53. Kallatt, S.; Umesh, G.; Bhat, N.; Majumdar, K. Photoresponse of Atomically Thin MoS2 Layers and Their Planar Heterojunctions. *Nanoscale* **2016**, *8*(33), 15213–15222.

54. Xu, Z.-Q.; Zhang, Y.; Wang, Z.; Shen, Y.; Huang, W.; Xia, X.; Yu, W.; Xue, Y.; Sun, L.; Zheng, C.; Lu, Y.; Liao, L.; Bao, Q. Atomically Thin Lateral p–n Junction Photodetector With Large Effective Detection Area. *2D Mater.* **2016**, *3*(4), 041001.
55. Wang, L.; Huang, L.; Tan, W. C.; Feng, X.; Chen, L.; Ang, K.-W. Pronounced Photovoltaic Effect in Electrically Tunable Lateral Black-Phosphorus Heterojunction Diode. *Adv. Electron. Mater.* **2018**, *4*(1), 1700442.
56. Duan, X.; Wang, C.; Shaw, J. C.; Cheng, R.; Chen, Y.; Li, H.; Wu, X.; Tang, Y.; Zhang, Q.; Pan, A.; Jiang, J.; Yu, R.; Huang, Y.; Duan, X. Lateral Epitaxial Growth of Two-Dimensional Layered Semiconductor Heterojunctions. *Nat. Nanotechnol.* **2014**, *9*(12), 1024–1030.
57. Levendorf, M. P.; Kim, C. J.; Brown, L.; Huang, P. Y.; Havener, R. W.; Muller, D. A.; Park, J. Graphene and Boron Nitride Lateral Heterostructures for Atomically Thin Circuitry. *Nature* **2012**, *488*(7413), 627–632.
58. Liu, Z.; Ma, L.; Shi, G.; Zhou, W.; Gong, Y.; Lei, S.; Yang, X.; Zhang, J.; Yu, J.; Hackenberg, K. P.; Babakhani, A.; Idrobo, J.-C.; Vajtai, R.; Lou, J.; Ajayan, P. M. In-Plane Heterostructures of Graphene and Hexagonal Boron Nitride With Controlled Domain Sizes. *Nat. Nanotechnol.* **2013**, *8*(2), 119–124.
59. Liu, L.; Park, J.; Siegel, D. A.; McCarty, K. F.; Clark, K. W.; Deng, W.; Basile, L.; Idrobo, J. C.; Li, A. P.; Gu, G. Heteroepitaxial Growth of Two-Dimensional Hexagonal Boron Nitride Templated by Graphene Edges. *Science* **2014**, *343*(6167), 163–167.
60. Huang, C.; Wu, S.; Sanchez, A. M.; Peters, J. J.; Beanland, R.; Ross, J. S.; Rivera, P.; Yao, W.; Cobden, D. H.; Xu, X. Lateral Heterojunctions Within Monolayer $MoSe_2$-WSe_2 Semiconductors. *Nat. Mater.* **2014**, *13*(12), 1096–1101.
61. Gong, Y.; Lin, J.; Wang, X.; Shi, G.; Lei, S.; Lin, Z.; Zou, X.; Ye, G.; Vajtai, R.; Yakobson, B. I.; Terrones, H.; Terrones, M.; Tay, B. K.; Lou, J.; Pantelides, S. T.; Liu, Z.; Zhou, W.; Ajayan, P. M. Vertical and In-Plane Heterostructures From WS_2/MoS_2 Monolayers. *Nat. Mater.* **2014**, *13*(12), 1135–1142.
62. Li, M. Y.; Shi, Y.; Cheng, C. C.; Lu, L. S.; Lin, Y. C.; Tang, H. L.; Tsai, M. L.; Chu, C. W.; Wei, K. H.; He, J. H.; Chang, W. H.; Suenaga, K.; Li, L. J. Epitaxial Growth of a Monolayer WSe_2-MoS_2 Lateral p-n Junction With an Atomically Sharp Interface. *Science* **2015**, *349*(6247), 524–528.
63. Tsai, M. L.; Li, M. Y.; Retamal, J. R. D.; Lam, K. T.; Lin, Y. C.; Suenaga, K.; Chen, L. J.; Liang, G.; Li, L. J.; He, J. H. Single Atomically Sharp Lateral Monolayer p-n Heterojunction Solar Cells with Extraordinarily High Power Conversion Efficiency. *Adv. Mater.* **2017**, *29*(32).
64. Zhang, X. Q.; Lin, C. H.; Tseng, Y. W.; Huang, K. H.; Lee, Y. H. Synthesis of Lateral Heterostructures of Semiconducting Atomic Layers. *Nano Lett.* **2015**, *15*(1), 410–415.
65. Li, H.; Li, P.; Huang, J. K.; Li, M. Y.; Yang, C. W.; Shi, Y.; Zhang, X. X.; Li, L. J. Laterally Stitched Heterostructures of Transition Metal Dichalcogenide: Chemical Vapor Deposition Growth on Lithographically Patterned Area. *ACS Nano* **2016**, *10*(11), 10516–10523.
66. Li, X.; Lin, M. W.; Lin, J.; Huang, B.; Puretzky, A. A.; Ma, C.; Wang, K.; Zhou, W.; Pantelides, S. T.; Chi, M.; Kravchenko, I.; Fowlkes, J.; Rouleau, C. M.; Geohegan, D. B.; Xiao, K. Two-Dimensional GaSe/$MoSe_2$ Misfit Bilayer Heterojunctions by van der Waals Epitaxy. *Sci. Adv.* **2016**, *2*(4), e1501882.
67. Chen, K.; Wan, X.; Wen, J.; Xie, W.; Kang, Z.; Zeng, X.; Chen, H.; Xu, J. B. Electronic Properties of MoS_2-WS_2 Heterostructures Synthesized With Two-Step Lateral Epitaxial Strategy. *ACS Nano* **2015**, *9*(10), 9868–9876.
68. Li, Z.; Zheng, J.; Zhang, Y.; Zheng, C.; Woon, W. Y.; Chuang, M. C.; Tsai, H. C.; Chen, C. H.; Davis, A.; Xu, Z. Q.; Lin, J.; Zhang, H.; Bao, Q. Synthesis of Ultrathin Composition Graded Doped Lateral WSe2/WS2 Heterostructures. *ACS Appl. Mater. Interfaces* **2017**, *9*(39), 34204–34212.

69. Novoselov, K. S.; Jiang, D.; Schedin, F.; Booth, T. J.; Khotkevich, V. V.; Morozov, S. V.; Geim, A. K. Two-Dimensional Atomic Crystals. *Proc. Natl. Acad. Sci. U. S. A.* **2005**, *102*(30), 10451–10453.
70. Duan, X.; Wang, C.; Pan, A.; Yu, R.; Duan, X. Two-Dimensional Transition Metal Dichalcogenides as Atomically Thin Semiconductors: Opportunities and Challenges. *Chem. Soc. Rev.* **2015**, *44*(24), 8859–8876.
71. Carvalho, A.; Wang, M.; Zhu, X.; Rodin, A. S.; Su, H.; Castro Neto, A. H. Phosphorene: From Theory to Applications. *Nat. Rev. Mater.* **2016**, *1*(11), 16061.
72. Castellanos-Gomez, A.; Buscema, M.; Molenaar, R.; Singh, V.; Janssen, L.; van der Zant, H. S. J.; Steele, G. A. Deterministic Transfer of Two-Dimensional Materials by All-Dry Viscoelastic Stamping. *2D Mater.* **2014**, *1*(1), 011002.
73. Zomer, P. J.; Dash, S. P.; Tombros, N.; van Wees, B. J. A Transfer Technique for High Mobility Graphene Devices on Commercially Available Hexagonal Boron Nitride. *Appl. Phys. Lett.* **2011**, *99*(23), 232104.
74. Novoselov, K. S.; Mishchenko, A.; Carvalho, A.; Castro Neto, A. H. 2D Materials and van der Waals Heterostructures. *Science* **2016**, *353*(6298), aac9439.
75. Ponomarenko, L. A.; Geim, A. K.; Zhukov, A. A.; Jalil, R.; Morozov, S. V.; Novoselov, K. S.; Grigorieva, I. V.; Hill, E. H.; Cheianov, V. V.; Fal'ko, V. I.; Watanabe, K.; Taniguchi, T.; Gorbachev, R. V. Tunable Metal–Insulator Transition in Double-Layer Graphene Heterostructures. *Nat. Phys.* **2011**, 7(12), 958–961.
76. Britnell, L.; Gorbachev, R. V.; Jalil, R.; Belle, B. D.; Schedin, F.; Mishchenko, A.; Georgiou, T.; Katsnelson, M. I.; Eaves, L.; Morozov, S. V.; Peres, N. M.; Leist, J.; Geim, A. K.; Novoselov, K. S.; Ponomarenko, L. A. Field-Effect Tunneling Transistor Based on Vertical Graphene Heterostructures. *Science* **2012**, *335*(6071), 947–950.
77. Haigh, S. J.; Gholinia, A.; Jalil, R.; Romani, S.; Britnell, L.; Elias, D. C.; Novoselov, K. S.; Ponomarenko, L. A.; Geim, A. K.; Gorbachev, R. Cross-Sectional Imaging of Individual Layers and Buried Interfaces of Graphene-Based Heterostructures and Superlattices. *Nat. Mater.* **2012**, *11*(9), 764–767.
78. Yu, W. J.; Liu, Y.; Zhou, H.; Yin, A.; Li, Z.; Huang, Y.; Duan, X. Highly Efficient Gate-Tunable Photocurrent Generation in Vertical Heterostructures Of Layered Materials. *Nat. Nanotechnol.* **2013**, *8*(12), 952–958.
79. Li, H.; Ye, L.; Xu, J. High-Performance Broadband Floating-Base Bipolar Phototransistor Based on WSe2/BP/MoS_2 Heterostructure. *ACS Photonics* **2017**, *4*(4), 823–829.
80. Ye, L.; Li, H.; Chen, Z.; Xu, J. Near-Infrared Photodetector Based on MoS_2/Black Phosphorus Heterojunction. *ACS Photonics* **2016**, *3*(4), 692–699.
81. Ye, L.; Wang, P.; Luo, W.; Gong, F.; Liao, L.; Liu, T.; Tong, L.; Zang, J.; Xu, J.; Hu, W. Highly Polarization Sensitive Infrared Photodetector Based on Black Phosphorus-on-WSe_2 Photogate Vertical Heterostructure. *Nano Energy* **2017**, *37*, 53–60.
82. Furchi, M. M.; Pospischil, A.; Libisch, F.; Burgdorfer, J.; Mueller, T. Photovoltaic Effect in an Electrically Tunable van der Waals Heterojunction. *Nano Lett.* **2014**, *14*(8), 4785–4791.
83. Cheng, R.; Li, D.; Zhou, H.; Wang, C.; Yin, A.; Jiang, S.; Liu, Y.; Chen, Y.; Huang, Y.; Duan, X. Electroluminescence and Photocurrent Generation From Atomically Sharp WSe_2/MoS_2 Heterojunction p-n Diodes. *Nano Lett.* **2014**, *14*(10), 5590–5597.
84. Lee, C. H.; Lee, G. H.; van der Zande, A. M.; Chen, W.; Li, Y.; Han, M.; Cui, X.; Arefe, G.; Nuckolls, C.; Heinz, T. F.; Guo, J.; Hone, J.; Kim, P. Atomically Thin p-n Junctions With van der Waals Heterointerfaces. *Nat. Nanotechnol.* **2014**, *9*(9), 676–681.
85. Komsa, H.-P.; Krasheninnikov, A. V. Electronic Structures and Optical Properties of Realistic Transition Metal Dichalcogenide Heterostructures From First Principles. *Phys. Rev. B* **2013**, *88*(8), 085318.
86. Kośmider, K.; Fernández-Rossier, J. Electronic Properties of the MoS_2-WS_2 Heterojunction. *Phys. Rev. B* **2013**, *87*(7), 075451.

87. Wong, J.; Jariwala, D.; Tagliabue, G.; Tat, K.; Davoyan, A. R.; Sherrott, M. C.; Atwater, H. A. High Photovoltaic Quantum Efficiency in Ultrathin van der Waals Heterostructures. *ACS Nano* **2017**, *11*(7), 7230–7240.
88. Long, M.; Liu, E.; Wang, P.; Gao, A.; Xia, H.; Luo, W.; Wang, B.; Zeng, J.; Fu, Y.; Xu, K.; Zhou, W.; Lv, Y.; Yao, S.; Lu, M.; Chen, Y.; Ni, Z.; You, Y.; Zhang, X.; Qin, S.; Shi, Y.; Hu, W.; Xing, D.; Miao, F. Broadband Photovoltaic Detectors Based on an Atomically Thin Heterostructure. *Nano Lett.* **2016**, *16*(4), 2254–2259.
89. Deng, Y.; Luo, Z.; Conrad, N. J.; Liu, H.; Gong, Y.; Najmaei, S.; Ajayan, P. M.; Lou, J.; Xu, X.; Ye, P. D. Black Phosphorus-Monolayer MoS_2 van der Waals Heterojunction p-n Diode. *ACS Nano* **2014**, *8*(8), 8292–8299.
90. Flöry, N.; Jain, A.; Bharadwaj, P.; Parzefall, M.; Taniguchi, T.; Watanabe, K.; Novotny, L. A $WSe_2/MoSe_2$ Heterostructure Photovoltaic Device. *Appl. Phys. Lett.* **2015**, *107*(12), 123106.
91. Pezeshki, A.; Shokouh, S. H.; Nazari, T.; Oh, K.; Im, S. Electric and Photovoltaic Behavior of a Few-Layer alpha-$MoTe_2/MoS_2$ Dichalcogenide Heterojunction. *Adv. Mater.* **2016**, *28*(16), 3216–3222.
92. Li, B.; Huang, L.; Zhong, M.; Li, Y.; Wang, Y.; Li, J.; Wei, Z. Direct Vapor Phase Growth and Optoelectronic Application of Large Band Offset SnS/MoS2 Vertical Bilayer Heterostructures With High Lattice Mismatch. *Adv. Electron. Mater.* **2016**, *2*(11), 1600298.
93. Yang, S.; Wang, C.; Ataca, C.; Li, Y.; Chen, H.; Cai, H.; Suslu, A.; Grossman, J. C.; Jiang, C.; Liu, Q.; Tongay, S. Self-Driven Photodetector and Ambipolar Transistor in Atomically Thin GaTe-MoS_2 p-n vdW Heterostructure. *ACS Appl. Mater. Interfaces* **2016**, *8*(4), 2533–2539.
94. Cho, A.-J.; Namgung, S. D.; Kim, H.; Kwon, J.-Y. Electric and Photovoltaic Characteristics of a Multi-Layer $ReS_2/ReSe_2$ Heterostructure. *APL Mater.* **2017**, *5*(7), 076101.
95. Sung, J. H.; Cha, S.; Heo, H.; Sim, S.; Kim, J.; Choi, H.; Jo, M.-H. Ultrafast Hot-Carrier Photovoltaics of Type-I Monolayer Heterojunctions in the Broad Spectral Ranges. *ACS Photonics* **2017**, *4*(3), 429–434.
96. Furchi, M. M.; Zechmeister, A. A.; Hoeller, F.; Wachter, S.; Pospischil, A.; Mueller, T. Photovoltaics in Van der Waals Heterostructures. *IEEE J. Sel. Top. Quantum Electron.* **2017**, *23*(1), 106–116.
97. Shastry, T. A.; Balla, I.; Bergeron, H.; Amsterdam, S. H.; Marks, T. J.; Hersam, M. C. Mutual Photoluminescence Quenching and Photovoltaic Effect in Large-Area Single-Layer MoS_2-Polymer Heterojunctions. *ACS Nano* **2016**, *10*(11), 10573–10580.
98. Jariwala, D.; Marks, T. J.; Hersam, M. C. Mixed-Dimensional van der Waals Heterostructures. *Nat. Mater.* **2017**, *16*(2), 170–181.
99. Ye, Y.; Dai, L. Graphene-Based Schottky Junction Solar Cells. *J. Mater. Chem.* **2012**, *22*(46), 24224.
100. Miao, X.; Tongay, S.; Petterson, M. K.; Berke, K.; Rinzler, A. G.; Appleton, B. R.; Hebard, A. F. High Efficiency Graphene Solar Cells by Chemical Doping. *Nano Lett.* **2012**, *12*(6), 2745–2750.
101. Li, X.; Zhu, H.; Wang, K.; Cao, A.; Wei, J.; Li, C.; Jia, Y.; Li, Z.; Li, X.; Wu, D. Graphene-on-Silicon Schottky Junction Solar Cells. *Adv. Mater.* **2010**, *22*(25), 2743–2748.
102. Shi, E.; Li, H.; Yang, L.; Zhang, L.; Li, Z.; Li, P.; Shang, Y.; Wu, S.; Li, X.; Wei, J.; Wang, K.; Zhu, H.; Wu, D.; Fang, Y.; Cao, A. Colloidal Antireflection Coating Improves Graphene-Silicon Solar Cells. *Nano Lett.* **2013**, *13*(4), 1776–1781.
103. Song, Y.; Li, X.; Mackin, C.; Zhang, X.; Fang, W.; Palacios, T.; Zhu, H.; Kong, J. Role of Interfacial Oxide in High-Efficiency Graphene-Silicon Schottky Barrier Solar Cells. *Nano Lett.* **2015**, *15*(3), 2104–2110.
104. Brus, V. V.; Gluba, M. A.; Zhang, X.; Hinrichs, K.; Rappich, J.; Nickel, N. H. Stability of Graphene-Silicon Heterostructure Solar Cells. *Phys. Status Solidi A* **2014**, *211*(4), 843–847.

105. Feng, T.; Xie, D.; Lin, Y.; Zang, Y.; Ren, T.; Song, R.; Zhao, H.; Tian, H.; Li, X.; Zhu, H.; Liu, L. Graphene Based Schottky Junction Solar Cells on Patterned Silicon-Pillar-Array Substrate. *Appl. Phys. Lett.* **2011**, *99*(23), 233505.
106. Lin, Y.; Li, X.; Xie, D.; Feng, T.; Chen, Y.; Song, R.; Tian, H.; Ren, T.; Zhong, M.; Wang, K.; Zhu, H. Graphene/Semiconductor Heterojunction Solar Cells With Modulated Antireflection and Graphene Work Function. *Energy Environ. Sci.* **2013**, *6*(1), 108–115.
107. Vazquez-Mena, O.; Bosco, J. P.; Ergen, O.; Rasool, H. I.; Fathalizadeh, A.; Tosun, M.; Crommie, M.; Javey, A.; Atwater, H. A.; Zettl, A. Performance Enhancement of a Graphene-Zinc Phosphide Solar Cell Using the Electric Field-Effect. *Nano Lett.* **2014**, *14*(8), 4280–4285.
108. Wang, P.; Li, X.; Xu, Z.; Wu, Z.; Zhang, S.; Xu, W.; Zhong, H.; Chen, H.; Li, E.; Luo, J.; Yu, Q.; Lin, S. Tunable Graphene/Indium Phosphide Heterostructure Solar Cells. *Nano Energy* **2015**, *13*, 509–517.
109. Li, X.; Chen, W.; Zhang, S.; Wu, Z.; Wang, P.; Xu, Z.; Chen, H.; Yin, W.; Zhong, H.; Lin, S. 18.5% Efficient graphene/GaAs van der Waals Heterostructure Solar Cell. *Nano Energy* **2015**, *16*, 310–319.
110. Wang, Y.; Zhang, Y.; Lu, Y.; Xu, W.; Mu, H.; Chen, C.; Qiao, H.; Song, J.; Li, S.; Sun, B.; Cheng, Y.-B.; Bao, Q. Hybrid Graphene-Perovskite Phototransistors With Ultrahigh Responsivity and Gain. *Adv. Opt. Mater.* **2015**, *3*(10), 1389–1396.
111. Lin, S.; Li, X.; Zhang, S.; Wang, P.; Xu, Z.; Zhong, H.; Wu, Z.; Chen, H. Graphene/CdTe Heterostructure Solar Cell and Its Enhancement With Photo-Induced Doping. *Appl. Phys. Lett.* **2015**, *107*(19), 191106.
112. Zhang, L.; Fan, L.; Li, Z.; Shi, E.; Li, X.; Li, H.; Ji, C.; Jia, Y.; Wei, J.; Wang, K.; Zhu, H.; Wu, D.; Cao, A. Graphene-CdSe Nanobelt Solar Cells With Tunable Configurations. *Nano Res.* **2011**, *4*(9), 891–900.
113. Ye, Y.; Gan, L.; Dai, L.; Dai, Y.; Guo, X.; Meng, H.; Yu, B.; Shi, Z.; Shang, K.; Qin, G. A Simple and Scalable Graphene Patterning Method and Its Application in CdSe Nanobelt/Graphene Schottky Junction Solar Cells. *Nanoscale* **2011**, *3*(4), 1477–1481.
114. Ye, Y.; Dai, Y.; Dai, L.; Shi, Z.; Liu, N.; Wang, F.; Fu, L.; Peng, R.; Wen, X.; Chen, Z.; Liu, Z.; Qin, G. High-Performance Single CdS Nanowire (Nanobelt) Schottky Junction Solar Cells With Au/Graphene Schottky Electrodes. *ACS Appl. Mater. Interfaces* **2010**, *2*(12), 3406–3410.
115. Lopez-Sanchez, O.; Alarcon Llado, E.; Koman, V.; Fontcuberta i Morral, A.; Radenovic, A.; Kis, A. Light Generation and Harvesting in a van der Waals Heterostructure. *ACS Nano* **2014**, *8*(3), 3042–3048.
116. Tsai, M. L.; Su, S. H.; Chang, J. K.; Tsai, D. S.; Chen, C. H.; Wu, C. I.; Li, L. J.; Chen, L. J.; He, J. H. Monolayer MoS_2 Heterojunction Solar Cells. *ACS Nano* **2014**, *8*(8), 8317–8322.
117. Gehring, P.; Urcuyo, R.; Duong, D. L.; Burghard, M.; Kern, K. Thin-Layer Black Phosphorus/GaAs Heterojunction p-n Diodes. *Appl. Phys. Lett.* **2015**, *106*(23), 233110.
118. Lin, S.; Li, X.; Wang, P.; Xu, Z.; Zhang, S.; Zhong, H.; Wu, Z.; Xu, W.; Chen, H. Interface Designed MoS_2/GaAs Heterostructure Solar Cell With Sandwich Stacked Hexagonal Boron Nitride. *Sci. Rep.* **2015**, *5*, 15103.
119. Lin, S.; Wang, P.; Li, X.; Wu, Z.; Xu, Z.; Zhang, S.; Xu, W. Gate Tunable Monolayer MoS_2/InP Heterostructure Solar Cells. *Appl. Phys. Lett.* **2015**, *107*(15), 153904.
120. Wang, P.; Lin, S.; Ding, G.; Li, X.; Wu, Z.; Zhang, S.; Xu, Z.; Xu, S.; Lu, Y.; Xu, W.; Zheng, Z. Enhanced Monolayer MoS2/InP Heterostructure Solar Cells by Graphene Quantum Dots. *Appl. Phys. Lett.* **2016**, *108*(16), 163901.

121. Chen, Z.; Liu, H.; Chen, X.; Chu, G.; Chu, S.; Zhang, H. Wafer-Size and Single-Crystal $MoSe_2$ Atomically Thin Films Grown on GaN Substrate for Light Emission and Harvesting. *ACS Appl. Mater. Interfaces* **2016**, *8*(31), 20267–20273.
122. Lin, S.-S.; Wu, Z.-Q.; Li, X.-Q.; Zhang, Y.-J.; Zhang, S.-J.; Wang, P.; Panneerselvam, R.; Li, J.-F. Stable 16.2% Efficient Surface Plasmon-Enhanced Graphene/GaAs Heterostructure Solar Cell. *Adv. Energy Mater.* **2016**, *6*(21), 1600822.
123. Hao, L. Z.; Gao, W.; Liu, Y. J.; Han, Z. D.; Xue, Q. Z.; Guo, W. Y.; Zhu, J.; Li, Y. R. High-Performance n-MoS_2/i-SiO_2/p-Si Heterojunction Solar Cells. *Nanoscale* **2015**, 7(18), 8304–8308.
124. Deckman, H. W.; Roxlo, C. B.; Yablonovitch, E. Maximum Statistical Increase of Optical Absorption in Textured Semiconductor Films. *Opt. Lett.* **1983**, *8*(9), 491.
125. Konstantatos, G.; Sargent, E. H. Nanostructured Materials for Photon Detection. *Nat. Nanotechnol.* **2010**, *5*(6), 391–400.
126. Polman, A. Applied Physics. Plasmonics Applied. *Science* **2008**, *322*(5903), 868–869.
127. Atwater, H. A.; Polman, A. Plasmonics for Improved Photovoltaic Devices. *Nat. Mater.* **2010**, *9*(3), 205–213.
128. Lin, J.; Li, H.; Zhang, H.; Chen, W. Plasmonic Enhancement of Photocurrent in MoS_2 Field-Effect-Transistor. *Appl. Phys. Lett.* **2013**, *102*(20), 203109.
129. Miao, J.; Hu, W.; Jing, Y.; Luo, W.; Liao, L.; Pan, A.; Wu, S.; Cheng, J.; Chen, X.; Lu, W. Surface Plasmon-Enhanced Photodetection in Few Layer MoS_2 Phototransistors With Au Nanostructure Arrays. *Small* **2015**, *11*(20), 2392–2398.
130. Butun, S.; Palacios, E.; Cain, J. D.; Liu, Z.; Dravid, V. P.; Aydin, K. Quantifying Plasmon-Enhanced Light Absorption in Monolayer WS_2 Films. *ACS Appl. Mater. Interfaces* **2017**, *9*(17), 15044–15051.
131. Petoukhoff, C. E.; Krishna, M. B.; Voiry, D.; Bozkurt, I.; Deckoff-Jones, S.; Chhowalla, M.; O'Carroll, D. M.; Dani, K. M. Ultrafast Charge Transfer and Enhanced Absorption in MoS_2-Organic van der Waals Heterojunctions Using Plasmonic Metasurfaces. *ACS Nano* **2016**, *10*(11), 9899–9908.
132. Jariwala, D.; Davoyan, A. R.; Tagliabue, G.; Sherrott, M. C.; Wong, J.; Atwater, H. A. Near-Unity Absorption in van der Waals Semiconductors for Ultrathin Optoelectronics. *Nano Lett.* **2016**, *16*(9), 5482–5487.
133. Piper, J. R.; Fan, S. Broadband Absorption Enhancement in Solar Cells with an Atomically Thin Active Layer. *ACS Photonics* **2016**, *3*(4), 571–577.
134. Wu, Y. B.; Yang, W.; Wang, T. B.; Deng, X. H.; Liu, J. T. Broadband Perfect Light Trapping in the Thinnest Monolayer Graphene-MoS_2 Photovoltaic Cell: The New Application of Spectrum-Splitting Structure. *Sci. Rep.* **2016**, *6*, 20955.
135. Feng, J.; Qian, X.; Huang, C.-W.; Li, J. Strain-Engineered Artificial Atom as a Broad-Spectrum Solar Energy Funnel. *Nat. Photonics* **2012**, *6*(12), 866–872.
136. Roldan, R.; Castellanos-Gomez, A.; Cappelluti, E.; Guinea, F. Strain Engineering in Semiconducting Two-Dimensional Crystals. *J. Phys. Condens. Matter* **2015**, *27*(31), 313201.
137. Çakır, D.; Sahin, H.; Peeters, F. M. Tuning of the Electronic and Optical Properties of Single-Layer Black Phosphorus by Strain. *Phys. Rev. B* **2014**, *90*(20), 205421.
138. Castellanos-Gomez, A.; Roldan, R.; Cappelluti, E.; Buscema, M.; Guinea, F.; van der Zant, H. S.; Steele, G. A. Local Strain Engineering in Atomically Thin MoS_2. *Nano Lett.* **2013**, *13*(11), 5361–5366.
139. Komsa, H.-P.; Krasheninnikov, A. V. Effects of Confinement and Environment on the Electronic Structure and Exciton Binding Energy of MoS_2 From First Principles. *Phys. Rev. B* **2012**, *86*(24), 241201.
140. Rand, B. P.; Burk, D. P.; Forrest, S. R. Offset Energies at Organic Semiconductor Heterojunctions and Their Influence on the Open-Circuit Voltage of Thin-Film Solar Cells. *Phys. Rev. B* **2007**, *75*(11), 115327.

141. Heremans, P.; Cheyns, D.; Rand, B. P. Strategies for Increasing the Efficiency of Heterojunction Organic Solar Cells: Material Selection and Device Architecture. *Acc. Chem. Res.* **2009**, *42*(11), 1740–1747.
142. Shi, H.; Yan, R.; Bertolazzi, S.; Brivio, J.; Gao, B.; Kis, A.; Jena, D.; Xing, H. G.; Huang, L. Exciton Dynamics in Suspended Monolayer and Few-Layer MoS_2 2D Crystals. *ACS Nano* **2013**, 7(2), 1072–1080.
143. Zhao, Y.; Xu, K.; Pan, F.; Zhou, C.; Zhou, F.; Chai, Y. Doping, Contact and Interface Engineering of Two-Dimensional Layered Transition Metal Dichalcogenides Transistors. *Adv. Funct. Mater.* **2017**, *27*(19), 1603484.
144. Du, Y.; Liu, H.; Deng, Y.; Ye, P. D. Device Perspective for Black Phosphorus Field-Effect Transistors: Contact Resistance, Ambipolar Behavior, and Scaling. *ACS Nano* **2014**, *8*(10), 10035–10042.
145. Kappera, R.; Voiry, D.; Yalcin, S. E.; Branch, B.; Gupta, G.; Mohite, A. D.; Chhowalla, M. Phase-Engineered Low-Resistance Contacts for Ultrathin MoS_2 Transistors. *Nat. Mater.* **2014**, *13*(12), 1128–1134.
146. Xu, Y.; Cheng, C.; Du, S.; Yang, J.; Yu, B.; Luo, J.; Yin, W.; Li, E.; Dong, S.; Ye, P.; Duan, X. Contacts Between Two- and Three-Dimensional Materials: Ohmic, Schottky, and p-n Heterojunctions. *ACS Nano* **2016**, *10*(5), 4895–4919.
147. Amani, M.; Lien, D. H.; Kiriya, D.; Xiao, J.; Azcatl, A.; Noh, J.; Madhvapathy, S. R.; Addou, R.; Kc, S.; Dubey, M.; Cho, K.; Wallace, R. M.; Lee, S. C.; He, J. H.; Ager, J. W., 3rd; Zhang, X.; Yablonovitch, E.; Javey, A. Near-Unity Photoluminescence Quantum Yield in MoS_2. *Science* **2015**, *350*(6264), 1065–1068.
148. Georgiou, T.; Jalil, R.; Belle, B. D.; Britnell, L.; Gorbachev, R. V.; Morozov, S. V.; Kim, Y. J.; Gholinia, A.; Haigh, S. J.; Makarovsky, O.; Eaves, L.; Ponomarenko, L. A.; Geim, A. K.; Novoselov, K. S.; Mishchenko, A. Vertical Field-Effect Transistor Based on Graphene-WS_2 Heterostructures for Flexible and Transparent Electronics. *Nat. Nanotechnol.* **2013**, *8*(2), 100–103.
149. Zhu, W.; Yogeesh, M. N.; Yang, S.; Aldave, S. H.; Kim, J. S.; Sonde, S.; Tao, L.; Lu, N.; Akinwande, D. Flexible Black Phosphorus Ambipolar Transistors, Circuits and AM Demodulator. *Nano Lett.* **2015**, *15*(3), 1883–1890.
150. Jang, H.; Park, Y. J.; Chen, X.; Das, T.; Kim, M. S.; Ahn, J. H. Graphene-Based Flexible and Stretchable Electronics. *Adv. Mater.* **2016**, *28*(22), 4184–4202.
151. Akinwande, D.; Petrone, N.; Hone, J. Two-Dimensional Flexible Nanoelectronics. *Nat. Commun.* **2014**, *5*, 5678.
152. Li, H.; Li, Y.; Aljarb, A.; Shi, Y.; Li, L. J. Epitaxial Growth of Two-Dimensional Layered Transition-Metal Dichalcogenides: Growth Mechanism, Controllability, and Scalability. *Chem. Rev.* **2017**. https://doi.org/1021/acs.chemrev.7b00212.
153. Tan, C.; Cao, X.; Wu, X. J.; He, Q.; Yang, J.; Zhang, X.; Chen, J.; Zhao, W.; Han, S.; Nam, G. H.; Sindoro, M.; Zhang, H. Recent Advances in Ultrathin Two-Dimensional Nanomaterials. *Chem. Rev.* **2017**, *117*(9), 6225–6331.
154. Yu, X.; Prevot, M. S.; Guijarro, N.; Sivula, K. Self-Assembled 2D WSe2 Thin Films for Photoelectrochemical Hydrogen Production. *Nat. Commun.* **2015**, *6*, 7596.
155. Jariwala, D.; Sangwan, V. K.; Lauhon, L. J.; Marks, T. J.; Hersam, M. C. Emerging Device Applications for Semiconducting Two-Dimensional Transition Metal Dichalcogenides. *ACS Nano* **2014**, *8*(2), 1102–1120.
156. Zhang, Z.; Chen, P.; Duan, X.; Zang, K.; Luo, J.; Duan, X. Robust Epitaxial Growth of Two-Dimensional Heterostructures, Multiheterostructures, and Superlattices. *Science* **2017**, *357*(6353), 788–792.
157. Rasmussen, F. A.; Thygesen, K. S. Computational 2D Materials Database: Electronic Structure of Transition-Metal Dichalcogenides and Oxides. *J. Phys. Chem. C* **2015**, *119*(23), 13169–13183.

CHAPTER 6

Black phosphorus: Light-matter interactions and potential applications

Jiong Yang, Yuerui Lu
Research School of Engineering, College of Engineering and Computer Science, Australian National University, Canberra, ACT, Australia

6.1 Introduction

Black phosphorus (BP) is a recently rediscovered two-dimensional (2D) material that has attracted tremendous attention owing to its unique anisotropic manner,[1–7] layer-dependent direct bandgaps,[8–10] and quasi-one-dimensional (1D) excitonic nature,[11–13] which are all in drastic contrast with the properties of other 2D materials, such as graphene[14] and transition metal dichalcogenide (TMD) semiconductors.[15–17] Monolayer BP has been of particular interest in investigating fundamental physics, such as 2D quantum confinement and many-body interactions,[11, 18] and exploring advanced photonic and optoelectronic applications. Light-matter interaction in BP is extraordinarily strong. The anisotropic nature of BP that originates from its puckering structure has been characterized by Raman spectroscopy.[2, 6, 13, 19–23] Exciton and trion properties in BP have also been characterized by photoluminescence (PL) spectroscopy.[10, 12, 24, 25] Regardless of the fact that it is unstable in ambient conditions and degrades quickly,[5, 9, 24, 26–29] BP owns one intriguing property, i.e., the layer-dependent direct bandgap in the infrared range.[9–11, 24, 30, 31] With this unique property, BP can bridge the zero-bandgap graphene[14, 32] and comparatively large-bandgap TMD semiconductors,[33–35] leading it to potential applications in the infrared regime, such as infrared photodetectors and absorbers.[21, 36–39] Besides, unlike TMD semiconductors that show an indirect-to-indirect bandgap transition only when the layer number is thinned down to monolayer,[33–35] BP is a direct bandgap semiconductor from monolayer to its bulk form,[5, 9, 30] resulting in high internal quantum efficiency and making it a promising candidate for future optoelectronic applications. In this chapter, we focus on the

2D Materials for Photonic and Optoelectronic Applications
https://doi.org/10.1016/B978-0-08-102637-3.00006-1

light-matter interactions in BP by analyzing its Raman and PL measurements, and we also discuss its potential photonic and optoelectronic applications.

6.2 Raman spectroscopy

Raman spectroscopy is a powerful tool to characterize the structural and electronic properties of few-layer BP. In particular, temperature-dependent Raman spectroscopy can help further understand the fine structure and properties of the material, such as atomic bonds, thermal expansion, and thermal conductivity.[9] Angle-dependent Raman measurements can quickly determine the crystalline orientation of a BP flake due to its strong anisotropic nature. Besides, Favron et al. also experimentally demonstrated that Raman spectroscopy could be used to identify the layer number of few-layer BP.[27] Different from the situation in TMD semiconductors, where the A_{1g} and $E_{2g}{}^{1}$ modes keep blue- and red-shifting with the layer number from bulk to monolayer, respectively,[40, 41] the $A_g{}^2$ mode exhibits redshift from bulk to bilayer BP and then blueshift when the layer number is further thinned down to monolayer. This peculiar trend may be caused by the combination of two out-of-phase monolayer infrared modes in bilayer BP that generates a new Raman-allowed mode near the $A_g{}^2$ frequency. Furthermore, the ratio between the intensity of $A_g{}^1$ and Si peak in Raman spectroscopy can also be used to roughly identify the layer number of BP samples.[26, 27]

6.2.1 Temperature-dependent Raman spectroscopy

Temperature-dependent Raman study for few-layer BP is important to further understand the fine structure and properties of the material, such as atomic bonds, thermal expansion, and thermal conductivity. The temperature dependence of the Raman spectra, measured at temperature ranging from 20°C to −160°C in a 5 L BP, is shown in Fig. 6.1. The decreasing temperature leads to the blueshift of the Raman phonon modes, $A_g{}^1$, B_{2g}, and $A_g{}^2$. The measured temperature dependence of the Raman mode frequency shift in few-layer BP can be characterized by a linear equation:

$$\omega = \omega_0 + \chi T \tag{6.1}$$

where ω_0 is the mode frequency at 0 K and χ is the first-order temperature coefficient. The measured χ values for modes $A_g{}^1$, B_{2g}, and $A_g{}^2$ in a 5 L BP are −0.023, −0.018, and −0.023 cm^{-1}/°C, respectively (Fig. 6.1B). The change of the Raman shift with temperature is determined by the anharmonic terms in the lattice potential energy, which is related to the anharmonic

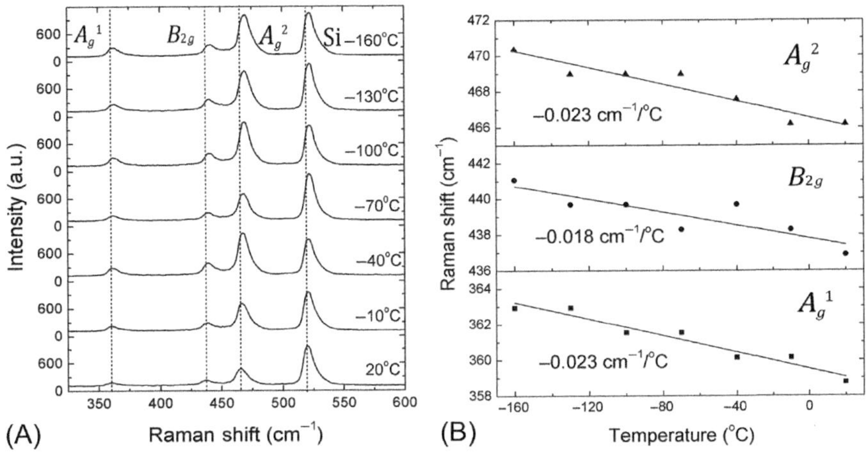

Fig. 6.1 Temperature-dependent Raman spectra of 5 L BP. (A) Raman spectra of 5 L BP at temperature ranging from 20°C to −160°C. (B) Temperature dependence of A_g^1, B_{2g}, and A_g^2 Raman peak positions. *Adapted with permission from Zhang, S.; Yang, J.; Xu, R.; Wang, F.; Li, W.; Ghufran, M.; Zhang, Y.-W.; Yu, Z.; Zhang, G.; Qin, Q.; Lu, Y., Extraordinary Photoluminescence and Strong Temperature/Angle-Dependent Raman Responses in Few-Layer Phosphorene. ACS Nano* ***2014****, 8(9), 9590-9596.*

potential constants, the phonon occupation number, and the thermal expansion of the crystal.[42] These measured χ values in 5 L BP are all larger in absolute value than those from both bilayer graphene (-0.0154 cm^{-1}/°C for G peak)[43] and few-layer MoS_2 (-0.0123 and -0.0132 cm^{-1}/°C for A_{1g} and E_{2g}^1 modes, respectively).[44] This indicates that phonon frequencies in few-layer BP are more sensitive to temperature modulation than those in graphene and MoS_2, which could be due to the superior mechanical flexibility of BP originating from its unique puckered crystal structure.[45]

6.2.2 Angle-dependent Raman spectroscopy

Compared with other 2D materials, BP shows strong anisotropic properties, allowing for various unique applications in optoelectronics. Normally, TEM is used to analyze the crystalline orientation of crystals. But the high-energy electrons in TEM are likely to introduce lots of defects in BP, which can expedite the reaction of BP with moisture and other possible reactants from the environment.[26] Thus, it is important to develop optical means to characterize the crystalline properties. It has been recently shown that second-harmonic generation is used to determine the crystalline orientation of MoS_2.[46] Both polarized reflection spectroscopy[1] and polarized IR spectroscopy[2] are capable of providing simple and nondestructive optical way to determine the crystalline orientation of BP. Angle-dependent Raman spectroscopy is an even simpler optical means to determine the crystalline orientation of BP. This process is critically important as it has minimal damage to BP and significantly shortened the time of characterization, allowing for more accurate analysis of intrinsic BP properties.

Linearly polarized Raman measurements were performed to quickly determine the crystalline orientation of a BP flake (15 L) (Fig. 6.2). The normally incident laser was in the z-direction. The polarization angle θ is relative to the zero-degree reference, which can be arbitrarily selected at the

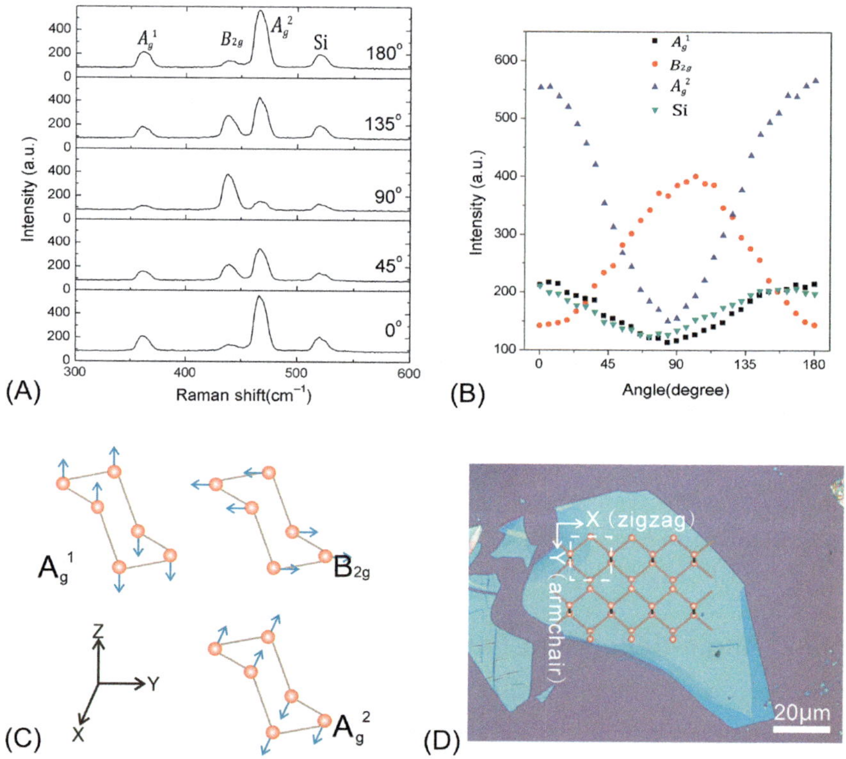

Fig. 6.2 BP crystalline orientation determination by polarization Raman spectra. (A) Raman spectra of 15 L BP under different polarization angles. (B) Polarization dependence of A_g^1, A_g^2, B_{2g} modes in 15 L BP and the Raman peaks in silicon. (C) Schematic plot showing vibration directions of A_g^1, A_g^2, B_{2g} Raman modes. (D) Crystalline orientation of the 15 L BP flake, determined by angle-dependent Raman measurement. *Adapted with permission from Zhang, S.; Yang, J.; Xu, R.; Wang, F.; Li, W.; Ghufran, M.; Zhang, Y.-W.; Yu, Z.; Zhang, G.; Qin, Q.; Lu, Y., Extraordinary Photoluminescence and Strong Temperature/Angle-Dependent Raman Responses in Few-Layer Phosphorene. ACS Nano* ***2014****, 8(9), 9590-9596.*

beginning. Raman intensities of B_{2g} and A_g^2 were observed to be significantly dependent on the polarization angle (Fig. 6.2A). The angle-dependent intensities of B_{2g} and A_g^2 modes both show an angle period of 180 degree and are out of phase (Fig. 6.2B). The Raman intensity of A_g^1 is less sensitive to the polarization angle (Fig. 6.2B). The measurement results can be perfectly explained by the vibration directions of these three Raman modes in crystalline BP. Fig. 6.2C shows a schematic of the vibration direction of phosphorus atoms in different Raman modes.[47] In A_g^2, B_{2g}, and A_g^1 vibrational modes, phosphorus atoms oscillate along the x (zigzag), y (armchair), and z (out-of-plane) directions, respectively. When laser polarization is parallel to x-direction (y-direction), the intensity of the A_g^2 mode will reach the maximum (minimum) value; meanwhile, the intensity of the B_{2g} mode will reach the minimum (maximum) value. Based on this polarization-dependent Raman measurement, the crystalline orientation of this 15 L BP flake was quickly determined, as indicated in Fig. 6.2D. This technique provides a fast and precise determination of crystalline orientation, without the need of complicated and high-resolution imaging systems, such as STM or TEM.

6.3 PL measurements

PL spectroscopy is a useful tool to characterize the electronic band structure of 2D materials, and it utilizes the interactions between laser photons and electrons/holes inside the measured material. Since the first rediscovery of BP, PL spectroscopy with different functionalities has been used to analyze it, including its optical bandgap, anisotropic nature, oxidation and degradation, and its exciton and trion dynamics.[9, 10, 12, 24, 25, 48, 49]

6.3.1 Layer-dependent PL

Because of the strongly layer-dependent peak energies and the direct bandgap nature of BP, layer number identification can be further confirmed by measuring their corresponding peak energies of PL emission (Fig. 6.3). Fig. 6.3A shows the normalized PL spectra of the mono- to five-layer BP samples.

The emission peak of the PL spectrum for a monolayer BP is at 711 nm, corresponding to a peak energy of 1.75 eV. This PL peak energy value was measured at −10°C, and it is expected not to vary too much at room temperature. Combining the results of previous work on few-layer BP (2 to 5 L)[9] with the results obtained from recent samples (1 to 5 L),[10] it can be clearly observed that the peak energy of PL emission shows unambiguous

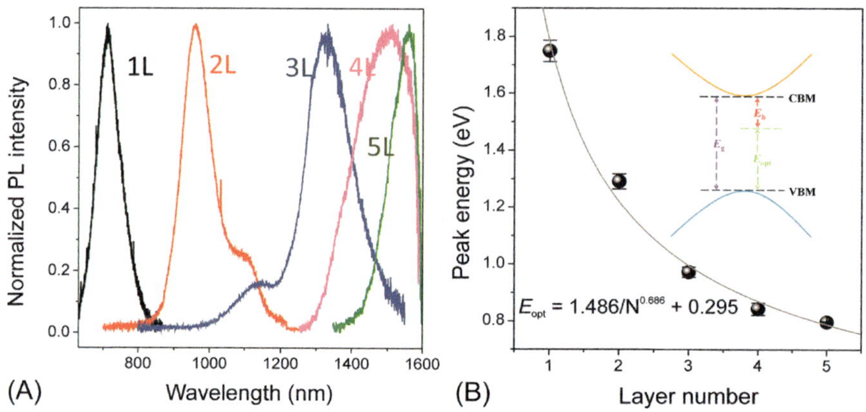

Fig. 6.3 PL characteristics of mono- and few-layer BP. (A) PL spectra of the mono- to five-layer BP samples. Each PL spectrum is normalized to its peak intensity and system background. (B) Evolution of PL peak energy with layer number of BP from experimental PL spectra, showing a rapid increase in peak energy as the layer number decreases. The solid gray line is the fitting curve $E_{opt} = 1.486/N^{0.686} + 0.295$, where E_{opt} is the optical gap in unit of eV and N is the layer number. Inset: energy diagram showing the electronic bandgap (E_g) and the exciton binding energy (E_b).

layer dependence (Fig. 6.3B). For each layer number, at least three samples were characterized; the measured peak energies for 1 to 5 L BP were 1.75 ± 0.04, 1.29 ± 0.03, 0.97 ± 0.02, 0.84 ± 0.02, and 0.80 ± 0.02 eV, respectively. The PL emission energy of BP samples with higher layer numbers (>5) is beyond the measurement wavelength range (up to 1600 nm) of a normal PL system.

The peak energy of PL emission, also termed as optical gap (E_{opt}), is the difference between electronic bandgap (E_a) and exciton binding energy (E_b) (Fig. 6.3B inset). Owing to the strong quantum confinement effect, free-standing monolayer BP is expected to have a large exciton binding energy of ~0.8 eV,[11, 18] whereas this value is expected to be only ~0.3 eV for monolayer BP on an SiO_2/Si substrate because of increased screening from the substrate. Using the measured electronic energy gap of ~2.05 eV from Pan et al.[50] and the measured optical gap of 1.75 eV in monolayer BP, the exciton binding energy of a monolayer BP on SiO_2/Si substrate was determined to be ~0.3 eV, which agrees very well with the prediction.[18] The optical gaps in BP increase rapidly with decreasing layer number because of the strong quantum confinement effect and the van der Waals interactions between neighboring sheets in few-layer BP.[11, 51] A power law form was used to fit the experimental data and obtain the fitting curve of

$$E_{opt} = 1.486/N^{0.686} + 0.295 \quad (6.2)$$

where E_{opt} is the optical gap in unit of eV and N is the layer number (Fig. 6.3B). The layer-dependent optical gaps, as indicated in Fig. 6.3B, agree very well with theoretical predictions.[8, 11] For bulk BP sample with large N value, its optical gap approaches the limit value of ~0.295 eV, which matches very well with the measured energy gap (~0.3 eV) of bulk BP.[1, 52] Previously, Ye et al.[5] observed bright exciton PL emission at ~1.45 eV from a monolayer BP that was coated with a layer of PMMA. This protective layer might introduce some defect states to the sample, which could change PL emission energy.

6.3.2 Power-dependent PL

The dynamics of excitons and trions have been of considerable interest for fundamental studies of many-body interactions, such as carrier multiplication and Wigner crystallization.[53] Monolayer BP, whose excitons are predicted to be confined in a quasi-1D space, provide an ideal platform for investigating remarkable exciton and trion dynamics in a reduced dimension. Exciton and trion dynamics have been successfully probed in a monolayer BP by controlling the photocarrier injection in power-dependent PL measurements (Fig. 6.4). The measured PL spectra exhibited two clear peaks with central wavelengths at ~705 nm (labeled "A") and ~760 nm (labeled "X"), whose intensities and peak positions are highly dependent on excitation power (Fig. 6.4A). After Lorentzian fitting of the measured PL spectra, the spectral components of these two peaks were extracted, indicated by red and blue curves in Fig. 6.4A. The intensity of higher-energy peak A increased almost linearly with laser power (Fig. 6.4B), while that of lower-energy peak X gradually saturated at a relatively high photocarrier injection. When the excitation power changed from 0.19 to 1.15 μW (Fig. 6.4C), the optical gap of A emission (E_A) monotonically decreased by ~15 meV, while that of X emission (E_X) increased by ~35 meV. The higher-energy peak A is attributed to exciton emission.

The origin of peak X is particularly interesting. Experimental observations do not support that peak X comes from the emission of localized excitons. First, the energy difference between PL emissions from the localized and free excitons is not sensitive to the relatively low density of photocarrier injection.[25, 54, 55] But in the experiment, this energy difference decreased significantly from ~150 to ~100 meV when the power of excitation laser increased from 0.19 to 1.15 μW. Furthermore, localized excitons are

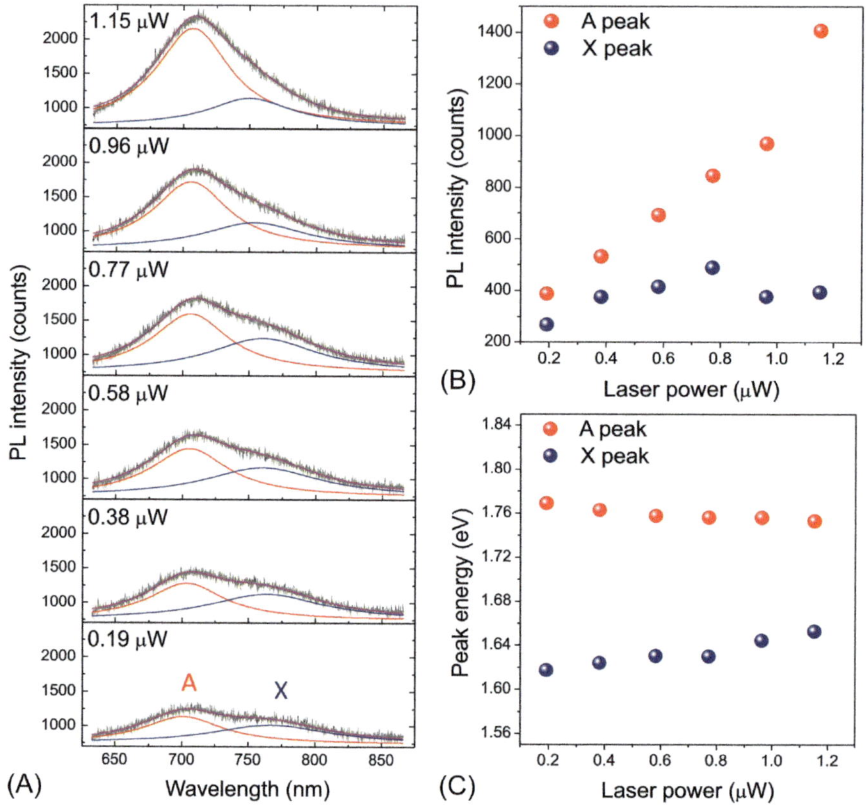

Fig. 6.4 Exciton and trion dynamics in monolayer phosphorene. (A) Measured PL spectra (solid *gray* lines) under various excitation laser power. PL spectra are fit to Lorentzians (solid *red* lines are exciton components, solid *blue* lines are trion components, and solid *pink* lines are the cumulative fitting results). (B) PL intensity of exciton (A) and trion (X) as a function of laser power. (C) PL peak energy of exciton (A) and trion (X) as a function of laser power.

expected to have a longer lifetime than free excitons because of the localization effect of squeezing an exciton in the zero-dimension-like state.[54, 56] However, time-resolved PL measurements revealed that the carrier lifetime from X state is less than that from A state. Therefore, the origin of X state can be attributed to trions. The trion peak intensity saturates at high photocarrier injection (Fig. 6.4B) because the free carriers coming from initial doping are almost depleted during the formation of trions. The trion density estimated from optical injection is comparable to the estimated initial doping level of the monolayer BP sample.

The difference between peak energies of exciton and trion PL emission is predicted to be[16, 57]:

$$E_A - E_X = E_{Tb} + \Delta E \tag{6.3}$$

where E_{Tb} is the binding energy of trions and ΔE is the average energy needed to add one carrier into the free carrier system. The energy difference $E_A - E_X$ represents the minimum energy for the removal of one electron (hole) from a negative (positive) trion. To convert a trion to an exciton, one of the two electrons (holes) in the negative (positive) trion will be unbound first ($+E_{Tb}$) and then added into the system ($+\Delta E$). During PL excitation, the same number of electrons and holes will be injected into the conduction band and valence band, respectively. At low excitation power (power close to zero), the system is in an equilibrium state, and ΔE is approximately equal to Fermi energy, E_F, that is determined at the initial doping level. At higher PL excitation power with more carrier injection, the initial doping becomes less important because of the enhanced screening effect from injected photocarriers.

In most conventional PL measurements with very high excitation power,[12, 17, 58] ΔE can be neglected and the trion binding energy can be approximately taken as the energy difference $E_A - E_X$. However, at a very low power range, the unique effect of ΔE cannot be observed. During the experiment, the start power excitation was set at 0.19 μW, which has an injected photocarrier concentration comparable to the initial doping level. With increasing excitation power, the energy difference $E_A - E_X$ decreased, indicating a reduction of ΔE. At a high excitation power (1.15 μW), the initial doping becomes less important and ΔE can be approximately neglected; the energy difference $E_A - E_X$ of ~100 meV at a high excitation power (1.15 μW) can be taken to be the upper limit of the trion binding energy in a monolayer BP. This value agrees well with the estimated trion binding energy in monolayer BP of $E_{Tb} \sim 0.3E_b$,[51] and the calculated exciton binding energy in a monolayer BP on SiO_2/Si substrate was $E_b \sim 0.3$ eV.[11] In a monolayer BP, both positive and negative trions are expected to have similar binding energies, since the effective masses of electrons and holes are almost equal.[8, 11]

By electrically injecting free carriers into a trilayer BP sample in a field-effect transistor (FET) structure, we have also observed the tuning of exciton and trion emission, and the measured trion binding energy in the trilayer BP was around 170 meV.

6.4 Potential applications

Due to its exceptionally unique properties compared to other 2D materials, few-layer BP has been proposed for various photonic and optoelectronic applications, including FETs,[1, 59, 60] photodetectors,[36, 61, 62] photovoltaic devices,[63–65] and saturable absorbers.[66–68] For example, Li et al. demonstrated the first FET device with few-layer BP and measured the carrier mobility to be $\sim$1000 cm^2 V^{-1} s^{-1} and the on/off ratio to be $\sim 10^5$.[1] This first demonstration of thin-layer BP integrated into FET devices showed BP's potential to be applied to the semiconductor industry. Youngblood et al. fabricated the waveguide-integrated BP photodetector with high responsivity, high response speed, and low dark current in the telecom band.[36] The measured intrinsic photoresponsivity can be up to 657 mA W^{-1} with a BP thickness of 100 nm, with the operating bandwidth exceeding 3 GHz. As BP is a direct bandgap semiconductor, it has been tested for photovoltaic applications. Buscema et al. fabricated a BP PN junction defined by local electrostatic doping and successfully observed the photovoltaic effect. One advantage of this BP photovoltaic device is that it realizes photoenergy harvesting in the near-IR range due to the intrinsic bandgap of few-layer BP. And lastly, BP has been intensively studied for saturable absorber applications for laser operation. Chen et al. demonstrated that mechanically exfoliated BP can be used as a new saturable absorber for Q-switching and mode-locking laser operation.[66] The aforementioned various applications of BP are merely examples of the broad-range IR to mid-IR applications that BP can be potentially integrated into. With the direct bandgap from monolayer to bulk and the strong anisotropic property, BP can certainly be extended to other relevant applications. It should be noted that most of the demonstrated applications are still in the laboratory prototype state; more research effort needs to be put in for field applications.

6.5 Conclusion

In this chapter, we have discussed light-matter interactions in BP by focusing on Raman and PL spectroscopy. Temperature- and angle-dependent Raman spectroscopy were used on few-layer BP samples to detect their fine structure and crystalline orientation, and the layer- and power-dependent PL measurements were used to determine the band structure and exciton and trion dynamics of thin-layer BP. By analyzing the different vibration modes in Raman spectra, the fine structure and other properties such as crystalline orientation can be further understood. Temperature-dependent Raman spectroscopy on thin-layer BP indicates that phonon frequencies in few-layer BP are more sensitive to temperature modulation than those in graphene and MoS_2, which could be due to the superior mechanical flexibility of BP originating from its unique puckered crystal structure. Angle-dependent Raman measurements give another quick and nondestructive way to determine the crystalline orientation of thin-layer BP. PL measurements can help understand the band structure of BP, and exciton and trion dynamics can be detected by optical or electrical tuning in PL measurements. In this chapter, few-layer BP was characterized by PL measurements, and its layer-dependent PL peaks clearly indicate the layer-dependent bandgaps. Besides, exciton and trion dynamics were detected by means of PL measurements with optical tuning, and trion binding energy was measured to be ~100 meV, larger than that of TMD materials, possibly due to the quasi-1D nature of exciton and trion in BP. In the last part of this chapter, we have briefly introduced the potential applications of BP. With its unique anisotropic properties, BP can be integrated into various photonic and optoelectronic applications.

References

1. Li, L.; Yu, Y.; Ye, G. J.; Ge, Q.; Ou, X.; Wu, H.; Feng, D.; Chen, X. H.; Zhang, Y. Black Phosphorus Field-Effect Transistors. *Nat. Nanotechnol.* **2014**, *9*(5), 372–377.
2. Xia, F.; Wang, H.; Jia, Y. Rediscovering Black Phosphorus as an Anisotropic Layered Material for Optoelectronics and Electronics. *Nat. Commun.* **2014**, *5*.
3. Fei, R.; Yang, L. Strain-Engineering the Anisotropic Electrical Conductance of Few-Layer Black Phosphorus. *Nano Lett.* **2014**, *14*(5), 2884–2889.
4. Na, J.; Lee, Y. T.; Lim, J. A.; Hwang, D. K.; Kim, G.-T.; Choi, W. K.; Song, Y.-W. Few-Layer Black Phosphorus Field-Effect Transistors With Reduced Current Fluctuation. *ACS Nano* **2014**, *8*(11), 11753–11762.
5. Wood, J. D.; Wells, S. A.; Jariwala, D.; Chen, K.-S.; Cho, E.; Sangwan, V. K.; Liu, X.; Lauhon, L. J.; Marks, T. J.; Hersam, M. C. Effective Passivation of Exfoliated Black Phosphorus Transistors against Ambient Degradation. *Nano Lett.* **2014**, *14*(12), 6964–6970.
6. Mao, N.; Tang, J.; Xie, L.; Wu, J.; Han, B.; Lin, J.; Deng, S.; Ji, W.; Xu, H.; Liu, K.; Tong, L.; Zhang, J. Optical Anisotropy of Black Phosphorus in the Visible Regime. *J. Am. Chem. Soc.* **2016**, *138*(1), 300–305.
7. Xu, R.; Yang, J.; Zhu, Y.; Yan, H.; Pei, J.; Myint, Y. W.; Zhang, S.; Lu, Y. Layer-Dependent Surface Potential of Phosphorene and Anisotropic/Layer-Dependent Charge Transfer in Phosphorene-Gold Hybrid Systems. *Nanoscale* **2016**, *8*(1), 129–135.
8. Qiao, J.; Kong, X.; Hu, Z.-X.; Yang, F.; Ji, W. High-Mobility Transport Anisotropy and Linear Dichroism in Few-Layer Black Phosphorus. *Nat. Commun.* **2014**, *5*.
9. Zhang, S.; Yang, J.; Xu, R.; Wang, F.; Li, W.; Ghufran, M.; Zhang, Y.-W.; Yu, Z.; Zhang, G.; Qin, Q.; Lu, Y. Extraordinary Photoluminescence and Strong

Temperature/Angle-Dependent Raman Responses in Few-Layer Phosphorene. *ACS Nano* **2014**, *8*(9), 9590–9596.

10. Yang, J.; Xu, R.; Pei, J.; Myint, Y. W.; Wang, F.; Wang, Z.; Zhang, S.; Yu, Z.; Lu, Y. Optical Tuning of Exciton and Trion Emissions in Monolayer Phosphorene. *Light Sci. Appl.* **2015**, *4,* e312.
11. Tran, V.; Soklaski, R.; Liang, Y.; Yang, L. Layer-Controlled Band Gap and Anisotropic Excitons in Few-Layer Black Phosphorus. *Phys. Rev. B* **2014**, *89*(23), 235319.
12. Xu, R.; Zhang, S.; Wang, F.; Yang, J.; Wang, Z.; Pei, J.; Myint, Y. W.; Xing, B.; Yu, Z.; Fu, L.; Qin, Q.; Lu, Y. Extraordinarily Bound Quasi-One-Dimensional Trions in Two-Dimensional Phosphorene Atomic Semiconductors. *ACS Nano* **2016**, *10*(2), 2046–2053.
13. Wang, X.; Jones, A. M.; Seyler, K. L.; Tran, V.; Jia, Y.; Zhao, H.; Wang, H.; Yang, L.; Xu, X.; Xia, F. Highly Anisotropic and Robust Excitons in Monolayer Black Phosphorus. *Nat. Nanotechnol.* **2015**, *10*(6), 517–521.
14. Geim, A. K.; Novoselov, K. S. The Rise of Graphene. *Nat. Mater.* **2007**, *6*(3), 183–191.
15. Radisavljevic, B.; Radenovic, A.; Brivio, J.; Giacometti, V.; Kis, A. Single-Layer MoS2 Transistors. *Nat. Nanotechnol.* **2011**, 6(3), 147–150.
16. Mak, K. F.; He, K.; Lee, C.; Lee, G. H.; Hone, J.; Heinz, T. F.; Shan, J. Tightly Bound Trions in Monolayer MoS2. *Nat. Mater.* **2013**, *12*(3), 207–211.
17. Ross, J. S.; Wu, S.; Yu, H.; Ghimire, N. J.; Jones, A. M.; Aivazian, G.; Yan, J.; Mandrus, D. G.; Xiao, D.; Yao, W.; Xu, X. Electrical Control of Neutral and Charged Excitons in a Monolayer Semiconductor. *Nat. Commun.* **2013**, *4,* 1474.
18. Rodin, A. S.; Carvalho, A.; Castro Neto, A. H. Excitons in Anisotropic Two-Dimensional Semiconducting Crystals. *Phys. Rev. B* **2014**, *90*(7), 075429.
19. Ling, X.; Liang, L.; Huang, S.; Puretzky, A. A.; Geohegan, D. B.; Sumpter, B. G.; Kong, J.; Meunier, V.; Dresselhaus, M. S. Low-Frequency Interlayer Breathing Modes in Few-Layer Black Phosphorus. *Nano Lett.* **2015**, *15*(6), 4080–4088.
20. Ribeiro, H. B.; Pimenta, M. A.; de Matos, C. J. S.; Moreira, R. L.; Rodin, A. S.; Zapata, J. D.; de Souza, E. A. T.; Castro Neto, A. H. Unusual Angular Dependence of the Raman Response in Black Phosphorus. *ACS Nano* **2015**, *9*(4), 4270–4276.
21. Viti, L.; Hu, J.; Coquillat, D.; Knap, W.; Tredicucci, A.; Politano, A.; Vitiello, M. S. Black Phosphorus Terahertz Photodetectors. *Adv. Mater.* **2015**, *27*(37), 5567–5572.
22. Yasaei, P.; Kumar, B.; Foroozan, T.; Wang, C.; Asadi, M.; Tuschel, D.; Indacochea, J. E.; Klie, R. F.; Salehi-Khojin, A. High-Quality Black Phosphorus Atomic Layers by Liquid-Phase Exfoliation. *Adv. Mater.* **2015**, *27*(11), 1887–1892.
23. Lu, W.; Ma, X.; Fei, Z.; Zhou, J.; Zhang, Z.; Jin, C.; Zhang, Z. Probing the Anisotropic Behaviors of Black Phosphorus by Transmission Electron Microscopy, Angular-Dependent Raman Spectra, and Electronic Transport Measurements. *Appl. Phys. Lett.* **2015**, *107*(2), 021906.
24. Pei, J.; Gai, X.; Yang, J.; Wang, X.; Yu, Z.; Choi, D.-Y.; Luther-Davies, B.; Lu, Y. Producing Air-Stable Monolayers of Phosphorene and Their Defect Engineering. *Nat. Commun.* **2016**, 7, 10450.
25. Xu, R.; Yang, J.; Myint, Y. W.; Pei, J.; Yan, H.; Wang, F.; Lu, Y. Exciton Brightening in Monolayer Phosphorene via Dimensionality Modification. *Adv. Mater.* **2016**, *28*(18), 3493–3498.
26. Andres, C.-G.; Leonardo, V.; Elsa, P.; Joshua, O. I.; Narasimha-Acharya, K. L.; Sofya, I. B.; Dirk, J. G.; Michele, B.; Gary, A. S.; Alvarez, J. V.; Henny, W. Z.; Palacios, J. J.; van der Zant, H. S. J. Isolation and Characterization of Few-Layer Black Phosphorus. *2D Mater.* **2014**, 1(2), 025001.
27. Favron, A.; Gaufres, E.; Fossard, F.; Phaneuf-Lheureux, A.-L.; Tang, N. Y. W.; Levesque, P. L.; Loiseau, A.; Leonelli, R.; Francoeur, S.; Martel, R. Photooxidation

and Quantum Confinement Effects in Exfoliated Black Phosphorus. *Nat. Mater.* **2015**, *14*(8), 826–832.
28. Kang, J.; Wood, J. D.; Wells, S. A.; Lee, J.-H.; Liu, X.; Chen, K.-S.; Hersam, M. C. Solvent Exfoliation of Electronic-Grade, Two-Dimensional Black Phosphorus. *ACS Nano* **2015**, *9*(4), 3596–3604.
29. Doganov, R. A.; O'Farrell, E. C. T.; Koenig, S. P.; Yeo, Y.; Ziletti, A.; Carvalho, A.; Campbell, D. K.; Coker, D. F.; Watanabe, K.; Taniguchi, T.; Neto, A. H. C.; Ozyilmaz, B. Transport Properties of Pristine Few-Layer Black Phosphorus by van der Waals Passivation in an Inert Atmosphere. *Nat. Commun.* **2015**, *6*.
30. Li, L.; Kim, J.; Jin, C.; Ye, G. J.; Qiu, D. Y.; da Jornada, F. H.; Shi, Z.; Chen, L.; Zhang, Z.; Yang, F.; Watanabe, K.; Taniguchi, T.; Ren, W.; Louie, S. G.; Chen, X. H.; Zhang, Y.; Wang, F. Direct Observation of the Layer-Dependent Electronic Structure in Phosphorene. *Nat. Nanotechnol.* **2016**, *12*, 21.
31. Cai, Y.; Zhang, G.; Zhang, Y.-W. Layer-Dependent Band Alignment and Work Function of Few-Layer Phosphorene. *Sci. Rep.* **2014**, *4*, 6677.
32. Novoselov, K. S.; Geim, A. K.; Morozov, S. V.; Jiang, D.; Katsnelson, M. I.; Grigorieva, I. V.; Dubonos, S. V.; Firsov, A. A. Two-Dimensional Gas of Massless Dirac Fermions in Graphene. *Nature* **2005**, *438*(7065), 197–200.
33. Mak, K. F.; Lee, C.; Hone, J.; Shan, J.; Heinz, T. F. Atomically Thin MoS_2: A New Direct-Gap Semiconductor. *Phys. Rev. Lett.* **2010**, *105*(13), 136805.
34. Zhang, Y.; Chang, T.-R.; Zhou, B.; Cui, Y.-T.; Yan, H.; Liu, Z.; Schmitt, F.; Lee, J.; Moore, R.; Chen, Y.; Lin, H.; Jeng, H.-T.; Mo, S.-K.; Hussain, Z.; Bansil, A.; Shen, Z.-X. Direct Observation of the Transition From Indirect to Direct Bandgap in Atomically Thin Epitaxial MoSe2. *Nat. Nanotechnol.* **2014**, *9*(2), 111–115.
35. Zhao, W.; Ghorannevis, Z.; Chu, L.; Toh, M.; Kloc, C.; Tan, P.-H.; Eda, G. Evolution of Electronic Structure in Atomically Thin Sheets of WS_2 and WSe_2. *ACS Nano* **2013**, 7(1), 791–797.
36. Youngblood, N.; Chen, C.; Koester, S. J.; Li, M. Waveguide-Integrated Black Phosphorus Photodetector With High Responsivity and Low Dark Current. *Nat. Photonics* **2015**, *9*(4), 247–252.
37. Zhang, B.; Lou, F.; Zhao, R.; He, J.; Li, J.; Su, X.; Ning, J.; Yang, K. Exfoliated Layers of Black Phosphorus as Saturable Absorber for Ultrafast Solid-State Laser. *Opt. Lett.* **2015**, *40*(16), 3691–3694.
38. Hanlon, D.; Backes, C.; Doherty, E.; Cucinotta, C. S.; Berner, N. C.; Boland, C.; Lee, K.; Harvey, A.; Lynch, P.; Gholamvand, Z.; Zhang, S.; Wang, K.; Moynihan, G.; Pokle, A.; Ramasse, Q. M.; McEvoy, N.; Blau, W. J.; Wang, J.; Abellan, G.; Hauke, F.; Hirsch, A.; Sanvito, S.; O'Regan, D. D.; Duesberg, G. S.; Nicolosi, V.; Coleman, J. N. Liquid Exfoliation of Solvent-Stabilized Few-Layer Black Phosphorus for Applications Beyond Electronics. *Nat. Commun.* **2015**, *6*.
39. Guo, Z.; Zhang, H.; Lu, S.; Wang, Z.; Tang, S.; Shao, J.; Sun, Z.; Xie, H.; Wang, H.; Yu, X.-F.; Chu, P. K. From Black Phosphorus to Phosphorene: Basic Solvent Exfoliation, Evolution of Raman Scattering, and Applications to Ultrafast Photonics. *Adv. Funct. Mater.* **2015**, *25*(45), 6996–7002.
40. Zhao, Y.; Luo, X.; Li, H.; Zhang, J.; Araujo, P. T.; Gan, C. K.; Wu, J.; Zhang, H.; Quek, S. Y.; Dresselhaus, M. S.; Xiong, Q. Interlayer Breathing and Shear Modes in Few-Trilayer MoS2 and WSe2. *Nano Lett.* **2013**, *13*(3), 1007–1015.
41. Li, S.-L.; Miyazaki, H.; Song, H.; Kuramochi, H.; Nakaharai, S.; Tsukagoshi, K. Quantitative Raman Spectrum and Reliable Thickness Identification for Atomic Layers on Insulating Substrates. *ACS Nano* **2012**, *6*(8), 7381–7388.
42. Postmus, C.; Ferraro, J. R.; Mitra, S. S. Pressure Dependence of Infrared Eigenfrequencies of KCl and KBr. *Phys. Rev.* **1968**, *174*(3), 983–987.

43. Calizo, I.; Balandin, A. A.; Bao, W.; Miao, F.; Lau, C. N. Temperature Dependence of the Raman Spectra of Graphene and Graphene Multilayers. *Nano Lett.* **2007**, 7(9), 2645–2649.
44. Sahoo, S.; Gaur, A. P. S.; Ahmadi, M.; Guinel, M. J. F.; Katiyar, R. S. Temperature-Dependent Raman Studies and Thermal Conductivity of Few-Layer MoS2. *J. Phys. Chem. C* **2013**, *117*(17), 9042–9047.
45. Wei, Q.; Peng, X. Superior Mechanical Flexibility of Phosphorene and Few-Layer Black Phosphorus. *Appl. Phys. Lett.* **2014**, *104*(25), 251915.
46. Yin, X.; Ye, Z.; Chenet, D. A.; Ye, Y.; O'Brien, K.; Hone, J. C.; Zhang, X. Edge Non-linear Optics on a MoS2 Atomic Monolayer. *Science* **2014**, *344*(6183), 488–490.
47. Sugai, S.; Shirotani, I. Raman and Infrared Reflection Spectroscopy in Black Phosphorus. *Solid State Commun.* **1985**, *53*(9), 753–755.
48. Liu, H.; Neal, A. T.; Zhu, Z.; Luo, Z.; Xu, X.; Tománek, D.; Ye, P. D. Phosphorene: An Unexplored 2D Semiconductor With a High Hole Mobility. *ACS Nano* **2014**, *8*(4), 4033–4041.
49. Lu, J.; Yang, J.; Carvalho, A.; Liu, H.; Lu, Y.; Sow, C. H. Light–Matter Interactions in Phosphorene. *Acc. Chem. Res.* **2016**, *49*(9), 1806–1815.
50. Liang, L.; Wang, J.; Lin, W.; Sumpter, B. G.; Meunier, V.; Pan, M. Electronic Bandgap and Edge Reconstruction in Phosphorene Materials. *Nano Lett.* **2014**, *14*(11), 6400–6406.
51. Thilagam, A. Two-Dimensional Charged-Exciton Complexes. *Phys. Rev. B* **1997**, *55*(12), 7804–7808.
52. Keyes, R. W. The Electrical Properties of Black Phosphorus. *Phys. Rev.* **1953**, *92*(3), 580–584.
53. Wigner, E. On the Interaction of Electrons in Metals. *Phys. Rev.* **1934**, *46*(11), 1002–1011.
54. Miyauchi, Y.; Iwamura, M.; Mouri, S.; Kawazoe, T.; Ohtsu, M.; Matsuda, K. Brightening of Excitons in Carbon Nanotubes on Dimensionality Modification. *Nat. Photonics* **2013**, 7(9), 715–719.
55. Iwamura, M.; Akizuki, N.; Miyauchi, Y.; Mouri, S.; Shaver, J.; Gao, Z.; Cognet, L.; Lounis, B.; Matsuda, K. Nonlinear Photoluminescence Spectroscopy of Carbon Nanotubes With Localized Exciton States. *ACS Nano* **2014**, *8*(11), 11254–11260.
56. Wang, H.; Zhang, C.; Chan, W.; Manolatou, C.; Tiwari, S.; Rana, F. Radiative Lifetimes of Excitons and Trions in Monolayers of the Metal Dichalcogenide MoS_2. *Phys. Rev. B* **2016**, *93*(4), 045407.
57. Huard, V.; Cox, R. T.; Saminadayar, K.; Arnoult, A.; Tatarenko, S. Bound States in Optical Absorption of Semiconductor Quantum Wells Containing a Two-Dimensional Electron Gas. *Phys. Rev. Lett.* **2000**, *84*(1), 187–190.
58. Kheng, K.; Cox, R. T.; d' Aubigné, M. Y.; Bassani, F.; Saminadayar, K.; Tatarenko, S. Observation of Negatively Charged Excitons X^- in Semiconductor Quantum Wells. *Phys. Rev. Lett.* **1993**, *71*(11), 1752–1755.
59. Avsar, A.; Vera-Marun, I. J.; Tan, J. Y.; Watanabe, K.; Taniguchi, T.; Castro Neto, A. H.; Özyilmaz, B. Air-Stable Transport in Graphene-Contacted, Fully Encapsulated Ultrathin Black Phosphorus-Based Field-Effect Transistors. *ACS Nano* **2015**, *9*(4), 4138–4145.
60. Buscema, M.; Groenendijk, D. J.; Blanter, S. I.; Steele, G. A.; van der Zant, H. S. J.; Castellanos-Gomez, A. Fast and Broadband Photoresponse of Few-Layer Black Phosphorus Field-Effect Transistors. *Nano Lett.* **2014**, *14*(6), 3347–3352.
61. Engel, M.; Steiner, M.; Avouris, P. Black Phosphorus Photodetector for Multispectral, High-Resolution Imaging. *Nano Lett.* **2014**, *14*(11), 6414–6417.
62. Yuan, H.; Liu, X.; Afshinmanesh, F.; Li, W.; Xu, G.; Sun, J.; Lian, B.; Curto, A. G.; Ye, G.; Hikita, Y.; Shen, Z.; Zhang, S.-C.; Chen, X.; Brongersma, M.;

Hwang, H. Y.; Cui, Y. Polarization-Sensitive Broadband Photodetector Using a Black Phosphorus Vertical p–n Junction. *Nat. Nanotechnol.* **2015**, *10*, 707.

63. Buscema, M.; Groenendijk, D. J.; Steele, G. A.; van der Zant, H. S. J.; Castellanos-Gomez, A. Photovoltaic Effect in Few-Layer Black Phosphorus PN Junctions Defined by Local Electrostatic Gating. *Nat. Commun.* **2014**, *5*, 4651.
64. Shenghuang, L.; Shenghua, L.; Zhibin, Y.; Yanyong, L.; Wai, N. T.; Zaiquan, X.; Qiaoliang, B.; Jianhua, H.; Chun-Sing, L.; Charles, S.; Feng, Y.; Ping, L. S. Solution-Processable Ultrathin Black Phosphorus as an Effective Electron Transport Layer in Organic Photovoltaics. *Adv. Funct. Mater.* **2016**, *26*(6), 864–871.
65. Yuanda, L.; Yongqing, C.; Gang, Z.; Yong-Wei, Z.; Kah-Wee, A. Al-Doped Black Phosphorus p–n Homojunction Diode for High Performance Photovoltaic. *Adv. Funct. Mater.* **2017**, *27*(7), 1604638.
66. Chen, Y.; Jiang, G.; Chen, S.; Guo, Z.; Yu, X.; Zhao, C.; Zhang, H.; Bao, Q.; Wen, S.; Tang, D.; Fan, D. Mechanically Exfoliated Black Phosphorus as a New Saturable Absorber for Both Q-Switching and Mode-Locking Laser Operation. *Opt. Express* **2015**, *23*(10), 12823–12833.
67. Sotor, J.; Sobon, G.; Macherzynski, W.; Paletko, P.; Abramski, K. M. Black Phosphorus Saturable Absorber for Ultrashort Pulse Generation. *Appl. Phys. Lett.* **2015**, *107*(5), 051108.
68. Luo, Z.-C.; Liu, M.; Guo, Z.-N.; Jiang, X.-F.; Luo, A.-P.; Zhao, C.-J.; Yu, X.-F.; Xu, W.-C.; Zhang, H. Microfiber-Based Few-Layer Black Phosphorus Saturable Absorber for Ultra-Fast Fiber Laser. *Opt. Express* **2015**, *23*(15), 20030–20039.

CHAPTER 7

Light-emitting devices

Qasim Khan[a], Sayed Ali Khan[a], Qiaoliang Bao[b]
[a]College of Electronic Science and Technology, Shenzhen University, Shenzhen, China
[b]Department of Materials Science and Engineering, ARC Centre of Excellence in Future Low-Energy Electronics Technologies (FLEET), Monash University, Clayton, VIC, Australia

7.1 Solid-state lighting: A breakthrough technology

Solid-state lighting (SSL) is a new technology used to generate artificial white light by using semiconductor diode that produces light in the visible range with the flowing of electric current inside the diode [electroluminescence (EL)]. The SSLs are mainly categorized into three major groups depending on the semiconducting materials used for its manufacturing: the polymer, organic, and inorganic light-emitting diodes (LEDs).

The SSL by using LEDs penetrated the global market, because it has mercury-free emission and also features outstanding characteristics, such as low energy consumption, high luminescence efficiency, high reliability, and superior long operational lifetime. The probable use of SSL is broad as it covers a wide range of lighting applications, such as residential lights, architecture lighting, medical lighting, traffic indicators, and sound-sensitive lights. These are the properties that surpassed both conventional lighting technologies (filament and fluorescent lamps) and is the reason for SSL to be considered as a promising alternative and subsequent successful commercialization. According to a report, published in 2015, 42 billion US dollars could be saved by 2025 by using new technologies that even improve 50% of the energy efficiency compared to the incandescent light bulb, which has an energy efficiency of 5% (Fig. 7.1). This is important because it led to a saving of 70 GW analogous to the power generated by nuclear power plants.[1]

A comparison of various technologies (incandescent, fluorescent, and solid-state lighting) was reported, clearly demonstrating that the LED technology surpasses the conventional lighting sources in terms of fundamental characteristics like energy efficiency, luminous efficacy (ratio of luminous flux to radiant flux), environmental friendly emission, and relatively long operational lifetime, which is illustrated in Table 7.1.

2D Materials for Photonic and Optoelectronic Applications
https://doi.org/10.1016/B978-0-08-102637-3.00007-3

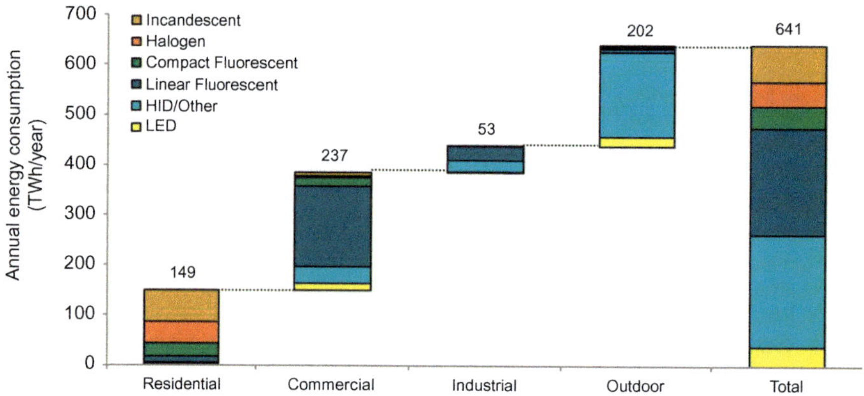

Fig. 7.1 Annual energy consumption broken down by sectors and lighting technology.[1]

Table 7.1 Comparison of various characteristics like energy efficiency, luminous efficacy (lumen/Watt), operational lifetime, production of heat, and emission of mercury of the commonly used light sources.[2–4]

Characteristics	Incandescent	Halogen	Compact fluorescent	High intensity discharge	LED
Energy efficiency	Very low	Low	High	High	Very high
Luminescence efficacy (Lumen\ Watt)	14	24	60–100	65–110	80–140
Operational lifetime (h)	1000	2000–3000	6000–10000	20,000	50,000
Heat production	Yes++	Yes++	Yes	Yes	No
Mercury emission	No	No	Yes	Yes	No

7.1.1 A short history of LED

An LED is an electrical device that has various layers of a single crystal semiconductor linked to the electrode, lying on a substrate. The fundamental LED consists of two layers of semiconducting material that form a p-n junction (Fig. 7.2), one is the n-doped having electrons in majority and the other is p-doped containing holes in majority. The semiconducting layers are doped by introducing a small amount of impurity atom having lower or

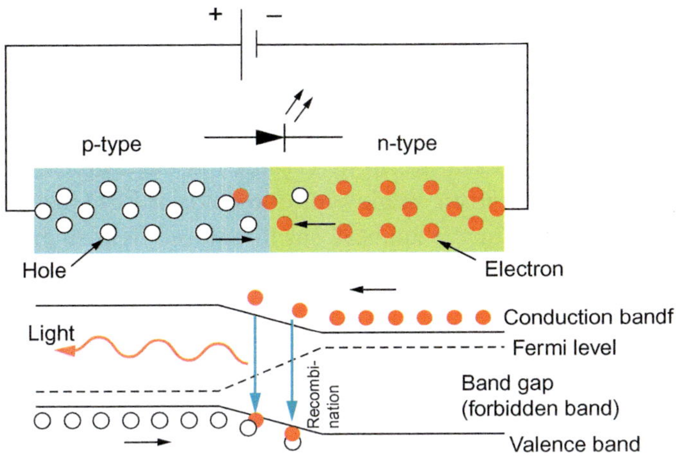

Fig. 7.2 Schematics of p-n junction for LEDs.

higher charge than the host atoms. When some tension (in the form of electric field) is applied to the diode the electrons and holes move to the junction, which results in radiative recombination in the area where they meet. The semiconducting materials that form the diode are characterized by the bandgap of the applied energy. The electrons and holes have various localized energy levels near the junctions. The electrons and holes lose their energy by releasing photons as a result of recombination and the released photons have energy according to the bandgap of the diode.

The first LED reported in the literature was invented by a Russian scientist Oleg Vladimirovich Losev in 1927.[5] After the end of the world wars, most of the attention was diverted to the infrared LEDs, prominently made from gallium arsenide (GaAs).[6] In 1962 an American scientist, Nick Holonyak Jr. developed, for the first time, a practical LED that has the emission in the visible region (red light).[7] Gallium arsenide phosphate (GaAsP) was later used for LEDs that emit a poor red light and work up to milli-Ampere current.[7] Finally in 1992 the high-brightness blue LED (GaN) was created by Shuji Nakamura of Nichia Corporation.[8] The present technology has improved to a great extent. The LEDs that work with nearly 1 A current are common nowadays. The development of gallium nitride (GaN) brought an outstanding scientific advancement in the field of LEDs. Before the development of GaN, the bright LEDs were only achievable for the emitting range of yellow—infrared. The discovery of GaN initiated the research to achieve blue, green, and white LEDs.

7.1.2 White light production

Unlike the conventional incandescent bulb and fluorescent lamp, LEDs emit a monochromatic light instead of white light according to the chemical compositions of the utilized semiconductors materials. On the basis of the emission color of the LED, it can be classified as blue, green, yellow, and red LEDs. The characteristic single-emitting color of LED makes the LEDs a potential candidate for the applications in colored light such as indicator lights. However, for the use of general lighting, it needs to emit the full spectrum of white light.

Various strategies to generate efficiently high-quality white light are as follows (Figs. 7.3 and 7.4):

1. Phosphor-converted LED (PC LED): here a monochromatic blue or ultraviolet (UV) LED is combined with different color phosphors to generate white light. The obtained LED is known as PC LEDs.
2. Red, green, and blue (RGB) combinations: here a number of monochromatic LEDs (RGB) are combined and mixed to generate white light. The white LED produced with this technique is called RGB LED.
3. Hybrid technique: a combination of monochromatic LEDs and PC LEDs are used in this method.

7.1.3 Organic LEDs

Organic LEDs are known as OLEDs (inorganic light-emitting diode are simply called LEDs). The organic materials used to generate artificial light were first established by Kallmann and coworkers in 1963 just after the

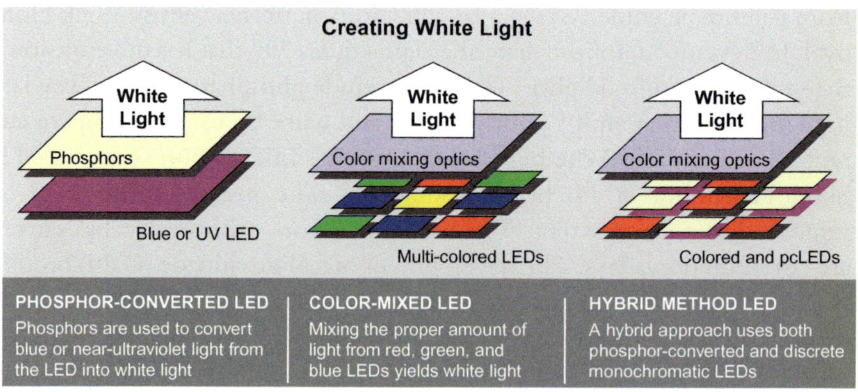

Fig. 7.3 Various approaches to produce high-quality artificial white light.[9]

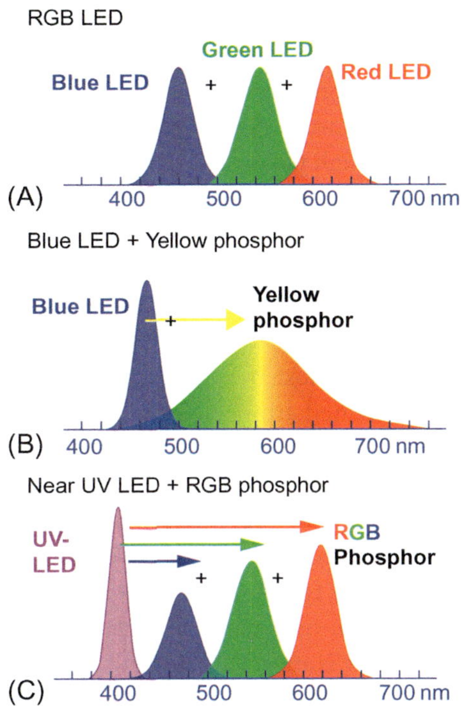

Fig. 7.4 Schematic illustration of generation of white light by using various method like: (A) RGB method (*red*+*green*+*blue* LEDs), (B) combination of *blue* LED and *yellow* phosphor (*blue*/UV LEDs+phosphors), and (C) combination of UV LED and various color phosphors.[10]

development of inorganic light-emitting materials, which were discovered in 1962.[11] Since the work of Tang and coworkers in 1987,[12] OLEDs based on a double-layer structure of organic materials gained significant interest for the preparation of flat panel displays and for the generation of artificial white light. This is due to several appealing characteristics such as ease of preparation, light weight, ultrathin structure, high flexibility, and ability to tune the device properties by adjusting the molecular structure of the organic materials.[12–19] Many studies have focused on developing comprehensive strategies to fabricate a wide range of colored OLEDs (blue, green, and red) based on molecular or polymeric thin films into matrix displays.[12] In 2013, Samsung Electronics illustrated the first curved television (TV) using OLEDs that have outstanding features like efficient color purity, outstanding contrast, and wide field of view. Two years later the same company launched a smartphone (Galaxy S_6 edge) by using the curved edge display with touch sensors to improve the design architecture of the device.

However, in spite of outstanding experimental research in the field of general lighting, OLEDs still lag behind inorganic LEDs in basic characteristics. Moreover, even though white OLEDs exhibit potential in back light unit of flat panel displays, a main disadvantage is the difficulty to acquire an efficient white light with high purity and long operational lifetime, thus narrowing their applications in the general lighting market. It has been found that phosphorescent emissive materials used to fabricate white OLEDs can harvest both singlet and triplet excitons, leading to enhanced internal quantum efficiency of 100%.[20–23] This was one of the most important progress. The phosphorescent emissive material, which is a platinum and iridium complex doped in organic host molecules, is used to create an emissive layer.[24] The energy obtained from electrical injection can be efficiently transferred from the singlet and triplet excitons of the organic host molecules to the triplet state of the doped phosphorescent material, which generates phosphorescence as a result of recombination and charge trapping. The internal quantum efficiency of the prepared doped OLEDs can be as high as 100%, which corresponds to a four-time enhancement to the fluorescent counterparts, since they can only harvest singlet excitons.

The device architecture is another issue in the path to commercialization of white OLEDs because the triplet state of the doped phosphorescent material emits a specific wavelength that corresponds to a specific color (blue, green, yellow, or red). It is very difficult to prepare a single molecule whose emission efficiently covers the entire visible region. In order to generate a high-quality white light, a combination of materials that emit various colors is required, thus complicating the device architecture design. Multiple phosphorescent dopants are added to achieve the goal. It is also worth noting that white light with higher photoluminescence quantum yield (PQLY) can be obtained by using multiple emissive layer OLEDs.[24, 25] In order to simplify the device structure, three phosphorescent emissive dopants are co-doped simultaneously in a single-molecule host with a wide bandgap. The present combination can decrease the thickness of the device hugely and can significantly enhance the efficiency because the OLED in this case consists of a single emissive layer instead of multiple emissive layers emitting various colors. However, the host with a broad energy gap usually have very short operational lifetime for the blue electro-phosphorescence, thus decreasing the total lifetime of the device. As illustrated in Fig. 7.5, the use of a single fluorescent emitting dopant efficiently controls all high-energy singlet excitons generated as a result of the electrification of the device for blue emission.[25] Remaining lower energy triplet excitons for green and red color

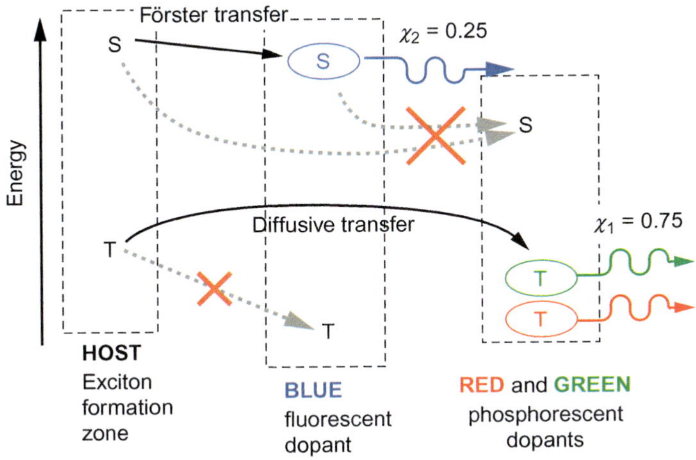

Fig. 7.5 Mechanism of energy transfer in the fluorescent/phosphorescent white OLEDs. *Reprinted by permission from Springer Nature: Sun, Y.; Giebink, N. C.; Kanno, H.; Ma, B.; Thompson, M. E.; Forrest, S. R., Management of Singlet and Triplet Excitons for Efficient White Organic Light-Emitting Devices. Nature* ***2006****, 440, 908, Copyright (2006)*

emission will be harvested by the phosphorescent emissive dopants.[26] This device structure can reach 100% internal quantum efficiency and have a long operational lifetime due to the highly stable blue emission of the blue fluorescent. In addition, external quantum efficiency (EQE) and device structure simplification can be further improved by using different host molecules and dopants, which in turn support commercialization of white OLEDs.

7.2 Light-emitting field-effect transistors based on two-dimensional materials

Light-emitting field-effect transistors (LEFETs or LETs—light-emitting transistors) are an emerging class of multifunctional optoelectronic devices that have been developed rapidly with the intention of simplifying circuitry, that is, they combine the light-emitting function of an OLED and the switching function of a field-effect transistor (FET) into a single device architecture.[27] The dual functionality of LETs has potential applications in the next generation of active matrix display technologies. It has been demonstrated that organic-based LETs have an efficiency that outperforms the equivalent LEDs. For example, devices with EQEs of 5% are demonstrated, which exceed the best achieved with OLEDs based on the same emitting layer and optimized transport layers (Fig. 7.6A and B).[27] The concept of using a *p-i-n* (p-doped/active-region/n-doped) semiconducting heterostructure in LETs provides a new approach to remarkably improve the device performance. These devices are 100 times more efficient than the equivalent OLED and 2 times more efficient than an optimized OLED with the same emitting layer. For the organic planer LETs, both electrons and holes are injected and transported in the light-emitting material, leading to maximum recombination and hence high EQE of 8.2% and high luminance efficiency of >28 cdA^{-1}.[28] The half-sphere on the light-emitting side of the LET device was introduced to optimize the light out-coupling efficiency. It is also predicted that in principle, EQEs of up to 21% can be achieved for an LET device.

Recently, large-area light-emitting transistors have been fabricated with unprecedented aperture (ratio between the emissive area to the entire area of the device) of 50%, easily surpassing the minimum requirement for

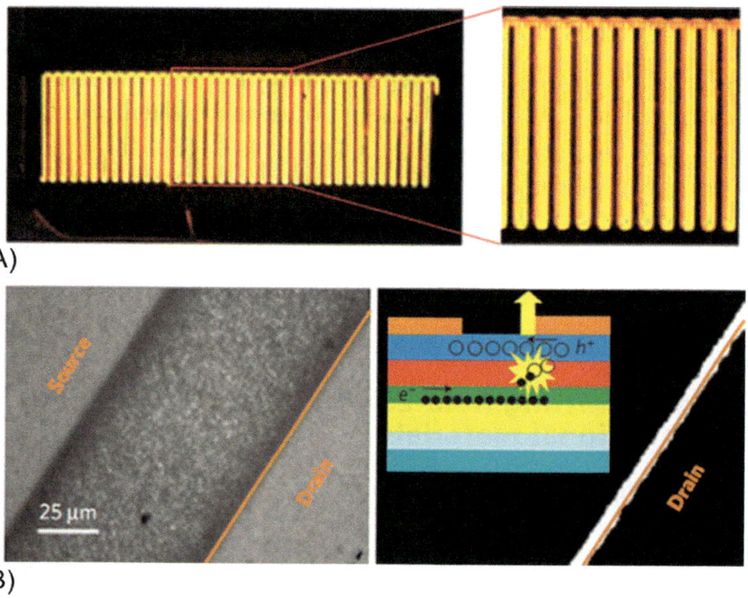

Fig. 7.6 (A-B) Optical micrographs of the lit trilayer OLET (organic LET) and its emission spectra.[26] (A) Optical micrograph of the interdigitated trilayer heterostructure OLET biased with $V_{ds}=V_{gs}=90$ V. Channel length and channel width are 150 μm and 20 cm, respectively. (B) Optical micrograph of the OLET channel when no bias is applied to the device, and when the applied bias is $V_{ds}=V_{gs}=90$ V. The schematic representation of the trilayer heterostructure OLET showing the expected location of the light-generation area is reported in the inset. (C–F) The transistor designs[28]:

Continued

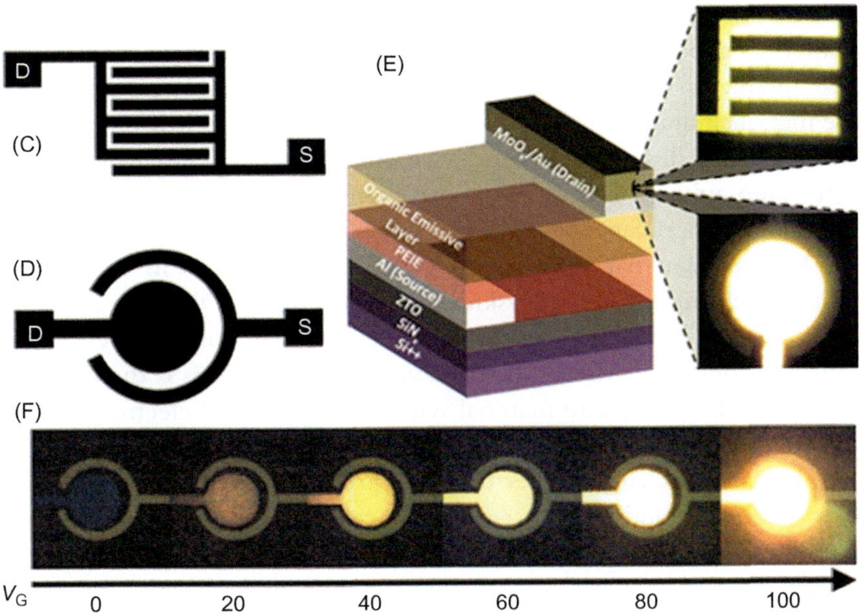

Fig. 7.6, cont'd (C) interdigitated pixel, (D) circular design for enhancing the aperture ratio, (E) the modified HLET (hybrid LET) architecture with a microscope image of the typical emission also shown for both the interdigitated and circular pixel, and (F) the area emission in this device is shown in series of images of the same device throughout a sweep of the gate voltage during a transfer measurement. *Panels (A) and (B) reprinted with permission from Capelli, R.; Toffanin, S.; Generali, G.; Usta, H.; Facchetti, A.; Muccini, M. Organic Light-Emitting Transistors With an Efficiency That Outperforms the Equivalent Light-Emitting Diodes. Nat. Mater.* ***2010****, 9(6), 496. Springer Nature. Copyright (2010). Panels (C)–(F) reprinted with permission from John Wiley and Sons, Copyright (2015).*

active-matrix organic light-emitting diode (AMOLED) displays (Fig. 7.6C–F).[29] The large aperture ratio of these LETs is independent of gate voltages and was achieved by engineering the charge transport and interface energies such that large area emission occurs through a semitransparent electrode. With this approach, pixels can be designed in any shape, for example, circular and square, and a larger aperture (66% increase) can be realized by increasing the drain to source area ratio. Hence, this strategy enables a new class of pixel design with exceptionally high electrical and optical performance, all while maintaining the fabrication advantage of solution processability.

The LETs reported to date have shown significant improvement in performance but are not yet suitable for display applications. The key problems of LETs thus far are their low EQEs at high brightness and poor ON/OFF ratio owing to the poor electrical and optical properties of the organic material. In order to use LETs for practical applications, certain prerequisites must be met, including (i) high EQE at high brightness,[30] (ii) low off-current to reduce power dissipation in the device, (iii) high switching capability (ON/OFF ratio) with acceptable temporal response (5 kHz being acceptable), and (iv) gate-voltage-dependent spectral tunability (multicolor emission).

It is known that the organic-inorganic perovskites in combination with purely inorganic perovskite quantum dots (PeQDs) are promising materials for optoelectronic devices due to their unique electrical and optical properties. One of the most attractive properties is the ambipolar nature of the perovskite material with nearly balanced electron and hole diffusion lengths.[31, 32] The long-range electron and hole diffusion lengths result in better charge carrier transport properties in a layer composed of perovskites. In addition, first-principle calculations for perovskite materials predict a hole mobility up to 3100 $cm^2 V^{-1} s^{-1}$ and electron mobility of 1500 $cm^2 V^{-1} s^{-1}$ with a concentration of 10^{16} cm^{-3}.

Heterostructure LEFET can be fabricated with solution-processed perovskites, a charge-transporting and light-emitting material. Thanks to their cost-effectiveness and ease of processing, hybrid perovskites have naturally attracted a vast interest for applications beyond photovoltaic energy conversion, such as water splitting, LEDs and tunable, electrically pumped lasers.[27, 33, 34] So far, transport parameters of perovskite materials were mostly deduced from the study of photovoltaic devices, which indicated ambipolar transport of holes and electrons within the perovskite active region, and long electron-hole pair diffusion length. As the perovskites have large and almost equal diffusion length for electrons and holes, they can be used as an electron-transport layer (ETL) as well as the hole-transport layer (HTL).[35]

7.2.1 Scope of the LEFET

Despite the rapid advancement of optoelectronic applications, a big gap remains in understanding the intrinsic transport properties of halide perovskites, namely charge carrier character, mobility, and charge transport mechanisms. To fill this gap, studies of basic FETs are urgently needed. Particularly, high photoluminescence (PL) efficiency and widely tunable bandgap from visible to infrared render perovskites are extremely attractive for the fabrication of solution-processable LEFETs, a device concept that may be integrated with heterogeneous optolectronic systems, such as flexible electroluminescent displays or electrically pumped lasers.

The realization of enhanced light emission is an important consequence from the LET device, which largely depends on carrier mobility. It has been demonstrated that the field-effect carrier mobility can be increased by almost two orders of magnitude below 200 K, affording immense potential for improving the performance of LETs.[36] The perovskite-based LET will pave the way to the realization of solution-processed hybrid perovskite light-emitting devices such as high brightness and full-color LETs.

7.2.2 Controllable synthesis of PeQDs and perovskite precursors

The PeQDs and two-dimensional (2D) perovskite precursors can be prepared by two-step solvothermal method.[37] In the first step, Cesium-oleate is prepared using the following procedures: Cs_2CO_3 is loaded into a three-neck flask along with octadecene and oleic acid (OA), dried for 1–2 h at elevated temperature (above 100°C), and then heated in a nitrogen (N_2) atmosphere to 150°C until all Cs_2CO_3 reacted with OA. Before injection into the flask, Cs-oleate has to be preheated to 100°C, as it precipitates out of 1-octadecene at room temperature.

In the second step, $CsPbX_3$ nanocrystals (NCs) are prepared using the following procedures: ODE and PbX_2 such as PbI_2, $PbBr_2$, $PbCl_2$, or their mixtures are loaded into three-neck flask and dried under vacuum for 1–2 h at 120°C. Dried oleylamine and dried OA are injected at 120°C under N_2. After the complete solubilization of PbX_2 salt, the temperature is adjusted for fine-tuning the size of the NCs while the Cs-oleate solution is quickly injected into the flask. After a few seconds, the reaction mixture is cooled by an ice-water bath. In order to prepare high-quality QDs with PL at certain wavelength, that is, 430 nm for blue, 540 nm for green, and 640 nm for red colors, the injection temperatures are adjusted between 120°C and 170°C.

Similarly, the organic-inorganic perovskite $CH_3NH_3PbX_3$ precursors can be prepared by following the two-step solvothermal methods described before. The perovskite $CH3NH_3PbX_3$ (X=I, Br, Cl) precursors are prepared by mixing CH_3NH_3X and PbX_2 in a 3:1 molar ratio for X=Br or Cl and 1:1 molar ratio for X=I in anhydrous *N*,*N*-dimethylformamide or γ-butyrolactone at 60°C for 12 h under nitrogen environment, followed by double filtering using a PTFF syringe filter (Whatman, 0.45 mm).

7.2.3 Characterization of PeQDs

The all-inorganic cesium lead halide PeQDs have been the new candidate for various optoelectronic devices.[30, 35] Routes have been developed for facile colloidal synthesis of monodisperse, 2–20 nm perovskite nanoparticles with cubic crystal structure. These $CsPbX_3$ PeQDs not only exhibit a bandgap that can be engineered by changing the composition but also give strong PL with a narrow spectral range [e.g., full width at half maximum (FWHM) of PL spectrum is as low as 12 nm]. Moreover, high quantum yields of up to 90% and wider color gamut covering up to 140% of the National Television System Committee (NTSC) color standard can be achieved. Based on our preliminary research, the absorbance, PL, transmission electron microscopic image, and CIE diagram of as-synthesized $CsPbI_3$ PeQDs are shown in Fig. 7.7.

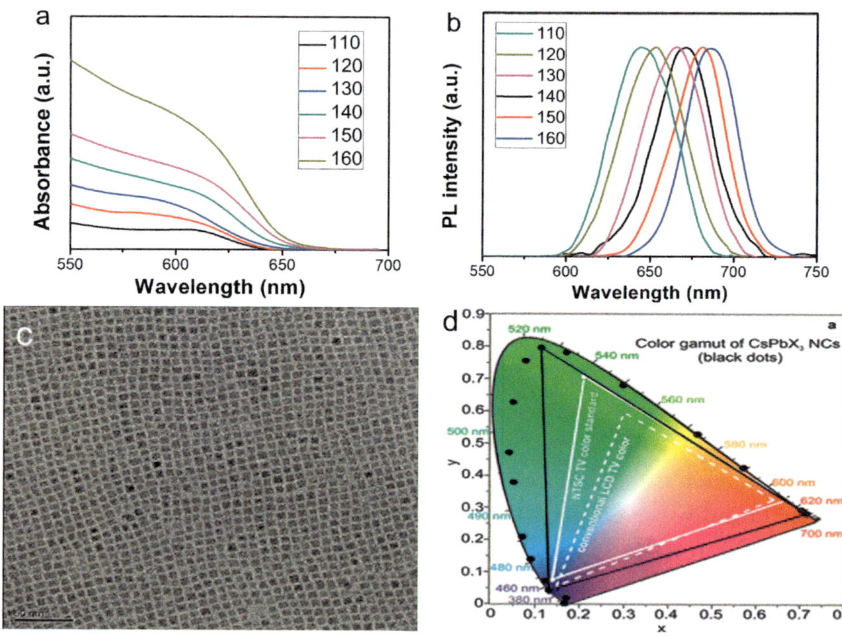

Fig. 7.7 (A) The UV-vis absorption spectra, (B) PL spectra of $CsPbI_3$ quantum-dots prepared at various temperatures, (C) TEM image of $CsPbI_3$ quantum dots prepared at 130°C, which indicates that the as-prepared QDs are monodisperse, and (D) the Commision Internationale de l'Eclairage diagram for the PeQDs. *Reprinted with permission from Muhieddine, K.; Ullah, M.; Maasoumi, F.; Burn, P. L.; Namdas, E. B., Hybrid Area-Emitting Transistors: Solution Processable and With High Aperture Ratios. Adv. Mater.* ***2015****, 27 (42), 6677-6682, ACS. Copyright (2015)*

The size and microstructure of PeQDs can be characterized by the high-resolution transmission electron microscopy (HRTEM). Optical characteristics can be analyzed by UV-visible and PL spectroscopies. X-ray photoelectron spectroscopy (XPS) and UV photoelectron spectroscopy (UPS) techniques can be used for studying surface characteristics. Photoluminance quantum yields (PLQYs) can be estimated according to standard procedure using appropriate dye molecules for blue, green, and red spectral regions (coumarine 343, fluorescein, and rhodamine 6G). PL lifetime measurements can be performed using a time-correlated single photon counting setup (TCSPC) in which SPC-130-EM counting module and BDL-488-SMN picosecond laser (Becker & Hickl) are equipped as excitation source. The pulse duration is 50 ps at a wavelength of 488 nm. Powder X-ray diffraction patterns (XRD) can be collected with STOE STADI P powder diffractometer to resolve the crystal structure and phase of PeQDs.

7.2.4 PeQLET device fabrication

The multicolor single device can be fabricated directly on the patterned ITO (indium tin oxide) substrate. The PMMA layer can be spin-coated as the gate-dielectric layer, and a low work function (LiF:Al) electrode can be deposited by thermal evaporation, followed by the spin-coating of three layers of red, green, and blue PeQDs in order. To enhance the carrier transport properties of these three layers, perovskite solution with different QDs is added, that is, $MAPbI_3$:$CsPbI_3$ for red, $MAPbBr_3$:$CsPbBr_3$ for green, and $MAPbCl_3$ for blue light-emitting materials. Further optimization can be carried out to achieve the optimum carrier transport in these layers and to achieve highest EQE and strongest brightness for each color. In addition to the material optimization, layer thickness and interface characteristics can be also optimized to achieve smooth and continuous surface coverage before the next layer deposition.

By combining the light-emitting device with the switching device (by adding gate terminal, Fig. 7.8), one can control the emissive zone inside the light-emitting device precisely. Under balanced carrier injection, gate-dependent EL can be observed from the transistor channel. In the gated device, the recombination zone position can be finely controlled by the gate voltage and polarity. To achieve simultaneous hole and electron injection in an LET, the local gate potential at drain and source electrodes must be larger than the threshold voltage of either of the charge carrier, that is, $|V_d| > |V_{th,h}|$ and $V_s > V_{th,e}$, or $V_d > V_{th,e}$ and $|V_s| > |V_{th,h}|$. Under this

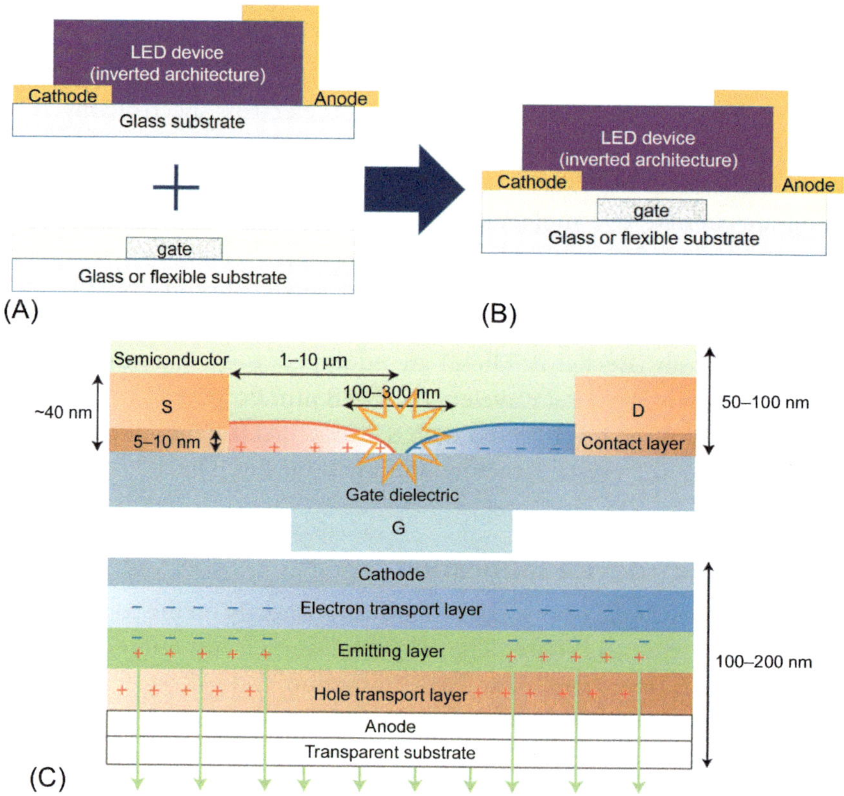

Fig. 7.8 (A-B) Schematic illustration of the integration of the LED device with the switching device, and (C) the operation mechanism of the electron-hole injection and recombination.

condition, drain source and gate voltages are tuned to control the injected current density of both carriers, which manipulate the spatial position of the emission zone as well as the EL intensity. Charges move vertically across the organic layers in the LED device as shown in Fig. 7.8C. The dimensions of the features of the device as well as the location of the recombination areas are shown in Fig. 7.8C. In LETs, the spatial separation between the exciton formation region and the metal electrodes prevents exciton-metal quenching and decreases the intensity of the electric field at the exciton location. These characteristics, in addition to the balanced electron and hole currents and the higher charge mobility, favor more-efficient light emission in LETs. In addition, the gate position is also responsible for tuning the position from near-to-drain to the center-of-channel.

7.2.5 Investigation of the mechanism in PeQDs-LETs

Currently, it is difficult for RGB-emitting phosphors comprising different materials to have luminous efficiencies of a multicolor display with balanced RGB components.[35] While there has been substantial research on patterned devices, one of the most feasible strategies for realizing RGB displays is color filtering of white light. Filtering, however, causes wastages up to 90% of the output light power to achieve color saturation, which requires that the display be operated at 10 times video brightness in order to meet the RGB color standard. This results in greater power consumption, faster pixel degradation, and shorter display lifetimes.[38]

The idea of spectral tunability was previously demonstrated with graphene. Being a tunable optical platform, graphene is a promising medium to achieve this goal (Fig. 7.9).[39] Bright spectrally tunable EL from blue (~450 nm) to red (~750 nm) at the graphene oxide-reduced-graphene oxide interface was observed. The EL results from the recombination of Poole-Frenkel emission ionized electrons at the localized energy levels arising from semi-reduced graphene oxide, and holes from the top of the π-band. Tuning of the emission wavelength is achieved by gate modulation of the participating localized energy levels. The demonstration of current-driven tunable LEDs not only represents a method for tuning the emission wavelength but also may find applications in high-quality displays. The device showed a high brightness of up to 6000 cdm^{-2}, with external quantum efficiency of around 1%. This mechanism also explains the observed spectral shift (Fig. 7.9B). Gating graphene lifts the chemical potential and thus the energy level of the lowest unoccupied discrete state; therefore, the excited electron energy is increased. Through this way, the EL peak can be adjusted within the whole PL range; and the relative emission intensity is determined by the distribution of density of states of the discrete energy levels. The EL emission can be continuously tuned from light blue to dark red by adjusting the Fermi levels, an exceptional achievement. The Fermi level and the doping level of the LET can therefore be modulated electrically instead of chemical doping.

7.3 Other optically pumped LED concepts

Quantum dots (QDs) and 2D materials have attracted great attention in recent decades. It is a field of high interest for both industry and academia due its outstanding characteristics like electrical, mechanical, and optical

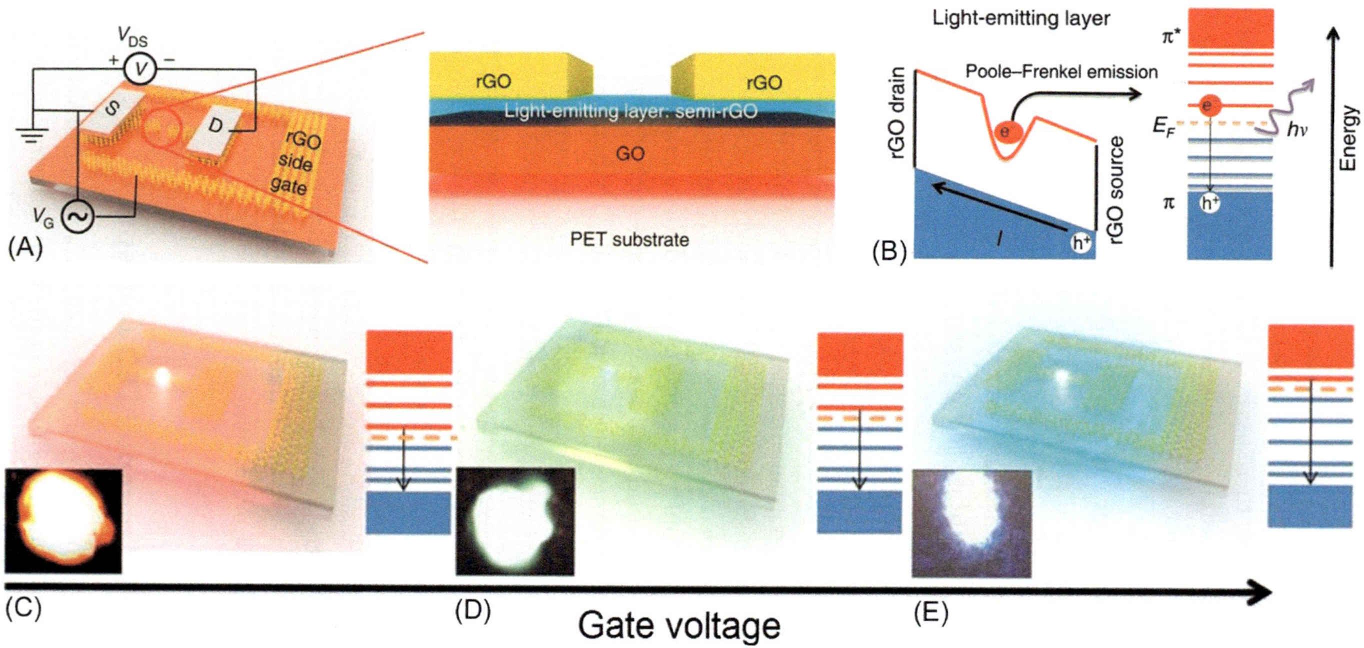

Fig. 7.9 Graphene multicolor LET and characterization. (A) Schematic representation of the LET device. A distinct semi-reduced GO (*blue*) at the interface between GO (*orange*) and rGO (*gold*) is responsible for light emission, (B) schematic representation of the charge injection process and light-emission mechanism, (C–E) schematic representation of the gate voltage-dependent EL. The Fermi level determines the lowest unoccupied energy state that mainly participates in the radiative recombination. Inset: corresponding emission images from a real device. *Adapted from Wang, X.; Tian, H.; Mohammad, M. A.; Li, C.; Wu, C.; Yang, Y.; Ren, T.-L., A Spectrally Tunable All-Graphene-Based Flexible Field-Effect Light-Emitting Device. Nat. Commun.* ***2015****, 6, 7767, under the Creative Commons Attribution 4.0 International License.*

properties compared to its bulk form.[40–44] Graphene is the first important character identified in 2D materials that has superior carrier mobility and broad absorption band.[42, 45] Furthermore, it has strong light-matter interaction, efficiently tunable charge carrier density, and broad-range light emission, making it a potential candidate for various optoelectronic applications such as light-emitting devices, solar cell, and photodetectors.[46–48] The most crucial fundamental demands of photodetector in terms of speed, sensitivity, and range of wavelength became possible with the discovery of graphene.

However, due to relatively low light absorption, that is, 2.3% per layer, photodetection with high response is a great challenge for graphene.[49] In order to resolve the problem another nanomaterials like QDs can be effectively combined with graphene by using hybrid architecture mechanism. The prepared combination of graphene and QDs greatly enhances the photosensitivity of the detector. On the other hand, 2D transition metal dichalcogenides (TMDs) such as WS_2, MoS_2, and $MoSe_2$ are semiconductors with tunable direct bandgap in the range 1.57–2 eV. Compared to the gapless graphene, TMDs are seen as a promising alternative in various optoelectronic and photonics applications.[43, 44, 50] The phenomenon of dielectric screening and quantum confinements in TMDs are responsible for the tightly packed electron-hole pairs named excitons, which are the key factor for the light absorption and emission spectra.[51, 52] Similar to other semiconductors, excitons in TMDs can be created with the injection of electrons and holes or with direct interaction of photon, which in turns leads to the emission of light under EL and/or photoluminescence.[52, 53] However, the photon-induced excitons in monolayer TMDs suffer from the recombination energy loss, which in turn results in PL with small quantum yields, thus narrowing down its potential in several applications of optically pumped light-emitting devices.[54]

In order to produce excitons in TMDs that will result in high quantum yields due to radiative recombination, a new and efficient technology will be need.

Optical emitter in the nanoscale range like fluorophores or semiconductor QDs are sensitive to their surrounding environment. So an optical excitation of nanocrystal QDs can be non-radiatively transferred to the surrounding environment by using various mechanism like Forster resonance energy transfer (FRET) and charge transfer. Compared to the charge transfer process, FRET is an efficient optical process for energy transfer through long ranges.[55] Thus a combination of the proximal nano-emitter, optically excited QDs with monolayer TMDs can bring great enhancement in PL of monolayer TMDs because of the near-field coupling, which enabled FRET from nano-emitter QDs to the TMDs. Another important phenomenon is the electrical control of FRET, which is a demand of various applications in the field of optoelectronics and photonics. The hybrid structure of nanocrystal QDs and TMDs provide the opportunity for electrically control FRET with electrostatic gating.[56, 57] In addition, the combination of nanostructure QDs with various wavelength emitting TMDs opens new research directions for white LED and LEFET. The heterostructure may even replace the traditional p-n junction, which is commonly formed due to electrostatic gating or chemical doping, to realise highly efficient state-of-the-art devices. The hybrid structure will provide a platform to combine various efficient 2D materials and QDs to prepare different types of optically pumped junctions.

7.3.1 Advantages and challenges

Heterostructures based on van der Waals coupling are used to combine the advantages of different TMDs materials, for example, responsivity in broadband region and strong light absorption. Furthermore, many optoelectronics devices like LEDs, FETs, and photodetectors became possible due to van der Waals heterostructures. In the case of efficient white light-emitting devices, white light generation from a single device simplify the device structure architecture. In addition, the considerable broad bandgap nanocrystal can work as a whispering gallery mode optical cavity for photoemission in the combined TMDs heterostructures, which may find applications in lasers.

1. The responsivity is an important parameter in considering an efficient photodetector. Remarkably high responsivity can be achieved by using the carriers propagating effect-dominated photodetector, but ultrahigh responsivity is achieved by sacrificing the response time. Therefore, it is a great challenge to balance these parameters for the appropriate applications.
2. Infrared photodetection is an important technology in the field of communications, military systems, and astronomy. Besides the traditionally and widely reported layered materials like graphene, non-layered materials like SnTe, $Pb_{1-x}Sn_xTe$, and $Pb_{1-x}Sn_xSe$ have narrow bandgaps and are considered as potential candidates for infrared photodetections. Thus efficient fabrication techniques of high-quality non-layered 2D materials must be developed. In conjunction, research in hybrid 2D non-layered materials is also required.

3. Remarkable EQE that is nearly 10% has been acquired in LED based on heterostructures so far. Despite this achievement, the EQE can be further enhanced by engineering the band structure with the help of van der Walls heterostructures. Also, it is crucial to increase the luminescence efficiency by exploiting novel, suitable materials and modifying the properties of TMDs materials.
4. Defects can significantly affect the corresponding properties of TMDs. These imperfections such as vacancies act as potential sources for the emission of light. Continuous efforts and developments are needed to analyze the characteristics of the various types of defects, highly efficient LEDs based on TMDs may be developed. It is also expected to have single photon emission from the various types of defects in layered 2D materials and its hybrid structures.

7.4 Conclusion

Decent lighting source are the main desire of human being from the ages. Although the application of the incandescent lamp is from the very beginning to the near past, the linear tube and compact fluorescent lamp have now replaced the conventional incandescent bulb due to the resent awareness regarding energy consumption and environmental issue. The fundamental characteristics of lighting technology is that energy efficiency has been solved with the development of compact fluorescent lamp but the emission of mercury vapor during continuous use created a new environmental problem. Materials science is involved in the challenge to search for alternate lighting source. Solid-state lighting (SSL) has many advantages over incandescent bulb and compact fluorescent lamp in terms of outstanding efficiency, mercury-free emission, and long operational lifetime.

Most white light is achieved by the combination of blue LED chip and yellow-emitting phosphor. But the light obtained from this combination cannot fulfil the requirements of high-quality warm white, due to the lack of red component. To obtain WLEDs (white LEDs) with high CRI (color-rendering index) as alternative materials for highly efficient RGB-emitting phosphors, QDs and 2D-TMDs must be produced. However, reabsorption phenomena in such tri-chromatic LEDs may also occur and cause a decrease in the device efficiency. Thus, a high stable and efficient light-emitting device needs to be fabricated. More specifically, the combination of QDs with 2D-TMDs in the form of hybrid heterostructure (QDs/2D-TMDs) emits a broader wavelength (cover the full region of visible light). In addition, the gate electrode is used to control the variation in color in the entire visible region by FRET. Controlling the color variation and modulation by using gate electrode is of tremendous potential in providing full spectrum device with enhanced electrical and optical properties that are not easily achievable with any other device architecture to date. The core of this chapter is to investigate highly efficient light-emitting devices that can provide a new design for lighting and display applications. This will be beneficial to the scientific community and industry for next-generation lighting source and display applications.

References

1. Nassau, K. *The Physics and Chemistry of Color: The Fifteen Causes of Color;* John Wiley & Sons, Inc., 2001.
2. Flesch, P. *Light and Light Sources;* Springer, 2006.
3. Khan, S. A.; Hao, Z.; Ji, W. W.; Khan, N. Z.; Abadikhah, H.; Hao, L.; Xu, X.; Agathopoulos, S.; Bao, Q. Crystal-Site Engineering for Developing Tunable Green Light Emitting Ba9Lu2Si6O24: Eu2+ Phosphors for Efficient White LEDs. *J. Alloys Compd.* **2018**, *767*, 374–381.
4. Braunstein, R. Radiative Transitions in Semiconductors. *Phys. Rev.* **1955**, *99*(6), 1892–1893.
5. Zheludev, N. The Life and Times of the LED—A 100-Year History. *Nat. Photonics* **2007**, *1*, 189.
6. Braunstein, R. Radiative Transitions in Semiconductors. *Phys. Rev.* **1955**, *99*(6), 1892–1893.
7. Holonyak, N.; Bevacqua, S. F. Coherent (Visible) Light Emission From Ga(As1 – xPx) Junctions. *Appl. Phys. Lett.* **1962**, *1*(4), 82–83.
8. Nakamura, S.; Mukai, T.; Senoh, M. Candela-Class High-Brightness InGaN/AlGaN Double-Heterostructure Blue-Light-Emitting Diodes. *Appl. Phys. Lett.* **1994**, *64*(13), 1687–1689.
9. US Department of Energy. *Solid-State Lighting: LED Basics;* 2012.
10. Han, J., Synthesis and Luminescence Properties of Rare Earth Activated Phosphors for Near UV-Emitting LEDs for Efficacious Generation of White Light. UC San Diego. ProQuest ID: Han_ucsd_0033D_13025. Merritt ID: ark:/20775/bb24587402. Retrieved from https://escholarship.org/uc/item/03t467f7, 2013.
11. Pope, M.; Kallmann, H. P.; Magnante, P. Electroluminescence in Organic Crystals. *J. Chem. Phys.* **1963**, *38*(8), 2042–2043.
12. Tang, C. W.; VanSlyke, S. A. Organic Electroluminescent Diodes. *Appl. Phys. Lett.* **1987**, *51*(12), 913–915.
13. So, F.; Kido, J.; Burrows, P. Organic Light-Emitting Devices for Solid-State Lighting. *MRS Bull.* **2011**, *33*(07), 663–669.
14. Wang, Z. B.; Helander, M. G.; Qiu, J.; Puzzo, D. P.; Greiner, M. T.; Hudson, Z. M.; Wang, S.; Liu, Z. W.; Lu, Z. H. Unlocking the Full Potential of Organic Light-Emitting Diodes on Flexible Plastic. *Nat. Photonics* **2011**, *5*(12), 753–757.
15. Wu, S.; Li, S.; Sun, Q.; Huang, C.; Fung, M. K. Highly Efficient White Organic Light-Emitting Diodes with Ultrathin Emissive Layers and a Spacer-Free Structure. *Sci. Rep.* **2016**, *6*, 25821.

16. Sun, Y.; Forrest, S. R. High-Efficiency White Organic Light Emitting Devices With Three Separate Phosphorescent Emission Layers. *Appl. Phys. Lett.* **2007**, *91*(26), 263503.
17. Ou, Q.-D.; Zhou, L.; Li, Y.-Q.; Shen, S.; Chen, J.-D.; Li, C.; Wang, Q.-K.; Lee, S.-T.; Tang, J.-X. Extremely Efficient White Organic Light-Emitting Diodes for General Lighting. *Adv. Funct. Mater.* **2014**, *24*(46), 7249–7256.
18. Han, T.-H.; Lee, Y.; Choi, M.-R.; Woo, S.-H.; Bae, S.-H.; Hong, B. H.; Ahn, J.-H.; Lee, T.-W. Extremely Efficient Flexible Organic Light-Emitting Diodes With Modified Graphene Anode. *Nat. Photonics* **2012**, *6*(2), 105–110.
19. Lee, J.; Chen, H. F.; Batagoda, T.; Coburn, C.; Djurovich, P. I.; Thompson, M. E.; Forrest, S. R. Deep Blue Phosphorescent Organic Light-Emitting Diodes With Very High Brightness and Efficiency. *Nat. Mater.* **2016**, *15*(1), 92–98.
20. Sun, Y.; Giebink, N. C.; Kanno, H.; Ma, B.; Thompson, M. E.; Forrest, S. R. Management of Singlet and Triplet Excitons for Efficient White Organic Light-Emitting Devices. *Nature* **2006**, *440*, 908.
21. Wang, Y.; Lu, Y.; Gao, B.; Wang, S.; Ding, J.; Wang, L.; Jing, X.; Wang, F. Self-Host Blue-Emitting Iridium Dendrimer Containing Bipolar Dendrons for Nondoped Electrophosphorescent Devices With Superior High-Brightness Performance. *ACS Appl. Mater. Interfaces* **2016**, *8*(43), 29600–29607.
22. Zhang, Q.; Tsang, D.; Kuwabara, H.; Hatae, Y.; Li, B.; Takahashi, T.; Lee, S. Y.; Yasuda, T.; Adachi, C. Nearly 100% Internal Quantum Efficiency in Undoped Electroluminescent Devices Employing Pure Organic Emitters. *Adv. Mater.* **2015**, *27*(12), 2096–2100.
23. Dias, F. B.; Bourdakos, K. N.; Jankus, V.; Moss, K. C.; Kamtekar, K. T.; Bhalla, V.; Santos, J.; Bryce, M. R.; Monkman, A. P. Triplet Harvesting With 100% Efficiency by Way Of Thermally Activated Delayed Fluorescence in Charge Transfer OLED Emitters. *Adv. Mater.* **2013**, *25*(27), 3707–3714.
24. Jiang, T.; Wang, F.; Tang, C.; Zhang, X.; Cao, X.; Tao, Y.; Huang, W. Carbazole/Phenylpyridine Hybrid Compound as Dual Role of Efficient Host and Ligand of Iridium Complex: Well Matching of Host-Dopant for Solution-Processed Green Phosphorescent OLEDs. *Dyes Pigments* **2018**, *150*, 130–138.
25. Jeong, H.; Shin, H.; Lee, J.; Kim, B.; Park, Y.-I.; Yook, K. S.; An, B.-K.; Park, J. Recent Progress in the Use of Fluorescent and Phosphorescent Organic Compounds for Organic Light-Emitting Diode Lighting. *J. Photon. Energy* **2015**, *5*(1), 057608.
26. Wang, S.; Wang, X.; Yao, B.; Zhang, B.; Ding, J.; Xie, Z.; Wang, L. Solution-Processed Phosphorescent Organic Light-Emitting Diodes With Ultralow Driving Voltage and Very High Power Efficiency. *Sci. Rep.* **2015**, *5*, 12487.
27. Capelli, R.; Toffanin, S.; Generali, G.; Usta, H.; Facchetti, A.; Muccini, M. Organic Light-Emitting Transistors With an Efficiency That Outperforms the Equivalent Light-Emitting Diodes. *Nat. Mater.* **2010**, *9*(6), 496.
28. Gwinner, M. C.; Kabra, D.; Roberts, M.; Brenner, T. J.; Wallikewitz, B. H.; McNeill, C. R.; Friend, R. H.; Sirringhaus, H. Highly Efficient Single-Layer Polymer Ambipolar Light-Emitting Field-Effect Transistors. *Adv. Mater.* **2012**, *24*(20), 2728–2734.
29. Muhieddine, K.; Ullah, M.; Maasoumi, F.; Burn, P. L.; Namdas, E. B. Hybrid Area-Emitting Transistors: Solution Processable and With High Aperture Ratios. *Adv. Mater.* **2015**, *27*(42), 6677–6682.
30. Protesescu, L.; Yakunin, S.; Bodnarchuk, M. I.; Krieg, F.; Caputo, R.; Hendon, C. H.; Yang, R. X.; Walsh, A.; Kovalenko, M. V. Nanocrystals of Cesium Lead Halide Perovskites (CsPbX3, X= Cl, Br, and I): Novel Optoelectronic Materials Showing Bright Emission With Wide Color Gamut. *Nano Lett.* **2015**, *15*(6), 3692–3696.

31. Xing, G.; Mathews, N.; Sun, S.; Lim, S. S.; Lam, Y. M.; Grätzel, M.; Mhaisalkar, S.; Sum, T. C. Long-Range Balanced Electron- and Hole-Transport Lengths in Organic-Inorganic CH3NH3PbI3. *Science* **2013**, *342*(6156), 344–347.
32. Samuel, D.; et al. Electron-Hole Diffusion Lengths Exceeding 1 Micrometer in an Organometal Trihalide Perovskite Absorber Stranks. *Snaith Henry J. Sci. (New York, N.Y.)* **2013**, *342*, 341–344.
33. Zhang, X.; Xu, B.; Zhang, J.; Gao, Y.; Zheng, Y.; Wang, K.; Sun, X. W. All-Inorganic Perovskite Nanocrystals for High-Efficiency Light Emitting Diodes: Dual-Phase CsPbBr3-CsPb2Br5 Composites. *Adv. Funct. Mater.* **2016**, *26*(25), 4595–4600.
34. Wang, Y.; Li, X.; Song, J.; Xiao, L.; Zeng, H.; Sun, H. All-Inorganic Colloidal Perovskite Quantum Dots: A New Class of Lasing Materials With Favorable Characteristics. *Adv. Mater.* **2015**, *27*(44), 7101–7108.
35. Zhang, F.; Zhong, H.; Chen, C.; Wu, X.-g.; Hu, X.; Huang, H.; Han, J.; Zou, B.; Dong, Y. Brightly Luminescent and Color-Tunable Colloidal CH3NH3PbX3 (X= Br, I, Cl) Quantum Dots: Potential Alternatives for Display Technology. *ACS Nano* **2015**, *9*(4), 4533–4542.
36. Chin, X. Y.; Cortecchia, D.; Yin, J.; Bruno, A.; Soci, C. Lead Iodide Perovskite Light-Emitting Field-Effect Transistor. *Nat. Commun.* **2015**, *6*, 7383.
37. Qasim, K.; Wang, B.; Zhang, Y.; Li, P.; Wang, Y.; Li, S.; Lee, S. T.; Liao, L. S.; Lei, W.; Bao, Q. Solution-Processed Extremely Efficient Multicolor Perovskite Light-Emitting Diodes Utilizing Doped Electron Transport Layer. *Adv. Funct. Mater.* **2017**, *27*(21), 1606874.
38. Kim, T.-H.; Cho, K.-S.; Lee, E. K.; Lee, S. J.; Chae, J.; Kim, J. W.; Kim, D. H.; Kwon, J.-Y.; Amaratunga, G.; Lee, S. Y. Full-Colour Quantum Dot Displays Fabricated by Transfer Printing. *Nat. Photonics* **2011**, *5*(3), 176.
39. Wang, X.; Tian, H.; Mohammad, M. A.; Li, C.; Wu, C.; Yang, Y.; Ren, T.-L. A Spectrally Tunable All-Graphene-Based Flexible Field-Effect Light-Emitting Device. *Nat. Commun.* **2015**, *6*, 7767.
40. Dai, X.; Zhang, Z.; Jin, Y.; Niu, Y.; Cao, H.; Liang, X.; Chen, L.; Wang, J.; Peng, X. Solution-Processed, High-Performance Light-Emitting Diodes Based on Quantum Dots. *Nature* **2014**, *515*(7525), 96.
41. Chen, W.; Wang, K.; Hao, J.; Wu, D.; Qin, J.; Dong, D.; Deng, J.; Li, Y.; Chen, Y.; Cao, W. High Efficiency and Color Rendering Quantum Dots White Light Emitting Diodes Optimized by Luminescent Microspheres Incorporating. *Nanophotonics* **2016**, *5*(4), 565–572.
42. Rathi, S.; Lee, I.; Lim, D.; Wang, J.; Ochiai, Y.; Aoki, N.; Watanabe, K.; Taniguchi, T.; Lee, G.-H.; Yu, Y.-J. Tunable Electrical and Optical Characteristics in Monolayer Graphene and Few-Layer MoS2 Heterostructure Devices. *Nano Lett.* **2015**, *15*(8), 5017–5024.
43. Mak, K. F.; Shan, J. Photonics and Optoelectronics of 2D Semiconductor Transition Metal Dichalcogenides. *Nat. Photonics* **2016**, *10*(4), 216.
44. Ahmed, S.; Yi, J. Two-Dimensional Transition Metal Dichalcogenides and Their Charge Carrier Mobilities in Field-Effect Transistors. *Nanomicro Lett.* **2017**, *9*(4), 50.
45. Novoselov, K. S.; Geim, A. K.; Morozov, S. V.; Jiang, D.; Zhang, Y.; Dubonos, S. V.; Grigorieva, I. V.; Firsov, A. A. Electric Field Effect in Atomically Thin Carbon Films. *Science* **2004**, *306*(5696), 666–669.
46. Lawton, L.; Mahlmeister, N.; Luxmoore, I. J.; Nash, G. R. Prospective for Graphene Based Thermal Mid-Infrared Light Emitting Devices. *AIP Adv.* **2014**, *4*(8), 087139.
47. Mahmoudi, T.; Wang, Y.; Hahn, Y.-B. Graphene and Its Derivatives for Solar Cells Application. *Nano Energy* **2018**, *47*, 51–65.
48. Koppens, F.; Mueller, T.; Avouris, P.; Ferrari, A.; Vitiello, M.; Polini, M. Photodetectors Based on Graphene, Other Two-Dimensional Materials and Hybrid Systems. *Nat. Nanotechnol.* **2014**, *9*(10), 780.

49. Nair, R. R.; Blake, P.; Grigorenko, A. N.; Novoselov, K. S.; Booth, T. J.; Stauber, T.; Peres, N. M.; Geim, A. K. Fine Structure Constant Defines Visual Transparency of Graphene. *Science* **2008**, *320*(5881), 1308.
50. Sun, Z.; Martinez, A.; Wang, F. Optical Modulators With 2D Layered Materials. *Nat. Photonics* **2016**, 10(4), 227–238.
51. Mak, K. F.; Lee, C.; Hone, J.; Shan, J.; Heinz, T. F. Atomically Thin MoS 2: A New Direct-Gap Semiconductor. *Phys. Rev. Lett.* **2010**, *105*(13), 136805.
52. Splendiani, A.; Sun, L.; Zhang, Y.; Li, T.; Kim, J.; Chim, C.-Y.; Galli, G.; Wang, F. Emerging Photoluminescence in Monolayer MoS_2. *Nano Lett.* **2010**, *10*(4), 1271–1275.
53. Sundaram, R.; Engel, M.; Lombardo, A.; Krupke, R.; Ferrari, A.; Avouris, P.; Steiner, M. Electroluminescence in Single Layer MoS2. *Nano Lett.* **2013**, *13*(4), 1416–1421.
54. Amani, M.; Lien, D.-H.; Kiriya, D.; Xiao, J.; Azcatl, A.; Noh, J.; Madhvapathy, S. R.; Addou, R.; Santosh, K.; Dubey, M. Near-Unity Photoluminescence Quantum Yield in MoS2. *Science* **2015**, *350*(6264), 1065–1068.
55. Swathi, R.; Sebastian, K. Long Range Resonance Energy Transfer From a Dye Molecule to Graphene Has (Distance)−4 Dependence. *J. Chem. Phys.* **2009**, *130*(8), 086101.
56. Newaz, A.; Prasai, D.; Ziegler, J.; Caudel, D.; Robinson, S.; Haglund, R., Jr.; Bolotin, K. Electrical Control of Optical Properties of Monolayer MoS2. *Solid State Commun.* **2013**, *155*, 49–52.
57. Mak, K. F.; He, K.; Lee, C.; Lee, G. H.; Hone, J.; Heinz, T. F.; Shan, J. Tightly Bound Trions in Monolayer MoS2. *Nat. Mater.* **2013**, *12*(3), 207–211.

CHAPTER 8

Flexible and stretchable inorganic electronics: Conductive materials, fabrication strategy, and applicable devices

Ge Shi, Sean E. Lowe, Yu Lin Zhong, Huijun Zhao
Centre for Clean Environment and Energy, School of Environment and Science, Griffith University, Gold Coast, QLD, Australia

8.1 Introduction

In the past two decades, the field of flexible/stretchable electronics has emerged with significant achievements. Flexible electronics have been used in a wide range of applications, e.g., strain sensors, artificial skins,[1] flexible touch screens,[2] transistors,[3] and energy harvesting and storage.[4] Thus, it is reasonable to believe that we will see commercial applications of flexible electronics in the near future.

However, scientists working in the field of stretchable electronics still face major challenges, particularly, in achieving high elastomeric deformation while maintaining desired electrical conductivity. Stretchability refers to the ability to conform to and recover from relative deformation, and high electrical performance is the ability to maintain electrical integrity under deformation. Ideally, both should be simultaneously achieved through integrating elastomeric substrates with electronic components. To meet the demands of high stretchability, electrical performance, and reliability, the following criteria are considered: (i) development of new materials capable of conducting electrons and altering shapes and (ii) design of new structures to accommodate large strain deformation without damaging the materials.

The convergence of material science, nanotechnology, and electrical engineering greatly aids the development of flexible/stretchable electrical devices. As a result, various methods for nanomaterial synthesis, design, and manufacturing have been created, to develop electronics exhibiting stretchability while featuring functional performances.

2D Materials for Photonic and Optoelectronic Applications
https://doi.org/10.1016/B978-0-08-102637-3.00008-5

The dramatic progress of nanomaterials has enabled researchers to fabricate flexible and conductive materials that can maintain their electrical properties under strain. These materials include nanosheets, nanotubes, and nanowires. Nanosheets, which are stable in single-atom or single-polyhedral thicknesses, exhibit completely different properties than their bulk peers. For example, graphene possesses many extraordinary attributes, e.g., its high intrinsic mobility (200,000 $cm^2\ v^{-1}\ s^{-1}$),[5] 2.3% absorption in the white light spectrum,[6] high specific surface area (2630 $m^2\ g^{-1}$),[5] and high Young's modulus (~1.0 TPa).[7] Nanotubes or nanowires provide intrinsic electrical and optical properties than their microscale grids. For instance, Ag nanowires have large failure strain (~20%) and the highest electrical performance (8 Ω/sq at a transmittance of 80%)[7] compared with other metal nanowires.[8]

The structural configurations of stretchable conductors can be classified as out-of-plane and in-plane structures. The out-of-plane structures are commonly created by transferring prepared conductive materials (e.g., films of carbon nanotube[9]) of specific geometry (e.g., a serpentine shape[10]) onto cured elastomeric substrates. This strategy is extensively used to form what it is defined as periodically structured materials (e.g., wavy structure[11]) by a process of prestrain-release-buckling, as reviewed in Fig. 8.2. The out-of-plane structures are generally held by the weak interactions between the conductor and the substrate—a simple adhesion of a device onto a cured substrate is difficult for robust applications. For in-plane structures, on the other hand, conductive electronics (e.g., 3D graphene foam[14]) are embedded into the uncured elastomer substrates; curing later on will form stretchable electronics, e.g., stretchable conductor[15]; this type of structures has an improved interaction.

The fabrication procedures of stretchable electronics can be divided as top-down and bottom-up strategies. The top-down strategy involves processing monolithic input material to create micro- or nanoscale morphologies of stretchable shapes, e.g., serpentine[16] and wavy[17] structures. It enables us to precisely create various complex patterning morphologies.[18] On the other hand, the bottom-up strategy offers facile fabrication of components by depositing raw materials to build the desired structure (e.g., 3D printing and other forms of additive manufacturing). This strategy easily allows a 3D structure[19] and transparent conductive films[20] to be built. Its advantages include controllable assembly and cost-effectiveness.

This chapter aims to comprehensively describe the development of flexible/stretchable electronics in healthcare, human-machine interactivity, energy conversion, and storage. First, we will describe advances in the synthesis and properties of nanomaterials that can ensure high performance and flexibility. Second, we introduce the major strategies and related manufacturing procedures to create stretchable structures and discuss their merits and limitations. Third, through reviewing the performance and structures of various electronics, we will attempt to show the relationship between mechanical deformation and overall performance. Last, we will present a brief discussion on research opportunities for future stretchable electrical conductors.

8.2 Conductive nanomaterials in stretchable electronics

The development of conductive but flexible materials is essential for flexible/stretchable electronics. Materials such as graphene,[20] carbon nanotubes (CNTs),[9] and metal nanoparticles/nanowires[21] are promising for a variety of applications, including flexible optical and electronic devices[22] and strain and biological sensors.[23] One common characteristic of these materials is their nanoscale thickness, because nanoscale thickness can minimize the strains induced by bending or stretching.[24]

8.2.1 Graphene

Graphene is composed of a single layer of carbon atoms arranged in a two-dimensional (2D) honeycomb lattice. Many extraordinary properties, such as its 2.3% light absorption,[25] large surface area (2630 $m^2\ g^{-1}$),[5a, 26] high Young's modulus (1 TPa),[27] and excellent electron mobility (200,000 $cm^2\ V^{-1}\ s^{-1}$),[5a] have been reported. Among the various synthesis methods,[28] only liquid-phase exfoliation of graphite and chemical vapor deposition (CVD) can control the layer number and guarantee structural integrity.

8.2.1.1 Liquid-phase exfoliation

Due to the demands placed on graphene used in stretchable electronics, graphene produced by liquid-phase exfoliation is uniquely suitable because of its (i) high structural integrity, (ii) high yield of single or few layers, (iii) high effectiveness in preventing aggregation, and (iv) high efficiency in terms of cost and time. Sonication methods have been widely utilized; thus, we will discuss them first.

Sonication can simultaneously achieve dispersion and exfoliation of graphite.[29] Sonication energy required for exfoliation is a function of the solvent-graphene interaction. Certain solvents, including benzyl benzoate, 1-methyl-2-pyrrolidinone (NMP), g-butyrolactone (GBL),

N,N-dimethylacetamide (DMA), *N*-vinyl-2-pyrrolidone (NVP), and *N,N*-dimethylformamide (DMF), are classified as good exfoliating solvents because their surface energies are similar with graphene.[30] In these solvents, graphene sheets are charged and generate strong electrostatic repulsion, which facilitates the disaggregation of layers and increases exfoliation efficiency. Other solvents, e.g., ethanol, acetone, and water, are poor solvents for graphite exfoliation because these have lower surface energies than graphene. Thus, to adjust the surface charge and increase exfoliation efficiency in these solvents, various surfactants, e.g., sodium dodecylbenzenesulfonate, sodium cholate, and polyvinylpyrrolidone, can be intercalated. After mixing the surfactants, solvent, and graphite together, these surfactants self-assemble on graphene surface.[31] The charge transfer between solvent molecules and surfactants will dramatically increase electrostatic repulsion of graphene.

Recently, some sonication-free methods have been reported. For example, Matsumoto et al. reported using microwave irradiation to exfoliating graphite in oligomeric ionic liquids, $IL2PF_6$ and $IL4PF_6$.[32] Oligomeric ionics can efficiently absorb microwaves and tightly adhere to graphite via a cation-π interaction. After 30 min of microwave irradiation, dilution, filtration, rinsing, and drying (Fig. 8.1A_1), the weight of the produced graphene is 93% of the weight of the initial graphite and consisted of 95% single layer (Fig. 8.1A_2). Graphene has a high structural integrity ($I_D/I_G \sim 0.14$, $C/O \sim 30$) and sheet resistance of 2.1 ± 0.9 kΩ/sq. Additionally, Paton et al. demonstrated a shear exfoliation method that can produce a large-scale defect-free graphene.[33] As shown in Fig. 8.1B_1, a mixer with a rotor/stator was applied to generate a high rotor speed (4500 rpm) and high shear rate ($>10^4$ s^{-1}) to exfoliate graphite. The solvent included NMP and sodium cholate, which has similar surface energy (69 mJm^{-2}) as that of graphene, such that exfoliation energy was minimized and shear exfoliation was effectively facilitated. The produced graphene had a thickness <15 nm and lateral dimensions of 300–800 nm (Fig. 8.1B_2). Although the shear exfoliated graphene had graphitic properties, their small lateral dimension (~300 nm) may affect their application in electronics because the small lateral dimension can lead to higher contact resistance.

Graphene platelets (GnPs) also may be used for developing stretchable electronics.[34] GnPs are fabricated by thermally expanding expandable graphite for a short time at 700°C. During thermal expansion, the intercalated solid chemical compounds react to generate a gas that will strongly overcome the van der Waals forces among the platelets to produce few-layer graphene.

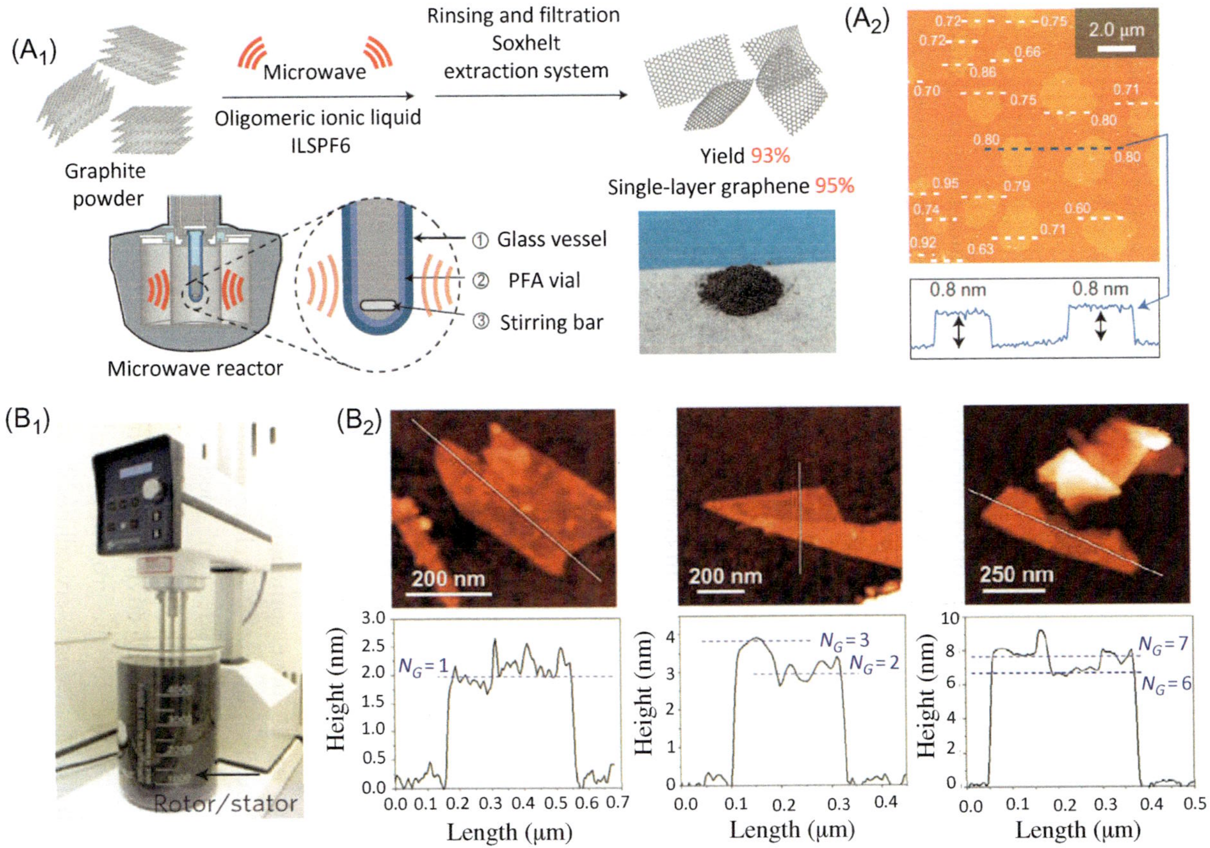

Fig. 8.1 (A_1) Schematic showing the experimental procedure of microwave irradiation. (A_2) AFM image of irradiated graphene sheets.[32] (B_1) Schematic showing shear exfoliation procedure, (B_2) AFM images and corresponding height profiles of shear exfoliated graphene.[33] *Panels (A1) and (A2) Copyright 2015, Nature Publishing Group. Panels (B1) and (B2) Copyright 2014, Nature Publishing Group.*

The thermal expansion will transform GIC flakes into worm-like fibers consisting of loosely joined thin graphene platelets. These GnPs have a lateral dimension >2 μm and thicknesses <15 nm. Superior to graphene oxide (GO), expansion along the graphite c-axis is generated by vaporizing intercalated compounds, rather than surface functional groups; so after expansion, the integrity of graphene is upheld. Therefore, GnPs have a high conductivity of ~1500 S/cm and a low I_D/I_G of 0.07.[35]

8.2.1.2 Chemical vapor deposition

The CVD[14] method is another effective way to produce large-area, high-quality graphene. During CVD, hydrocarbon precursor gases (e.g., methane,[36] acetylene,[37] and methanol[38]) are decomposed as carbon radicals in transition metals. Then, carbon atoms precipitate out to form single- or few-layer graphene on a metal surface at a certain temperature. The metal substrate works as a catalyst that determines the reaction rate and the quality of resulting graphene. Ni and Cu are the most common catalysts used. Due to the high solubility of carbon in Ni, the Ni-catalyzed graphene tends to be few-layer or multilayer.[39] In contrast, benefiting from ultralow carbon solubility in Cu, the precipitation rate of carbon can be controlled and thus large-area, uniform monolayer graphene can be grown on Cu surfaces.[40] Apart from Cu, recently, hexagonal boron nitride (h-BN) has been considered as a good substrate candidate, due to its insulating properties, chemical inertness, and small lattice misfit.[41]

Although high-quality graphene has been synthesized via CVD, even at sizes up to 30 in.,[42] there are still several areas for further research. First, CVD-derived graphene film is polycrystalline and thus contains many grain boundaries that disrupt electronic connectivity and structural continuity.[43] Thus, controllable fabrication of single-crystal graphene would be an important step for the development of flexible electronics. One of the strategies is to reduce the number of nucleation sites before the nuclei grow to flakes, also called the "seed selection" method.[44] Graphene nanoparticles initially are formed on oxidized Cu foil; some of them are then reduced during annealing in an Ar/H_2 environment. The remaining graphene nanoparticles become "seeds" for the growth of further hexagonol flakes.[45] Apart from oxidized Cu foil, single-crystalline Ni (111),[46] Cu(111),[47] Ge(110),[48] binary metal alloys,[49] and liquid metals[50] were used to synthesize single-crystalline graphene. The successful growth of single-crystal graphene can be basically attributed to (i) the elimination of grain boundaries of metal substrate to ensure uniformity of single-crystal structure of graphene and (ii) the high mobility of metal atoms that can eliminate the formation of nucleation vacancies or defects within graphene.

Second, the transfer of graphene from the substrate to the device is also worth investigating, because it may produce defects, such as wrinkles, or leave impurities.[51] To solve this problem, electrochemical methods were applied to intercalate gas bubbles between graphene and the metal substrate, resulting in easy removal of graphene.[52] Plasma treatment-generated gas bubbles were also applied to speed the etching of Cu foil in dilute etchant.[53] Although research on CVD graphene has made dramatic progress, reliable synthesis and clean transfer of single-crystalline graphene are still great challenges. The mechanism of single-crystalline graphene synthesis on various substrates is still not understood completely.

Third, three-dimensional (3D) macroporous graphene, graphene foam, has been synthesized from CVD graphene.[54] During synthesis, graphene was precipitated on the scaffold surface, such as a nickel foam,[14] and grew as a foam. Because of the difference between the thermal expansion coefficients of nickel and graphene, ripples and wrinkles formed on graphene foam.[14] Interestingly, these ripples and wrinkles could provide piezoresistivity to the foam. After incorporating the graphene foam into a liquid PDMS matrix and curing, the composite exhibited piezoresistivity, which indicates that when strain was applied to these graphene foam/PDMS composites, the resistance of the defects changed immediately, resulting in a high gauge factor of the sensor.[55]

8.2.2 Carbon nanotubes

CNTs can be viewed as hollow cylinders of rolled-up graphene sheets having a hexagonal lattice of sp^2-bonded carbon atoms. These can be synthesized using various methods, such as arc discharge,[56] laser ablation,[57] and CVD.[58] CNTs can be divided into single-walled (SCNTs), double-walled (DCNTs),[59] and multi-walled (MCNTs)[60] structures. The electronic properties of CNTs heavily depend on their graphene lattice orientation, also called chiral index (n, m), and thus CNTs can be divided as metallic ($m=n$ or $m-n$ is a multiple of 3) or semiconductive (all other cases).

After synthesis, roughly one-third of CNTs are metallic and others are semiconducting[61]; so a direct application of the mixture of metallic and semiconducting CNTs could degrade device performance. The alignment of CNTs can enhance the stretchability of electrical conductors.[62] In addition, the purity of semiconductive CNTs is essential to determine the performance of transistors.

8.2.2.1 CNTs for stretchable conductors and sensors

CNTs have a high aspect ratio ($>10^6$) and strong tensile strength ($\sim$100 GPa).[63] The intertwining and curvilinear configurations contribute to the flexibility of CNT films. These curvilinear structures can be straightened to accommodate tensile strain and maintain connection. Alignment can enhance the conductivity of CNT films in the uniaxial direction. Two strategies are commonly applied to align CNTs: (i) using a prestrain-release-buckling process[64] and (ii) aligned placement of vertical CNT forests onto the substrates.[65]

The prestrain-release-buckling process can generate aligned and wavy structures of CNTs. This process includes three steps: (i) straining an elastomer substrate, (ii) depositing or coating CNT films onto the prestrained elastomer substrates, and (iii) releasing the prestrain and shaping the CNT film into a wavy shape (Fig. 8.2A). During prestrain, the intertwining of CNTs is eliminated and all CNTs are oriented to the prestrain direction, resulting in alignment. Under stretching, the wavy structures of aligned CNTs are opened to accommodate strain, and thus contact between CNTs is retained. On the other hand, CNTs without prestrain will be straightened and lose contact between each other.[12] A stretchable conductor consisting of aligned wavy CNTs can exhibit much less resistance changes ($<6\%$) than flat conductors ($\sim 43\%$) with 100% stretching.[12] The prestrain method has been widely used to produce aligned and wavy structure of CNTs, which have been used in various stretchable electronics.[11b, 66]

In the second strategy, vertically aligned thin SWCNT films were utilized as a sensor.[67] SWCNTs were first grown by CVD, and then, these were laid side by side onto a flat PDMS substrate, aligned perpendicularly to the strain direction (Fig. 8.2B). After sealing with PDMS, the strain sensor exhibited excellent durability and stability. With stretching, some CNTs straightened and formed a mesh-like network, resulting in stretchability (Fig. 8.2B). Moreover, 3D "accordion-like" CNT forests (Fig. 8.2C_1) provided strong stretchability to CNT/polyurethane composites[13] (Fig. 8.2C_2). First, CNTs were grown on silicon wafers. Next, polyurethane polymer penetrated the forest to form a 3D "accordion-like" structure (Fig. 8.2C_3). Within a stretching cycling, skeletons of the "accordion-like" forest will open and close (Fig. 8.2C_4), which results in an extremely high failure strain (over 1400%) and high stability after 100 cycles of twisting or bending.

In addition, a method combining the two strategies provides both high alignment and stretchability of the CNT strain sensor.[68] First, a vertically aligned CNT array was grown by CVD. Second, an elastic Ecoflex strip was stretched to 100% as a prestrained substrate. Last, the aligned CNT

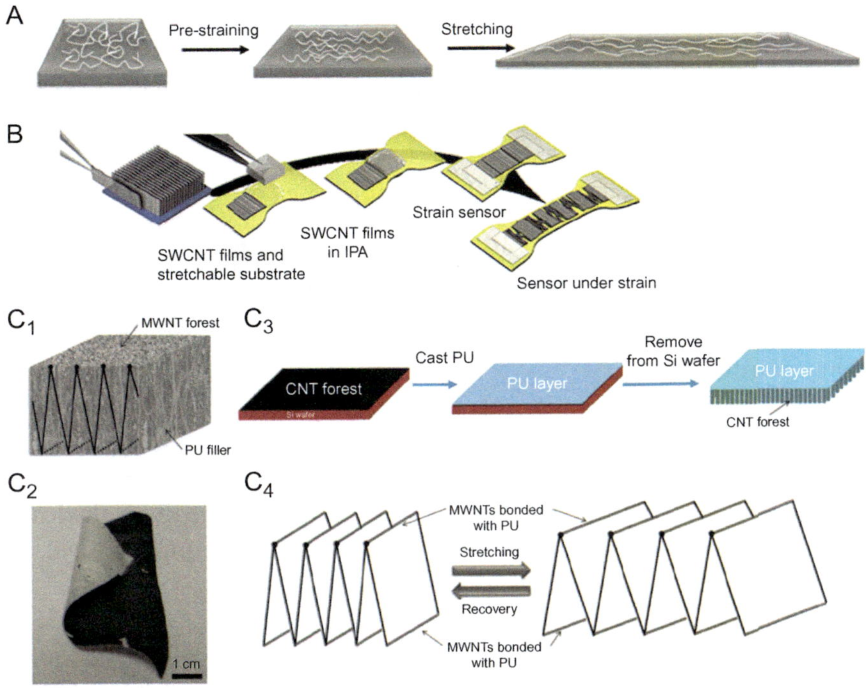

Fig. 8.2 (A) Sketch of a prestrained and stretched CNT network from 0% to 100% strain. (B) Schematic showing production of a SWCNT strain sensor. (C_1) Aligned MWNT forest interconnected with PU. (C_2) Flexible CNT/PU conductor. (C_3) Fabrication of the CNT/PU conductor. (C_4) Opening and closing of the 3D accordion structure. *Panel (A) reproduced with permission from Lin, L.; Liu, S.; Fu, S.; Zhang, S.; Deng, H.; Fu, Q. Fabrication of Highly Stretchable Conductors via Morphological Control of Carbon Nanotube Network. Small* ***2013****, 9, 3620–3629. Copyright 2013, Wiley-VCH. Panel (B) Copyright 2011, Nature Publishing Group. Panels (C_1–C_4) reproduced with permission from Shin, M. K.; Oh, J.; Lima, M.; Kozlov, M. E.; Kim, S. J.; Baughman, R. H. Elastomeric Conductive Composites Based on Carbon Nanotube Forests. Adv. Mater.* ***2010****, 22, 2663–2667. Copyright 2010, Wiley-VCH.*

array was attached onto the substrate by dry spinning. After releasing, the array was transformed to a wavy structure. The resistance of the array only increased by a factor of 2 until 400% strain was obtained, because (i) the prestrain-generated wavy structure can be straightened to maintain connection until stretching to 100%, and (ii) the alignment can ensure the connection of CNTs even until 900% strain.

8.2.2.2 Solution-based sorting of semiconducting CNTs

As mentioned previously, after synthesis, roughly one-third of CNTs are metallic and others are semiconducting[61]; so a direct application of the mixture of metallic and semiconducting CNTs could degrade device performance. Metallic CNTs (m-CNTs) have a great potential for electrical conductors, because of their low bandgap (<0.1 eV[62]). On the other hand, semiconducting CNTs are promising channel materials in field-effect transistors (FETs) because of their high bandgap (>0.5 eV[69]), on/off ratio (~10^4 to 10^6),[63] and high mobility (10^3 to 10^5 $cm^2\ V^{-1}\ s^{-1}$).[70]

To improve the purity of CNTs, solution-based methods have been proposed to separate SCNTs. The mechanism of such methods is using polymer dispersants that exhibit selective interactions with certain lattice orientations of SWNTs. Recently developed solution-based methods include gel chromatography,[71] conjugated polymer wrapping chromatography,[72] and ultracentrifugation.[73]

Gel chromatography uses columns to selectively adsorb semiconducting SWCNTs dispersed in a sodium dodecyl sulfate aqueous solution. These columns contain an allyl dextran-based size-exclusion gel that specifically interacts with semiconducting SWCNTs, but m-CNTs exhibit the lowest level of interactions with the gel. As a result, semiconducting SWCNTs are sorted and finally collected as unbound nanotubes.[74]

During chromatography, conjugated polymers selectively wrap the semiconducting SWCNTs, while the unwrapped SWCNTs and bundles remain in the sediment. The selective dispersion of semiconducting SWCNTs by conjugated polymers is a result of the differences in the polarizabilities between metallic SWCNTs (7.0×10^5 Å^2) and semiconducting SWCNTs (140 Å^2). When conjugated polymers, e.g., poly(3-dodecylthiophene), are wrapped with SWCNTs, complexes containing metallic SWCNTs have a higher affinity to aggregate with each other. As a result, the bundled metallic SWCNT-polymer complexes as sediment after centrifugation, while the semiconducting SWCNT-polymer complex remains in the suspension.[75]

Ultracentrifugation was also utilized to sort semiconducting SCNTs with the assistance of polymer surfactants.[76] For instance, Derenskyi et al. used poly(9,9-di-n-dodecylfluorenyl-2,7-diyl) (PF12) and poly(3-dodecylthiophene-2,5-diyl) (P3DDT), both of which have alkyl side chains with 12 carbon atoms, as polymer surfactants.[76] PF12 and P3DDT attached on the surface of semiconducting SWCNTs through π-interaction during ultrasonication. Using ultracentrifugation, contaminants, bundles, and undispersed metallic species were removed.

8.2.3 Silver nanowires

Despite the exciting progress of carbon-based stretchable conductors, their sheet resistance (ca. 400 Ω/sq[77]) is still higher than that of metal networks (ca. 20 Ω/sq[78]). For the development of current-driven devices, such as organic LEDs and solar cells, metal nanomaterials have become new candidates to provide low sheet resistance (<100 Ω/sq).[79]

8.2.3.1 Synthesis of silver nanowires

Due to their high electrical conductivity, ductility, and relatively easy synthesis, silver nanowires (AgNWs) are seen as a candidate to replace indium tin oxide for flexible electronic devices. Currently, polyol synthesis is the most widely used method to synthesize AgNWs. This reaction includes reducing metallic precursors (e.g., $AgNO_3$) with a polyol (e.g., ethylene glycol) in the presence of polyvinylpyrrolidone. At a high temperature, ethylene glycol ($HOCH_2CH_2OH$) is oxidized to aldehyde (CH_3CHO), which additionally reduces the silver ions to metallic silver atoms, as shown by the following reactions:

$$\text{(i)}\ 2HOCH_2CH_2OH \rightarrow 2CH_3CHO + 2H_2O\ \text{and}$$

$$\text{(ii)}\ 2Ag^+ + 2CH_3CHO \rightarrow CH_3COCOCH_3 + 2Ag + 2H^+$$

Then, the Ag atoms nucleate as seeds with nanocrystals and grow into nanorods.[80] Polyvinylpyrrolidone is then added and it adheres to the sides of the nanorods, whose ends are still open and absorb more atoms to grow. Meanwhile, various nucleating agents, e.g., HNO_3[81] or NaCl,[82] must be added to prevent the silver seeds from agglomerating into spheroids.

AgNWs are typically grown with diameters of 30–200 nm and lengths of 1–25 μm.[83] In a percolation network, contact resistance between nanowires is the main contributor of the main resistance. Thus, effort has been devoted to produce AgNWs with greater lengths. In 2008, Korte et al. reported a growth method for the synthesis of AgNWs with a length of 10–50 μm and uniform diameters of ~100 nm, where CuCl or $CuCl_2$ were added as nucleating agents to control the formation rate of Ag seeds and the growth rate of AgNWs.[84] In a further development, it was shown that a repeated reduction of $AgNO_3$ generated AgNWs having an average length and diameter of 96.1 μm and 160 nm, respectively.[21]

8.2.3.2 Enhancement of the conductivity of AgNW networks

Although AgNWs have attracted much research interest, mainly due to their high conductivity and mechanical ductility, there are two issues that can degrade AgNW properties in applications: (i) contact resistance between nanowires and (ii) poor adhesion between the nanowires and plastic substrates. The proposed solutions to these problems are introduced in this section.

Plasmonic techniques have been applied to weld AgNW junctions (Fig. 8.3A_1).[85] Plasmonic welding stimulates the mobility of atoms in nanowire junctions and results in epitaxial recrystallization of atoms (Fig. 8.3A_2). Contact resistance at junctions was reduced from ~5000 to ~2 kΩ after 60 s of welding.

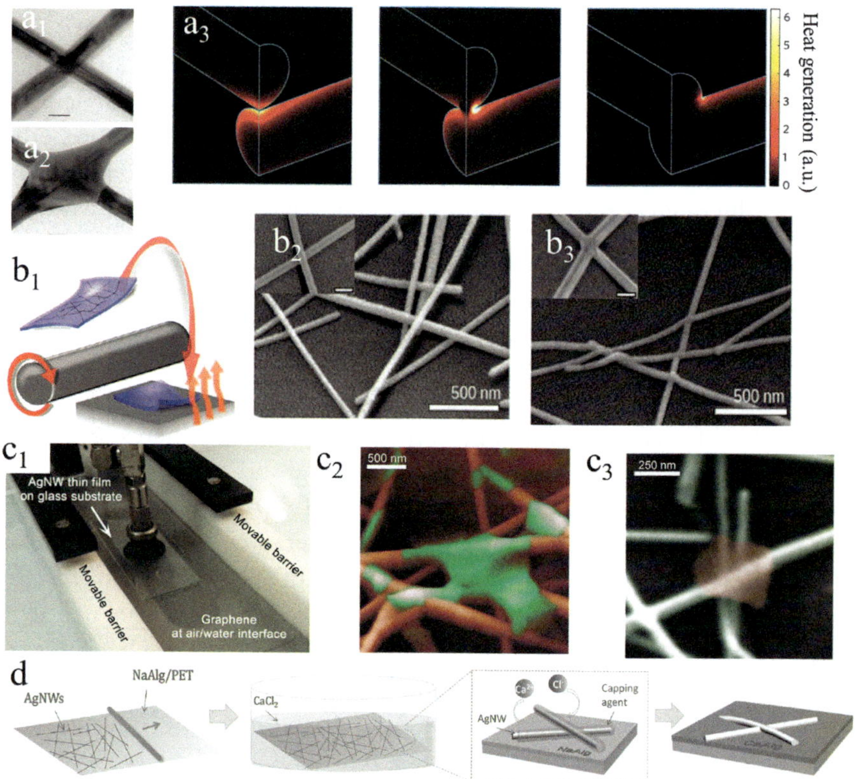

Fig. 8.3 (A_1 and A_2) TEM micrographs of as-made AgNW junction and (A_3) FEM simulation of optical heat generation at AgNW junctions during nanowelding in which heat generation occurs primarily in the bottom nanowire within a few nanometers of the gap.[85] (B_1) Schematic of hot rolling of AgNW networks on PET substrate with a steel rod. (B_2) AgNWs without rolling. (B_3) AgNWs heat-rolled at 165°C; inset, magnification of nanowire junctions.[86] (C_1) A photograph of LS horizontal depositing graphene film that forms on water-air interface onto a glass substrate containing AgNWs. (C_2) False color AFM height micrograph. (C_3) SEM micrograph of graphene-wrapped AgNW junctions.[87] (D) Schematic of the fabrication of flexible AgNW/CaAlg/PET film.[88] *Panels (A_1–A_3) Copyright 2012, Nature Publishing Group. Panels (B_1–B_3) Copyright 2013, American Chemical Society. Panels (C_1–C_3) Copyright 2014, Wiley-VCH. Panel (D) Copyright 2015, Wiley-VCH.*

In addition, hot rolling has been shown to be effective at improving the metallic contact (Fig. 8.3B_1). AgNWs were flattened and welded together at their intersection joints through hot rolling with a pressure 50 psi at 165°C (Fig. 8.3B_2 and B_3), resulting in a decrease in sheet resistance from ~5800 to ~17.5 Ω/sq.[86] These controllable welding processes could ensure welding effect and prevent damage to the substrate; so it has a potential to be applied in the fabrication of solar cells, light-emitting diodes, and touch screen display.

Apart from welding, bridging AgNWs by adding conductive materials is another strategy to reduce contact resistance. Selective nucleation and growth of silver nanoparticles at silver nanowire junctions has been reported. The chemically reduced silver atoms can nucleate, deposit, and grow at the junction region, which finally fill the gap between AgNWs.[89] The silver conductive nanonetwork after the treatment could achieve an optical transmittance of 89.4% and a sheet resistance of 14.9 Ω/sq.

Graphene is also used to bridge the gap between nanowires. For instance, Jurewicz el at. used graphene Langmuir films to cover the spray-coated AgNW films to boost the conductivity of a transparent film.[87] First, Langmuir films of graphene formed at air-water interfaces with 25 mN m^{-1} surface pressure. Next, a substrate containing AgNW networks was applied to fabricate a Langmuir-Schaefer (LS) horizontal deposition film, resulting in the transfer of graphene monolayers onto the AgNW network (Fig. 8.3C_1). This method decreased surface resistance of the film from 600 kΩ/sq at T=97% to 200 Ω/sq at T=93% because the deposited graphene sheets wrapped the nanowire junctions (Fig. 8.3C_2 and C_3) and reduced the wire-wire junction resistance.

Other materials used to bridge AgNWs include Au nanofilms, which were formed by galvanic displacement and can decrease junction resistance from 1 GΩ to 450 Ω[7]; GO sheets, which were wrapped with AgNWs to decrease sheet resistance from 1045 to 26 Ω/sq[90]; and SCNTs, which were coated on AgNWs and generated a sheet resistance of 203 Ω/sq of the film under bending.[91] Among these bridging materials, high-conductivity metals may be more effective than carbon-based materials. This is because (i) these metals, including Cu, Ag, and Au, have lower resistivity than carbon-based materials, and (ii) after coating, these metal particles or nanofilms can form intermetallic bonds with AgNWs, which is stronger than the van der Waals force between AgNWs and carbon-based materials.

Although the junction treatments have the capability to reduce resistance, weak adhesion between AgNWs and elastic substrates is still a problem. Under stretching or bending, these AgNWs may be easily wiped off from their substrate. Some treatment may be necessary to create strong adhesion with substrates.

$CaCl_2$ sintering was reported as a method to enhance adhesion between AgNWs and an elastic substrate.[88] Fig. 8.3D schematically describes the fabrication of a conductive, transparent, and flexible AgNW/CaAlg/PET film bonded with CaAlg gel. Sodium alginate (NaAlg) solution was first spin-coated onto a PET substrate. Subsequently, AgNW solution was coated onto the NaAlg/PET film, and then the film was immersed into $CaCl_2$ solution. During immersion, the chloride ion, Cl^-, detaches the capping agent (polyvinylpyrrolidone, added during AgNW synthesis) from the surface of AgNW. Meanwhile, metallic Ca^{2+} ions react with NaAlg to form water-resistant alginate gel (CaAlg) to bond AgNWs tightly on the substrate. As a result, the sheet resistances of flexible AgNW/PET films can be reduced from 300 to 30.6 Ω/sq with a transmittance ~92% and can be maintained nearly at a constant value during 1000 cyclic bendings. In addition, a UV-curable adhesive photopolymer can be used as a polymer substrate to form strong interactions with AgNWs, resulting in a low sheet resistance increase of 9.5% at 1 Ω/sq after 2000 cyclic bendings.[92] Silane modification[93] and deposition of GO[94] were also reported to enhance bonding between AgNWs and the substrates.

8.2.4 Gold nanoparticles and nanowires

Gold presents a natural material of flexible metal and low resistance. With a rapid development of metal synthetic and microelectronics systems, various Au nanomaterials, including nanoparticles, nanowires, and nanomesh, have attracted great interest in terms of the development of flexible/stretchable electronic devices.[95]

8.2.4.1 Gold nanoparticles

Gold nanoparticles are produced by chemical reduction of chloroauric acid ($HAuCl_4$), in a process known as the Turkevich method.[96] The reductant, e.g., triisopropylsilane (TIPS), can reduce Au^{3+} ions to Au^+ ions. Then, a disproportionation reaction occurs to convert 3 Au^+ ions to Au^{3+} and 2 Au^0 atoms. The Au^0 atoms act as nucleation seed and bind with dissolved Au^+ ions to grow gold nanoparticles.[97]

After incorporation of AuNPs into an elastic matrix, these particles have a great capacity for reversible nanoscale restructuring. When the matrix is deformed, nanoparticles can rearrange themselves to maintain electrical connection. Flexible electrode arrays (Fig. 8.4A_1) were fabricated by inkjet

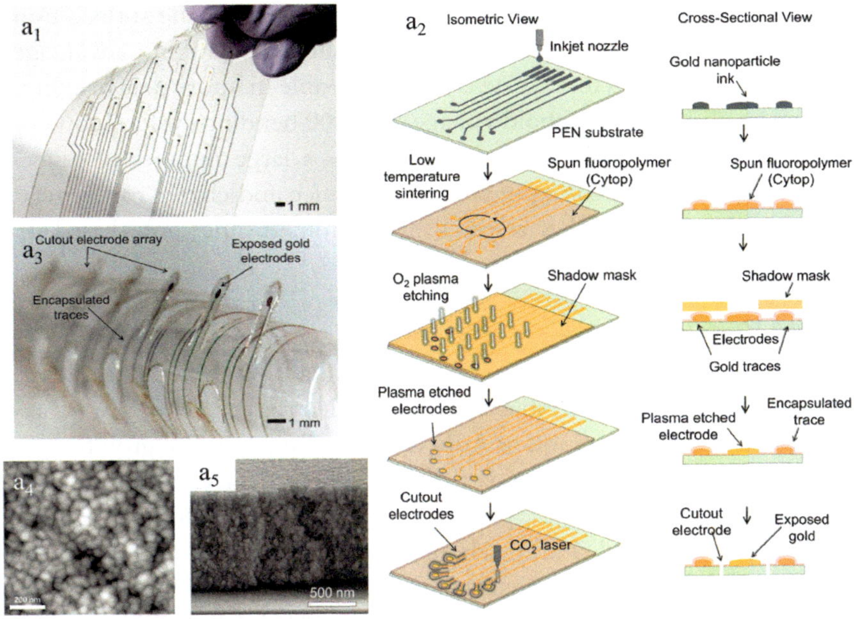

Fig. 8.4 (A_1) Inkjet-printed flexible gold electrode array. (A_2) Fabrication process of the printed electrode array and with cross-sectional views. (A_3) Cutout electrodes after being bent to a 5-mm radius. (A_4) AFM micrograph of the surface of printed AuNP line. (A_5) SEM micrograph of the cross-section of printed AuNP line.[98] *Copyright 2015, Wiley-VCH.*

printing AuNP ink onto a polyethylene naphthalate substrate.[98] In brief, the fabrication procedure included (i) AuNP ink in a petroleum hydrocarbon solvent printed to produce conductive interconnections, (ii) fluoropolymer layer spin-coated to encapsulate gold interconnections, (iii) shadow-masked O_2 plasma treatment to etch the fluoropolymer layer to produce exposed electrode, and (iv) using a CO_2 laser to cut around the electrodes to create freestanding flaps (Fig. 8.4A_2). The cutout electrodes (Fig. 8.4A_3) demonstrated higher mechanical flexibility than the array without flexure cuts, because these can extend like freestanding flaps. The resulting electrodes have very low surface resistance of 0.30–0.34 Ω/sq due to the highly packed structure of AuNPs (Fig. 8.4A_4 and A_5). No significant change in resistance during bending and twisting tests was observed. Moreover, benefiting from a similar highly packed structure, an Au/Pu film had a conductivity as high as 6800 and 2400 S/cm until extension strain reached 110%.[99]

The AuNP film produced by LS deposition also served as gate electrodes in flexible FETs.[100] A 2D AuNP monolayer film is initially formed on the surface of water. Subsequently, the layer is lifted from the surface using a PDMS pad and transferred to a designed substrate. Taking advantage of well-ordered AuNPs in the gate layer, the flexible memory device exhibited almost no change in threshold voltage after 500 bending cycles. However, a nanoparticle-based conductive film contains a large number of junctions between particles, resulting in contact resistance and charge carrier scattering. For this reason, other research has examined other gold-based structures, including gold nanowires.[101]

8.2.4.2 Gold nanowires

Gold nanowires (AuNWs) have been adopted as an important candidate for the development of flexible/stretchable electronics. There are various methods for the preparation of AuNWs, including template growth,[102] oriented attachment of Au nanoparticles,[103] and seed-mediated growth.[104]

For example, Feng et al. produced single-crystalline, ultrathin AuNWs based on a modified Turkevich method.[104a] In brief, a gold precursor ($HAuCl_4$) was gradually reduced by triisopropylsilane (TIPS) in the presence of oleylamine. After reduction, oleylamine-gold halides (AuCl) were rapidly formed. Then, the polymeric strands of AuCl complex are grown along the [111] direction. The polymer can be described as a linear chain composed of a AuCl ion-rich backbone surrounded by oleylamines. After aging the solution at room temperature for 2 days, the solution turned from yellow to dark red (Fig. 8.5A_1 and A_2), indicating complete reduction of the Au^+-amine complex into oleylamine-capped AuNW dispersion in hexane (Fig. 8.5A_3). Electron microscopy studies showed each nanowire is single crystalline growing along the [111] direction (Fig. 8.5A_4) with an inter-fringe distance of ~0.23 nm and a width of ~2.5 nm. The nanowires have a diameter of ~2.5 nm and length in tens of micrometers.[105]

Using long AuNWs, Cheng's group fabricated a nanomembrane via the Langmuir Blodgett technique. The nanomembrane has an optical transmittance of 90%–97% in the spectral window of 300 – 1100 nm, an electrical resistance of ~1142 kΩ/sq, and a breaking strength of ~14 Nm^{-1}.[105] Using this material, a wearable pressure sensor (Fig. 8.5B_1) was fabricated by sandwiching an AuNW-impregnated tissue paper between a PDMS sheet and an electrode array, as shown in Fig. 8.5B_2.[101] This device can exhibit strong piezoresistivity, as shown in Fig. 8.5B_3: under an external pressure, a small compressive deformation of the tissue paper enabled more AuNWs to be in contact with electrodes, leading to higher current. Upon unloading, both PDMS and tissue paper recovered to their original

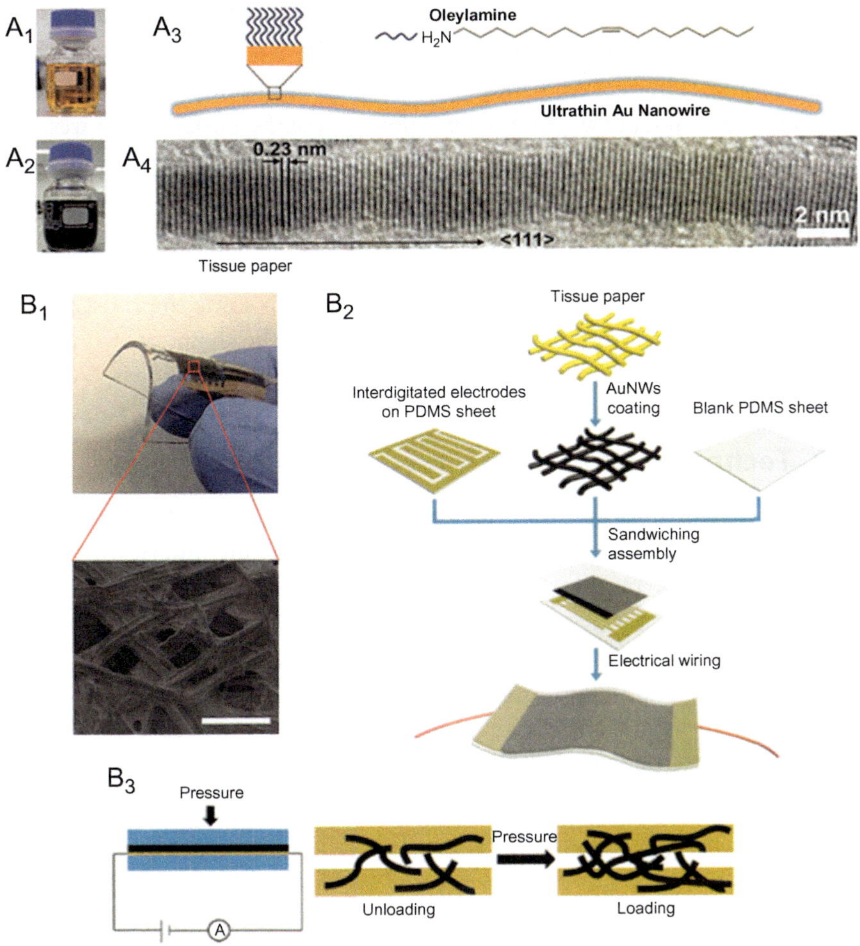

Fig. 8.5 Photographs of AuNW suspension at (A_1) initial and (A_2) final stage of reaction. (A_3) Schematic of oleylamine-capped AuNWs. (A_4) High-resolution TEM micrograph of an individual single-crystalline ultrathin AuNW.[105] (B_1) Photograph showing the bendability of the sensor and SEM micrograph of AuNW-coated tissue fibers. Scale bar, 100 μm. (B_2) Schematic showing the fabrication of a flexible AuNW sensor and (B_3) the piezoresistivity mechanism. *Panel (A_1–A_4) Copyright 2013, Wiley-VCH. Panels (B_1–B_3) reproduced with permission from Gong, S.; Schwalb, W.; Wang, Y.; Chen, Y.; Tang, Y.; Si, J.; Shirinzadeh, B.; Cheng, W. A Wearable and Highly Sensitive Pressure Sensor With Ultrathin Gold Nanowires. Nat. Commun.* ***2014****, 5, 3132. Copyright 2014, Nature Publishing Group.*

shapes, reducing the amount of AuNWs bridging the finger electrode pairs, leading to a decrease in the current. Because long AuNWs possess low contact resistance and high flexibility, these can be also applied in other flexible/stretchable electronics, e.g., strain sensors[106] and photodetectors.[107]

Admittedly, some challenges remain. One of the difficulties is to reduce impurities and improve the yield of gold nanomaterial-forming reactions. For example, spherical Au nanoparticles always exist when using oleylamine to synthesize Au nanowires, reducing the yield of the target nanowires. Second, large-scale production is also hard to achieve because of the complicated synthesis processes. Third, it is still difficult to characterize nanowires (~2 nm) and nanosheets (thickness <5 nm), because under electron beam irradiation, these tend to melt or undergo phase transformation. Despite these limitations, Au nanomaterials remain some the most promising to service in stretchable electronics with the help of frontiers of technology.

8.3 Techniques for realizing stretchability

It is often critical to assemble nanomaterials into particular structures that facilitate stretchability, because these materials generally inherit low stretchability, such as <1% for silicon,[108] <10% for CNT and for graphene.[27] Currently, three kinds of morphologies have been extensively utilized to accommodate deformations: (i) wave or spring-based configurations, (ii) serpentine-based configuration, and (iii) fiber-based configurations.[109]

8.3.1 Techniques for fabricating wavy structural configurations

The use of a wavy structural configuration has become an effective strategy to achieve mechanical stretchability. It can be generated by a prestrain-release-buckling process, as shown in Fig. 8.6A. This configuration can accommodate large tensile and compressive strains by changing wave amplitude and wavelength. The strategy has been applied in fabricating AgNWs/PDMS,[113] CNT/PDMS,[114] and Si/PDMS conductors.[11a] The maximum applied strain, ε_{max}, can be represented as:

$$\varepsilon_{max} = \varepsilon_{pre} + \varepsilon_{fracture} \tag{8.1}$$

where ε_{pre} is the prestrain and is equal to the strain the electronics can be extended, and $\varepsilon_{fracture}$ is the fracture strain of the material. A theoretical calculation of wave amplitudes and wavelengths during prestrain was reviewed in Refs. 109, 115.

Strain distributions of electrodes with different wavy configurations were analyzed by finite element modeling,[111] and the results are shown in Fig. 8.6B. When a flat electronic is stretched, large stresses will be induced, resulting in cracks (Fig. 8.6B$_1$). On the other hand, wavy structures can

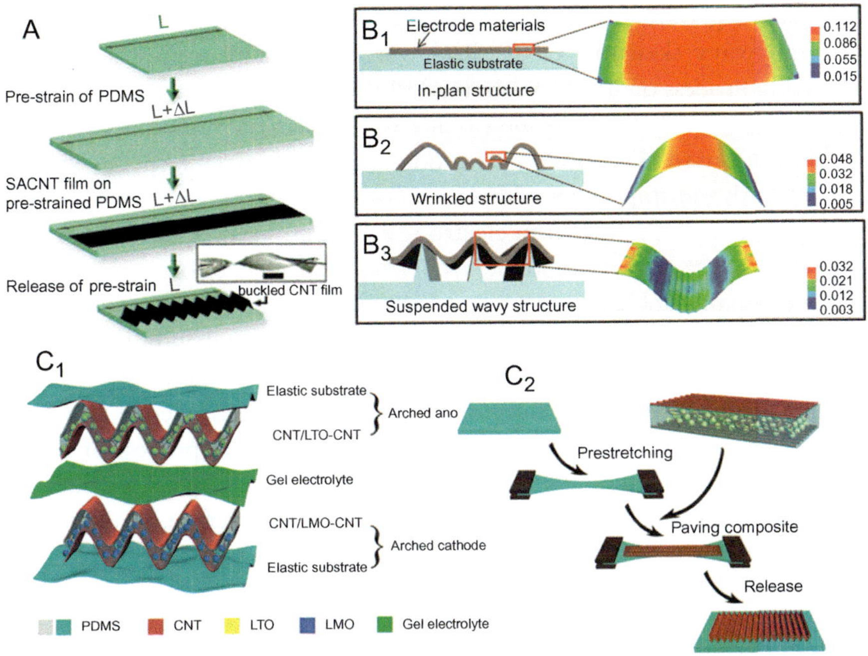

Fig. 8.6 (A) Prestrain-release-buckling process used to fabricate a flexible conductor containing wavy superaligned CNTs; inset, image showing the flexibility of the conductor.[110] (B) Electrodes and their strain distribution for (B_1) flat structure, (B_2) wavy structure adhered on the substrate, and (B_3) suspended wavy structure.[111] (C_1) Schematic structure of a multilayered lithium-ion battery using a wavy electrode. (C_2) The prestrain-release-buckling process to fabricate the battery.[112] *Panel (A) Copyright 2015, Royal Society of Chemistry. Panels (B_1–B_3) and (C_1, C_2) Copyright 2015 Wiley-VCH.*

release stress via changes in wave amplitude. However, if a wavy structure is adhered to a stretchable substrate, the adhered portion will continue to suffer strain, which can limit the stretchability of the device (Fig. 8.6B_2). If a wavy structure could be suspended instead of adhering it on a stretchable substrate directly, stress in the electrode material could be reduced, resulting in enhanced stretchability (Fig. 8.6B_3).[111] Notably, the suspended arrangement can enhance stretchability, but how to avoid delamination from the substrate is still a problem. Therefore, a balanced arrangement of providing both adhesion to the substrate and enough space to change wave amplitude could be the best strategy to achieve high stretchability and reliability.

Taking advantages of the wavy configuration, stretchable electrodes that can bear a 400% strain and exhibit stable electrochemical performance in lithium-ion batteries were developed.[112] The multilayer wavy structure is shown in Fig. 8.6C_1, in which two aligned CNT films act as a current collector and wavy skeleton. CNT films were formed by a prestrain-release-buckling process on 450% prestrained PDMS substrates (Fig. 8.6C_2). Inside the electrode, lithium manganese oxide and lithium titanium oxide nanoparticles were used as cathode and anode active materials, respectively. The wavelength and amplitudes of the wavy configuration change during stretching, greatly reducing strain in electrode materials. Consequently, resistance increases <20% when the strain is <400%, and the electrochemical performance remains stable after 500 stretching cycles with a strain of 400%.

Although the wavy structure can realize high stretchability, there exist some drawbacks: (i) The wavy structure can only enhance stretchability in one direction, so sophisticated engineering design is needed to achieve multi-, even omnidirectional, stretchability. (ii) In the process of release, it is still practically difficult to control buckling, especially with a high pre-strain condition. Therefore, engineering of wavy structure has a space for improvement in the future.

8.3.2 Techniques for fabricating serpentine interconnect structural configuration

Serpentine configurations have been introduced for embedding metal nanowires into elastic substrates. Distinguished from the "wavy" structure in which strain is modulated through changing wave amplitude direction, the serpentine configuration accommodates strain by simultaneously changing both the wavelength and amplitudes of bound nanowire.

The geometrical parameters of a periodic serpentine ribbon are illustrated in Fig. 8.7A_1: the serpentine loop is defined by the ribbon width w, arc radius R, arc angle α, and arm length l. The end-to-end distance S of a serpentine unit can be defined as:

$$S = 4\left(R\cos\alpha - \frac{l}{2}\sin\alpha\right) \tag{8.2}$$

where R is the arc radius, α is the arc angle, and l is the straight part length.

When the ribbon is stretched along the horizontal direction, the arc segment of serpentine will be elongated, and the straight segment will be compressed, resulting in a tensile displacement, u_0. The applied strain, ε_{app}, is defined as:

$$\varepsilon_{app} = 2u_0/S \tag{8.3}$$

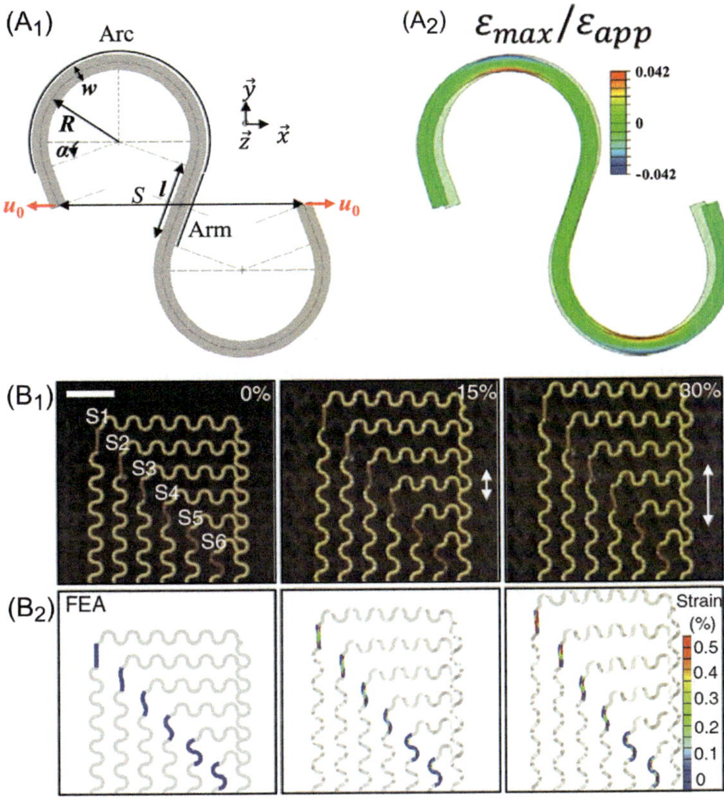

Fig. 8.7 (A_1) Geometric parameters and boundary conditions of a freestanding, thick serpentine unit cell. (A_2) Finite element models of strain distribution in a serpentine unit cell with maximum strain at the inner crest of the arc.[116] (B_1) Photographs of SiNR strain gauges under different strains (0%, 15%, and 30%) and (B_2) the corresponding finite element analysis results. Scale bar, 1 mm.[117] *Panels (A_1) and (A_2) Copyright 2015, Elsevier Ltd. Panels (B_1) and (B_2) Copyright 2014, Nature Publishing Group.*

where u_0 is the tensile displacement at each end, and S is the end-to-end distance of a serpentine unit.

The stretchability of a serpentine unit is defined as the critical strain beyond the intrinsic rupture strain of the material.[118]

Strain distribution in a representative serpentine unit cell is given by Fig. 8.7A_2. It is obvious that the maximum strain always occurs at the inner crest of the arc. The mechanism of a serpentine under stretching is described.[116]

The effect of strains on different serpentine configurations was examined by applying a range of strains (0%, 15%, and 30%) to silicon nanoribbon (SiNR)-based strain gauges.[117] These strain gauges consisted of six kinds of serpentine configurations with curvatures of 0 (S1), 1.94 (S2),

4.74 (S3), 7.4 (S4), 9.75 (S5), and 10 mm^{-1} (S6) (Fig. 8.7B_1). According to finite element analysis results, as the applied strain increases, the strain gauges with small SiNR curvatures experience considerably greater strains compared with those with larger curvatures (Fig. 8.7B_2). The serpentine designs with larger curvatures relieve more strain than smaller ones. The SiNR S6 sustains strains up to ~30%, whereas S1 fractures at ~10% strain. This result could indicate that the serpentine configuration with larger curvature enables the material to endure larger deformation and thus is suitable for the production of stretchable conductors.

Some stretchable devices were fabricated based on the serpentine structure. For example, to maximize the system-level softness and stretchability of a stretchable heater, a serpentine structural mesh of AgNWs was formed as shown in Fig. 8.8A_1.[119] AgNWs were uniformly mixed within polystyrene-butadiene-styrene elastomer. The mixture was then cast on a serpentine-patterned PDMS mold to form the heater. After adding copper wires and conductive paste, it was further encapsulated by two polystyrene-butadiene-styrene encapsulation layers that were also formed by patterning polystyrene-butadiene-styrene solution in a PDMS mold (Fig. 8.8A_2). Unlike film-type encapsulation, the serpentine type polystyrene-butadiene-styrene encapsulation could induce less stress during stretching and thus enhance stretchability. As a result, the stretchable heater can endure 5000 cycles of ~30% tensile stretching without observable change in resistance or joule-heating characteristics. The serpentine type encapsulation design may be more widely utilized in wearable devices in the future because it can form more stable and conformal contact with skin than simple film-like devices.

Serpentine electrodes (Fig. 8.8B_1) were produced by E-beam deposition of gold atoms through a shadow mask (Fig. 8.8B_2).[120] These gold-patterned electrodes were additionally coated on collagen film and used as electrodes, interconnections, localized heaters, or strain gauges. Because of the serpentine structure, serpentine resistors exhibited stable resistance (>99%) (Fig. 8.8B_3).

Moreover, serpentine mesh layouts of electrodes and interconnects were applied in multifunctional devices to protect the functioning of rigid active components, such as LEDs, photodetectors, electrocardiogram sensors, electromyographic sensors, etc., under deformation[121] (Fig. 8.8C_1). Under stretching, these serpentine layouts are first twisted and extended out of the plane of the substrate. Then, these will change shape to accommodate tensile strain (Fig. 8.8C_2)[18] and protect rigid active components. The serpentine mesh-shaped interconnection provided strong flexibility to the multifunctional devices that can accommodate various mechanical motions of the human skin[122] (Fig. 8.8C_3).

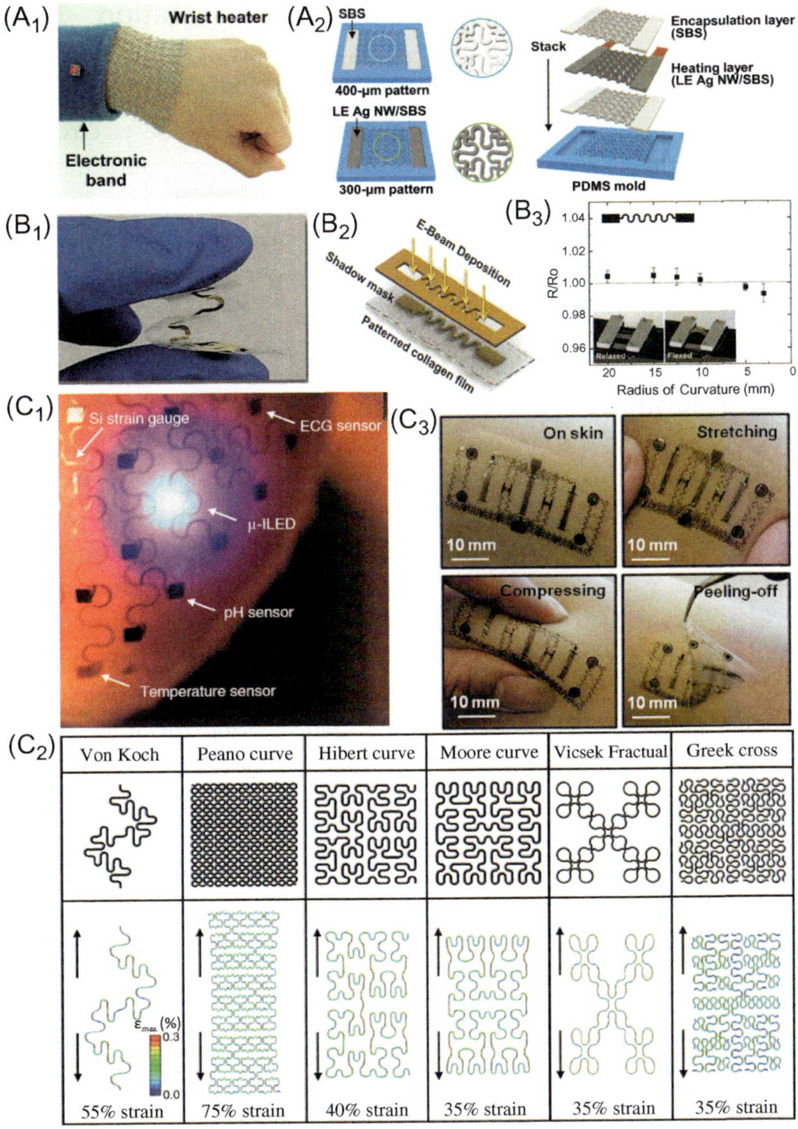

Fig. 8.8 (A_1) Photograph of stretchable heater and (A_2) schematic of heater fabrication.[119] (B_1) E-beam deposition of gold atoms to form serpentine electrodes, (B_2) flexible gold electrodes and (B_3) resistance change versus radius of curvature.[120] (C_1) 3D multifunctional integumentary membrane with ECG sensor, strain gauge, LED, pH sensor, and temperature sensor, enveloping a rabbit heart.[121] (C_2) Models of serpentine layouts and finite element analysis images of each structure under stretching.[18] (C_3) Photographs of a multifunctional device mounted on the forearm, under stretching, compression, and peeling off.[122] *Panels (A_1) and (A_2) Copyright 2015, American Chemical Society. Panels $(B_1–B_3)$ and (C_3) Copyright 2015, Wiley-VCH. Panel (C_1) and (C_2) Copyright 2014, Nature Publishing Group.*

8.3.3 Techniques for fabricating fiber configuration

In the past decade, by integrating conductive materials into conventional fiber structures, new types of stretchable/wearable electronics have emerged, such as strain sensors, energy storage devices, electronic skin, etc.[123] Generally, these fiber-structured electronics are fabricated by two strategies: (i) processing conductive materials into a fiber shape and then weaving them into textile structures and (ii) utilizing un-conductive fiber materials as substrates and integrating or depositing stretchable conductive materials into or onto the substrates.

8.3.3.1 Weaving conductive materials to produce stretchable fiber

Vertical aligned CNT arrays have the merits of being lightweight, having high specific strength, and the ability to create an ideal mesoporous structure for high electrolyte accessibility. Thus, CNT fibers woven from CNT arrays have become popular substrates for constructing fiber electrodes or conductors.[124] A spring-like yarn electrode can be formed by spinning aligned MCNT fibers.[125] First, aligned MCNTs were spun into a CNT fiber. Next, these parallel aligned CNT fibers were twisted as a large spring-like yarn, which consisted of coiled loops of CNT fibers along the yarn axis (Fig. 8.9A_1 and A_2). In each coil loop, CNTs were highly aligned along the helical direction (Fig. 8.9A_3). The spring-like coil structure provides enough space for fiber movement, so the spring-like yarn can be stretched to 100% strain without breaking individual CNTs (Fig. 8.9A_4). According to the fabrication process as shown in Fig. 8.9B, aligned CNT yarns were also woven into textiles and further electrodeposited with PANI to obtain electrodes for supercapacitors.[126] The resulting textile supercapacitor is able to maintain its specific capacitance of 272.7 F g^{-1} at 96.4% after 200 cyclic bendings.

AgNWs are also used as a starting material to obtain highly conductive stretchable fibers. It is hard to weave AgNWs directly; so AgNWs need to be integrated into an elastic matrix and then woven into fibers. Recently, highly stretchable AgNWs-AgNPs (Ag nanoparticles) embedded into styrene-butadiene-styrene (SBS) elastomeric fibers were reported. These can be used as stretchable conductors and electrodes in strain sensors.[127] The fabrication procedure is shown in Fig. 8.9C. First, AgNW-mixed SBS solution was injected into polyvinyl alcohol (PVA) liquid and coagulated as fibers, which was twisted by a wet spinning method. Next, the AgNP precursor was absorbed onto the AgNW-mixed SBS fiber, and then converted into AgNPs directly inside the fiber. The composite fiber did not

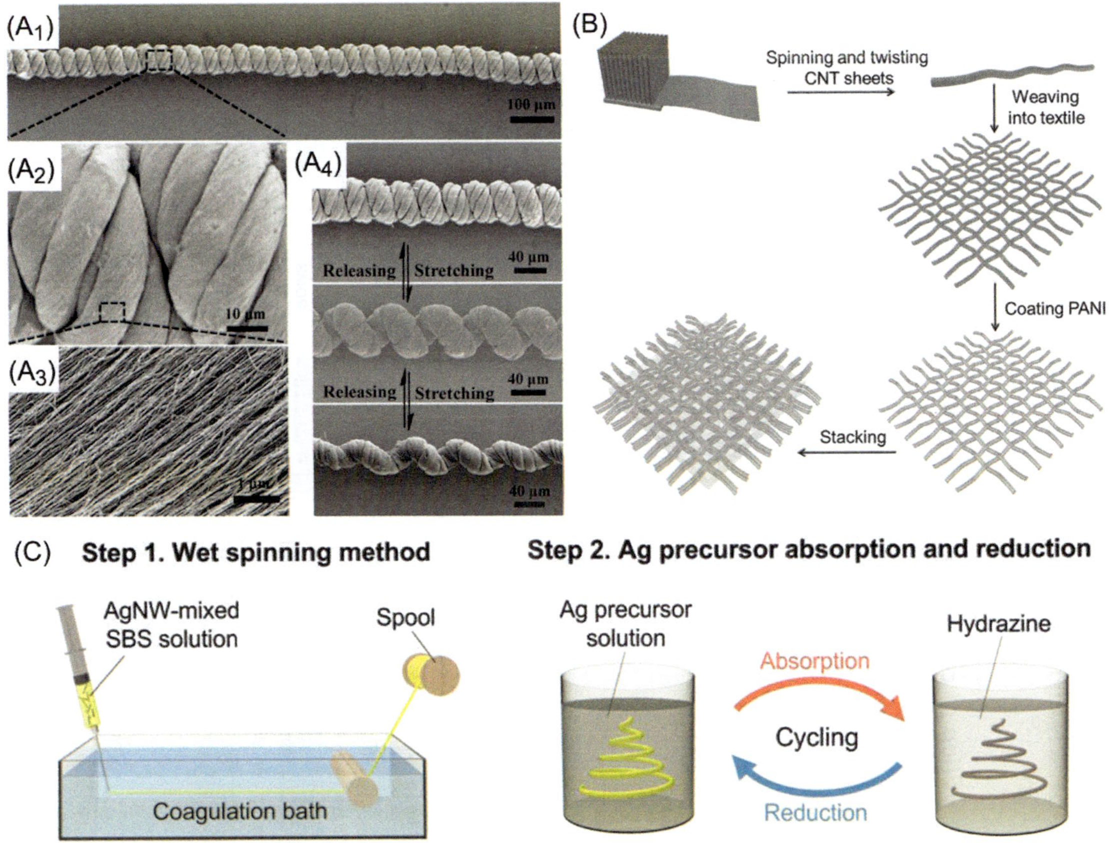

Fig. 8.9 SEM micrographs (A_1 and A_2) of spring-like yarn composed of (A_3) aligned CNT fibers. (A_4) Morphology change of a yarn during a stretching release process with a strain of 100%. (B) Schematic showing the fabrication of a textile supercapacitor based on CNT/PANI composite fibers.[126] (C) Schematic of the fabrication of a highly stretchable conductive device consisting of AgNW-AgNP-embedded SBS fibers.[127] *Panels (A_1–A_4) reproduced with permission from Zhang, Y.; Bai, W.; Cheng, X.; Ren, J.; Weng, W.; Chen, P.; Fang, X.; Zhang, Z.; Peng, H. Flexible and Stretchable Lithium-Ion Batteries and Supercapacitors Based on Electrically Conducting Carbon Nanotube Fiber Springs. Angew. Chem.* ***2014****, 126, 14792–14796. Copyright 2014, Wiley-VCH. Panel (B) Copyright 2014, Wiley-VCH. Panel (C) Copyright 2015, Wiley-VCH.*

lose its electrical connections until 220% stretching, because the AgNWs can bridge the cracks in the AgNP network.

In addition, flower-shaped Ag nanoparticles (Ag nanoflowers) and PU were employed to produce stretchable conductive fibers using a scalable wet spinning process. The resulting Ag nanoflowers/PU composite fibers exhibited extremely high conductivity of 41,245 S/cm and the lowest rupture strain ~90%.[128]

8.3.3.2 Deposition of conductive materials as sheath in sheath-core fibers

The coupling of elastic textile materials as core and high conductive nanomaterials as sheath can combine the positive attribute of each material, realizing high conductivity, stretchability, and cost-effectiveness. For example, CNT has been deposited onto elastic fiber substrates to produce a material with stretchability and conductivity for wearable electronics.[124a] Peng's group has produced a superelastic fiber-shaped supercapacitor according to the process in Fig. 8.10A.[129] Initially, prestrained elastic Ecoflex fiber

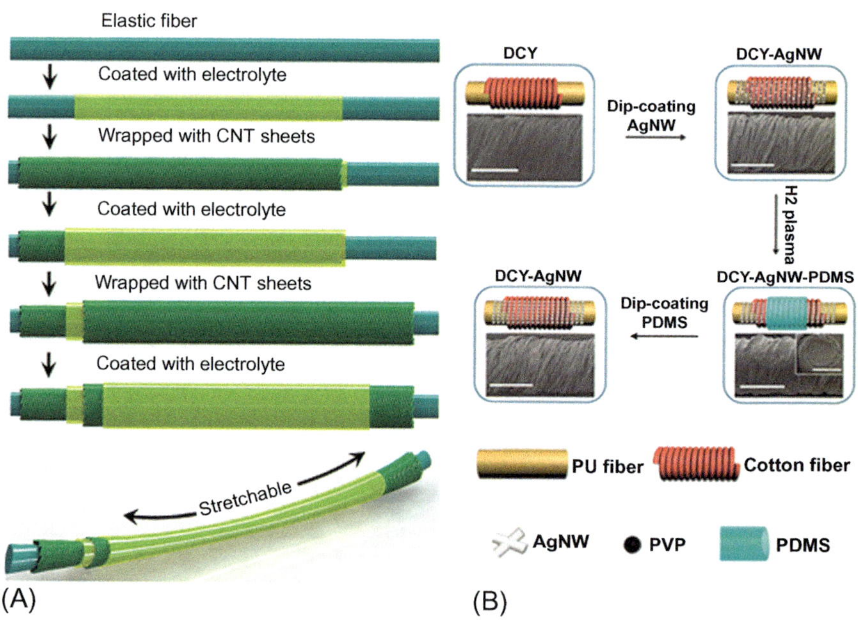

Fig. 8.10 (A) Illustration of the fabrication of a highly stretchable fiber-shaped supercapacitor with core-sheath structure.[129] (B) Fabrication of a AgNW/yarn fiber. *Panel (A) Copyright 2013, Wiley-VCH. Panel (B) reproduced with permission from Cheng, Y.; Wang, R.; Sun, J.; Gao, L. Highly Conductive and Ultrastretchable Electric Circuits from Covered Yarns and Silver Nanowires. ACS Nano* ***2015****, 9, 3887–3895. Copyright 2015. American Chemical Society.*

was used as core and CNTs and polyaniline were deposited onto the CNTs as sheath. These polyaniline CNT/Ecoflex fibers served as the inner electrode that was coated with a thin layer of PVA gel electrolyte. Next, the CNT/PANI layer was deposited on the PVA electrolyte as the outer electrode. The second PVA electrolyte was finally coated to obtain the final fiber-shaped supercapacitor. During stretching, these CNTs will first separate following the straightening of Ecoflex core, then slide away with and rearrange themselves to accommodate strain. By virtue of the highly stretchable structure, the PANI/CNT/Ecoflex fiber is able to maintain illumination of an LED under a dramatic stretching strain of 400%.[130]

Liu et al. wrapped the CNTs along the rubber core coaxial direction; after the release of a high prestrain (1400%), periodic hierarchical buckling in two dimensions of the CNT sheath will be generated. Due to the wavy buckled sheath and elastic core, during stretching, the wavy structures inside the sheath will be stretched, sheath coils will be separated, and individual nanotubes or nanofibers will rearrange themselves. Due to the unique coiling of these CNTs onto the rubber core, the resistance of the fiber only reduced by 5.01% with 3000% strain.[131]

Aside from CNTs, AgNWs can be wrapped with elastomeric fibers.[128] A conductive and stretchable composite AgNW/yarn fiber was prepared by the procedure shown in Fig. 8.10B.[15] The substrate is a double-covered yarn (DCY) consisting of PU fiber as core and cotton yarn that forms a spiral configuration onto the PU fiber. The PU fiber can provide superb elasticity, and the spiral-shaped cotton yarn can provide a large prestrain (>500%), high surface area, and hygroscopicity to absorb AgNWs. Then, AgNWs were dip-coated onto the DCY surface and formed a thin sheath. After encapsulation with a PDMS layer, the AgNW/yarn fiber was finally produced. The encapsulation of PDMS was able to prohibit buckling of AgNWs during deformation of the fiber. Because of the high elasticity of PU fiber, the spiral configuration of cotton yarn, and the high stretchability of AgNWs, the DCY/AgNW fibers can maintain conductivities with 500% strain and 1000 cycles of 200% strain, although the conductivity did experience a gradual decline.

8.4 Application of stretchable electronics

The applications of conventional electronics are sometimes restricted due to their rigid and brittle nature. In daily life, electronics with high flexibility and deformability are highly desirable. Stretchable electronics could simultaneously satisfy the demand of electronic functions and mechanical deformation.

This section presents some typical examples of stretchable electronics, including stretchable LEDs, strain sensors, artificial skin, supercapacitors, and FETs.

8.4.1 Stretchable LED arrays

Stretchable LED arrays could be one the most visible and convenient applications of stretchable electronics. For example, coating PDMS on an AgNW film results in an AgNW/PDMS conductor.[113] Through repetitive stretching/releasing cycles, a periodic wavy pattern appeared on the surface of the conductor (Fig. 8.11A_1). Using two patterned AgNW/PDMS conductors, a simple LED circuit, as shown in Fig. 8.11A_2, was fabricated. The LED was able to retain its illumination intensity until the AgNW/PDMS conductors were stretched to a strain of 50% (Fig. 8.11A_3) or folded, because (i) the wavy structure can accommodate stretching strain, as mentioned in Section 8.3.1, and (ii) these AgNWs can slide against each other and be entangled, creating microdeformation in order to maintain percolation networks.

Apart from film-based conductors, AgNW/yarn fibers fabricated by the procedure shown in Fig. 8.10C were utilized as a conductor to connect 4×3 LEDs.[15] As shown in Fig. 8.11B, all LEDs can function with bending and 30% stretching strain. The success of this stretchable LED array is mainly determined by three factors: (i) using flexible conductive materials, AgNWs,

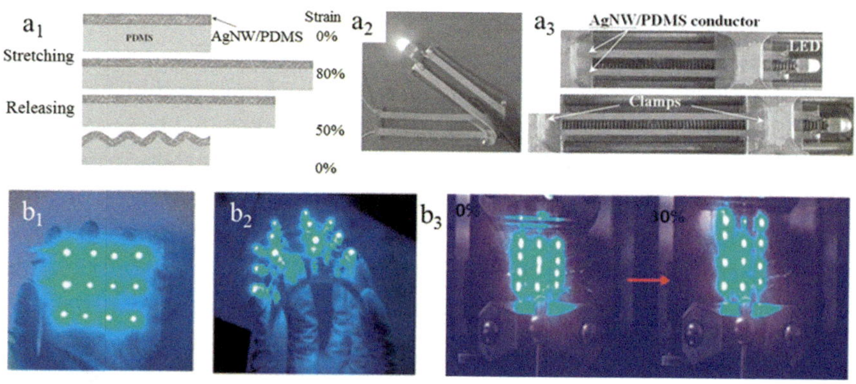

Fig. 8.11 (A_1) Stretching-releasing to form a wavy structure on the AgNW/PDMS conductor surface. Photographs of the LED circuit under (A_2) bending and (A_3) 0% and 30% tensile strain. (B_1) Illuminated stretchable LED arrays using AgNW/yarn fibers as interconnections; the LED arrays kept their illumination intensity when (B_2) bent and (B_3) stretched to a strain of 30%.[15] *Part (A) reproduced with permission from Xu, F.; Zhu, Y. Highly Conductive and Stretchable Silver Nanowire Conductors. Adv. Mater.* ***2012****, 24, 5117–5122. Copyright 2013, Wiley-VCH. Panels (B_1–B_3) Copyright 2015, American Chemical Society.*

and stretchable core materials (PU fiber and cotton yarn), (ii) producing a unique stretchable core-sheath fiber structure, and (iii) fabricating stable connections between stretchable conductors and LEDs, which was formed with silver paste.

For the development of LED arrays, realizing the circuit design could be a challenge, because conventional print or filtration methods could cause limited resolution and severe constraints on pattern configuration. Recently, precise circuits were completed by a method combining pattern printing and AgNW deposition.[132] As shown in Fig. 8.12A_1, first, a desired circuit pattern of wax is printed onto PVDF membranes. Second, through filtration, AgNWs were deposited onto the circuit part of the membrane that was not covered by wax. Third, the AgNW-based circuit was transferred onto a PDMS substrate. The pattern was fully transferred without any visible defects, and the whole substrate was soft and can be repeatedly stretched (Fig. 8.12A_2 and A_3). Using this method, two kinds of AgNW conductors are accomplished by transferring two AgNW circuits onto PDMS substrates (Fig. 8.12A_4). Last, 5×5 LEDs are mounted onto the conductors to make an LED array (Fig. 8.12A_5). The LED arrays can be stretched to about 20% strain (Fig. 8.12A_6) and wrapped around cylinders without losing functionality.

Other than AgNWs, SCNT gel was used as an electrode material. SWNTs were patterned on a PDMS substrate by printing the gel through shadow masks to form stretchable conductors (Fig. 8.12B_1).[133] After integrating a driving cell active matrix and organic LEDs with printed stretchable conductors, a 16×16-pixel stretchable active-matrix organic LED display was fabricated. This LED display can maintain illumination when the device was either spread over an arbitrary curved surface, bent, or crumpled (Fig. 8.12B_2).

8.4.2 Wearable strain sensor

Wearable strain sensors have extensive applications as a human motion detector in medical contexts (e.g., rehabilitation assistance and personal health monitoring[134]), human-machine interfaces (HMI, e.g., performance monitoring and receiving data of human emotions[135]), and entertainment (e.g. wireless controllers in immersive games or for drones[136]). There are four demands that should be met for wearable strain sensors: (i) the ability to measure large strain (>20%) without failure, (ii) high signal-to-noise ratio, (iii) large gauge factor (>10) in order to perceive strain precisely, and (iv) fast response time (<50 ms) to ensure timely signal acquisition.[101, 106]

According to sensing mechanisms, wearable strain sensors are categorized into two types: piezopotential and piezoresistive. The basic structure

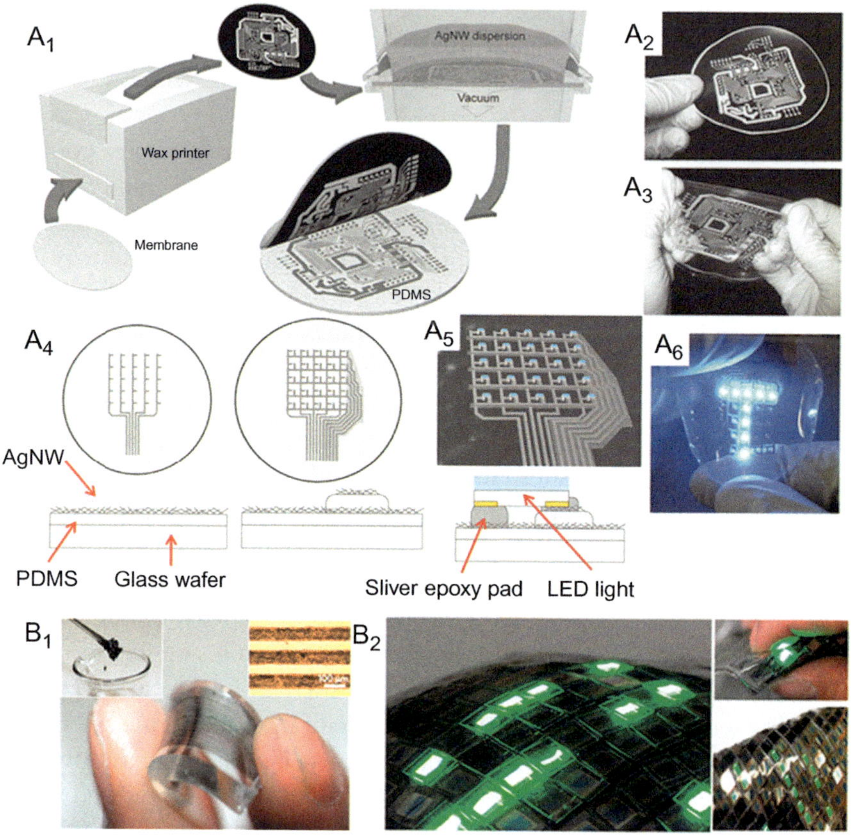

Fig. 8.12 (A_1) Schematic of a printing circuit on PVDF membrane, AgNW dispersion, and transferring the AgNW circuit onto a PDMS substrate. (A_2) Example of a wax patterned circuit on a PVDF membrane. (A_3) The transferred AgNW circuit on a stretchable PDMS substrate; schematic description of the structure of (A_4) AgNW conductors and (A_5) LED arrays. Top images are the surface view and bottom images are the cross-sectional view. (A_6) Photograph of LED arrays that can maintain illumination under stretching strain of 20%. (B_1) Printed CNT arrays on a PDMS sheet; the inset images show SWNTs dispersed in paste and a micrograph of printed elastic conductors with a linewidth of 100 μm. (B_2) Photographs of a stretchable LED display that can be spread over arbitrary curved surface and can be folded without degradation of its illumination.[133] *Panels (A_1–A_6) reproduced with permission from Tybrandt, K.; Vörös, J. Fast and Efficient Fabrication of Intrinsically Stretchable Multilayer Circuit Boards by Wax Pattern Assisted Filtration. Small* ***2016****, 12, 180–184. Copyright 2015, Wiley-VCH. Panels (B_1) and (B_2) Copyright 2009, Nature Publishing Group.*

of a piezopotential-type sensor is a capacitor that consists of two electrodes to generate a capacitance:

$$C = \varepsilon A / d \tag{8.4}$$

where ε is the dielectric constant, A is the area of the electrodes, and d is distance between the two electrodes. When strain is applied to the sensor, A and d will change, resulting in piezopotential performance.

The piezoresistive strain sensors work through a disconnection mechanism between conductive nanomaterial fillers. Without loading, these nanomaterial fillers stack together as a conductive film and connected to form overlap. When a strain is applied, the connection between these nanomaterials changes, resulting in a resistance change. The sensor sensitivity is evaluated by a gauge factor (*GF*), which can be defined by:

$$GF = (\Delta R/R_0)/\varepsilon \tag{8.5}$$

where ΔR is the resistance change with strain, R_0 is the resistance prior to straining, and ε is the applied strain.

Piezoresistivity of the sensor can be mainly attributed to four effects: (i) the area of overlap is reduced until microcracks form, (ii) tunneling resistance changes between two adjacent nanomaterials due to changes in microcracks, (iii) influence of geometrical changes of conducive films, and (iv) piezoresistivity of individual nanosheets or nanoparticles due to deformation upon loading.

The relative change of resistance ($\Delta R/R_0$) can be expressed as:

$$\Delta R/R_0 = (1 + 2\nu)\varepsilon + \Delta\rho/\rho_0, \tag{8.6}$$

where ν is Poisson's ratio, $\Delta\rho$ is resistivity variation, $(1+2\nu)\varepsilon$ refers to the influence of geometrical changes, and $\Delta\rho/\rho_0$ indicates the piezoresistivity of GnPs.

The key challenge with these types of strain sensors is realizing high piezoresistive performance while avoiding electrical failure under strain. To overcome this challenge, a fragmented CVD graphene foam (FGF) was integrated into a highly sensitive strain sensor.[55b] Briefly, the sensor was fabricated by coating FGF on polyimide tape, encapsulating with PDMS (Fig. 8.13A). Because of the interconnection among FGF pieces, under slight deformation (strain <0.8%), these pieces will reorient themselves in the direction of applied strain, resulting in a GF of 21. Due to its thinness (0.55 mm) and encapsulation with PDMS, the sensor can be attached on human skin to monitor pulse signals from the radial artery of the wrist or a bending signal from the elbow and finger. In addition to graphene, AuNW strain sensors also exhibit strong performance, including the ability to maintain stretchability and durability for up to 5000 cycles of 30% strain tensile.[106] The bundling morphologies of AuNWs inside the strain sensor provide GF of 9.9 for low strain (0%–5%) and 6.9 for large strain

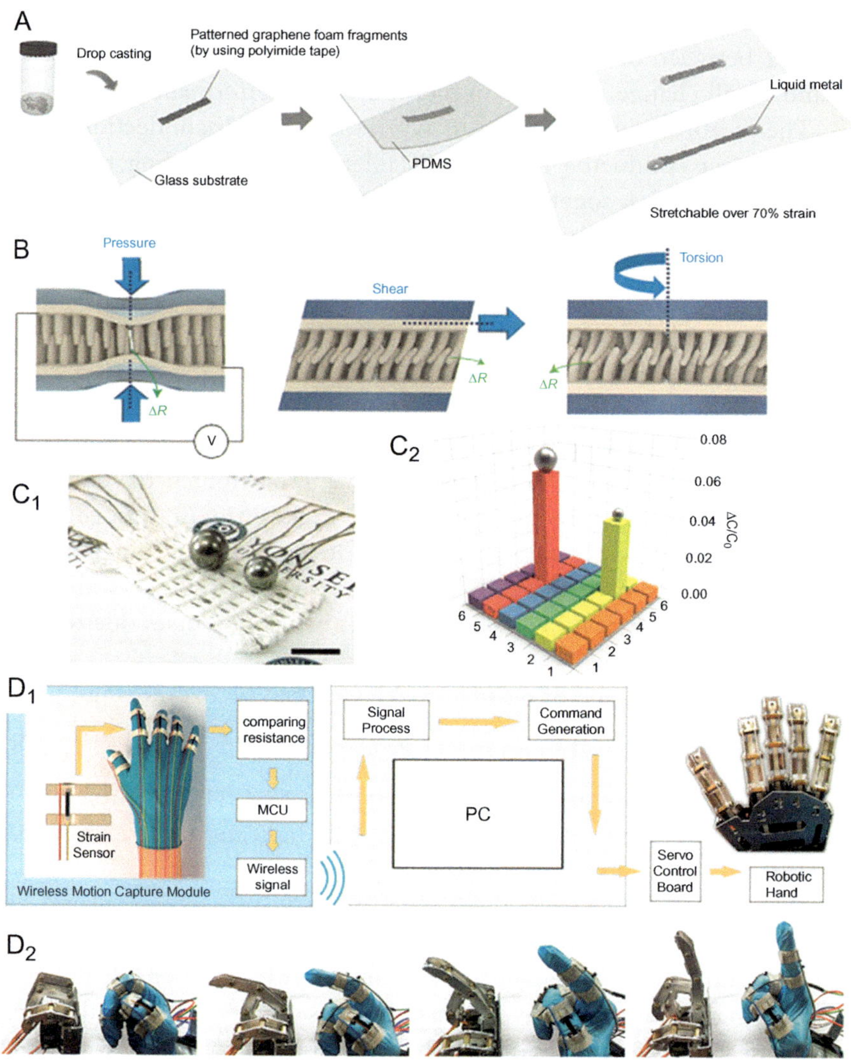

Fig. 8.13 (A) Fabrication of a strain sensor incorporated with fragmented CVD graphene foam.[55b] (B) Sensing mechanism of the strain sensor based on interlocked nanohairs.[137] (C_1) Pixelated pressure sensor array holding two beads, scale bar, 1 cm. (C_2) Pressure distribution of the two beads detected by the pressure sensor array.[136] (D_1) Illustration of an HMI system of motion capture and robotic hand control. (D_2) Robotic hand mimicking finger bending.[124b] *Panels (A) and (C_1, C_2) Copyright 2015, Wiley-VCH. Panel (B) Copyright 2012, Nature Publishing Group. Panels (D_1) and (D_2) Copyright 2015, Nature Publishing Group.*

(<50%), so the strain sensor is able to detect various human motions from forearm muscle motions to wrist pulses.

Pang et al. created a flexible and highly functional strain sensor that consists of several vertically interlocked polymer nanohairs (Fig. 8.13B).[137] These polymer nanohairs were coated with platinum to endow them with conductivity. Owing to the unique interlocking between nanowires, under different loadings, resistance between the interlocked structures is changed, resulting in different piezoresistive performance, e.g., GFs of 11.45 for pressure, 0.75 for shear, and 8.53 for torsion.

Apart from attaching the strain sensor on skin to detect human motions, strain sensor arrays can be used to detect any type of motion/pressure. For instance, a pixelated strain sensor array of 6×6 units has been produced by weaving AgNP-coated polystyrene-butadiene-styrene fiber[136]; when two beads were respectively located on two sensor pixels in the array (Fig. 8.13C_1), the loaded sensor pixels generated distinct electrical signals compared to the unloaded ones. By analyzing the electrical performance of all the sensor pixels together, the pressure location can be easily detected, as shown in Fig. 8.13C_2. Thus, using strain sensor arrays fabricated from, e.g., patterned graphene[138] or AgNW sensors,[139] the spatial distribution of external pressure can be detected. The resolution of these arrays highly depends on the number of sensors/pixels. To achieve a high resolution, a large number of sensors are required in an array. Therefore, in the future, the use of nanomaterials and micromanufacturing techniques may become increasingly important to maximize resolution.

HMI is another novel application for strain sensors. The function of HMIs includes capturing human motions and controlling the motion of machines. The whole system is shown schematically in Fig. 8.13D_1.[124b] First, human motion generates a resistance change that is captured by a microcontroller (MCU). Second, a wireless signal is transferred to a PC through Bluetooth. Last, the signal is processed to generate a command to finally control a robotic hand through a servo control board. This robotic hand in the HMI system can precisely mimic finger bending (Fig. 8.13D_2). Other HMI systems were produced with strain sensors developed with AuNW ink,[140] graphene platelet film,[141] and AgNW/graphene film.[142]

8.4.3 Artificial electronic skin

Electronic skin (e-skin) is an artificial skin that mimics human skin using electronic sensors to monitor the sense of pressure, temperature, and vibration. E-skin has found many potential applications, such as wearable devices,[143] artificial prosthetics,[144] health monitoring,[145] and smart robots.[146] In an e-skin, tactile sensing and stretchability are both important because tactile sensing of the skin allows the body to communicate with the outside environment, whereas stretchability of the skin enables free movements. Generally, according to the mechanism of converting external stimuli into electronic signals, the recently emerged e-skin could be divided as piezoresistive e-skin and piezopotential e-skin.

8.4.3.1 Piezoresistive e-skin

The tactile sensing mechanism of piezoresistive e-skin is such that when external stimuli occur, the contact condition among conductive materials in the sensor will change, resulting in resistance change. Through the analysis of resistance changes, information about the stimuli can be identified.

Chou et al. created a bio-inspired stretchable e-skin having tactile sensing and tactile color-changing abilities. The essential components of the e-skin are a piezoresistive pressure sensor and stretchable organic electrochromic devices (Fig. 8.14A).[147] The tunable resistive pressure sensor includes an array of PDMS micropyramids that were coated with SWNTs everywhere expect for the top part. Electrochromic devices were fabricated by coating an organic electrochromic polymer, poly(3-hexylthiophene-2,5-diyl) (P3HT), onto a SWNT-coated PDMS substrate. As pressure increases, increasing area of the SWNT-coated pyramid comes into contact with the counterelectrode sheet, resulting in a progressive resistance decrease (Fig. 8.14B). Meanwhile, the color of the electrochromic device will change, corresponding to various pressure intensities (Fig. 8.14C). The PDMS micropyramid arrays were partially coated with SWNTs to ensure high sensitivity of the e-skin. The electrochromic aspect, where the color can be changed through molecular rearrangement, give a chameleon-like property to the e-skin. However, the high price of P3HT, e.g., \$637 per gram from Sigma-Aldrich, may affect their popularity in real applications.

8.4.3.2 Piezopotential e-skin

The piezopotential e-skin employs a capacitance generated between two layers of electrodes. When an external stimulus occurs, the electrode area and distance between two electrodes will change, resulting in piezopotential performance.

A sandwich structured PDMS/Ag/Ecoflex/Ag/PDMS capacitor array provides an example of a piezopotential e-skin (Fig. 8.15A).[148]

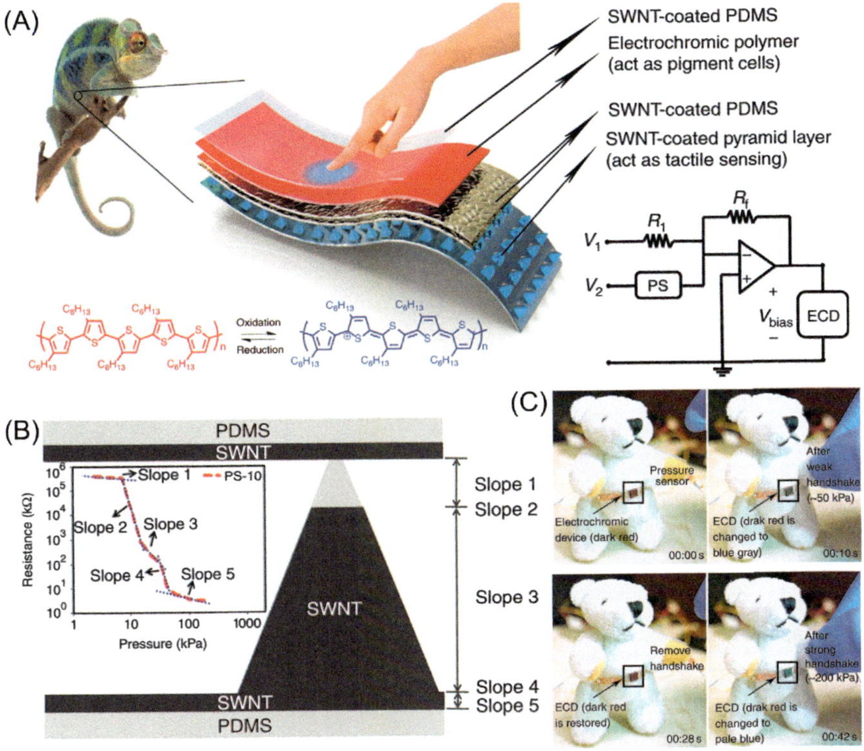

Fig. 8.14 (A) Schematic of the concept of a chameleon-inspired e-skin; bottom image shows the structures of both neutral and oxidized states of the electrochromic polymer, P3HT. (B) Schematic of SWNT-coated pyramidal microstructured PDMS; inset shows the resistance response of a sensor under pressure of 10 kPa, also called PS-10. (C) Photographs of a teddy bear wearing an e-skin that changes color in response to tactile stimuli. *Reproduced with permission from Chou, H.-H.; Nguyen, A.; Chortos, A.; To, J. W.; Lu, C.; Mei, J.; Kurosawa, T.; Bae, W.-G.; Tok, J. B.-H.; Bao, Z., A Chameleon-Inspired Stretchable Electronic Skin With Interactive Colour Changing Controlled by Tactile Sensing. Nat. Commun.* ***2015****, 6, 8011. Copyright 2015, Nature Publishing group.*

Here, serpentine Ag electrodes were separated with an Ecoflex elastic dielectric layer. The overall sensor capacitance C_s includes fringe capacitance C_f, plate capacitance C_p, and dielectric overlay fringe capacitance C_s. Contacts between the sensor and human body will change C_f, due to the interaction between fringe electric field of the sensor and the human body. Under pressure or stretching, C_p and C_s will be altered due to the distance change between electrodes (Fig. 8.15B). The piezopotential e-skin can detect a minimum pressure of 6 Pa due to the array's structure. However, probably due to the sandwich structure, the external stimuli must transfer from PDMS

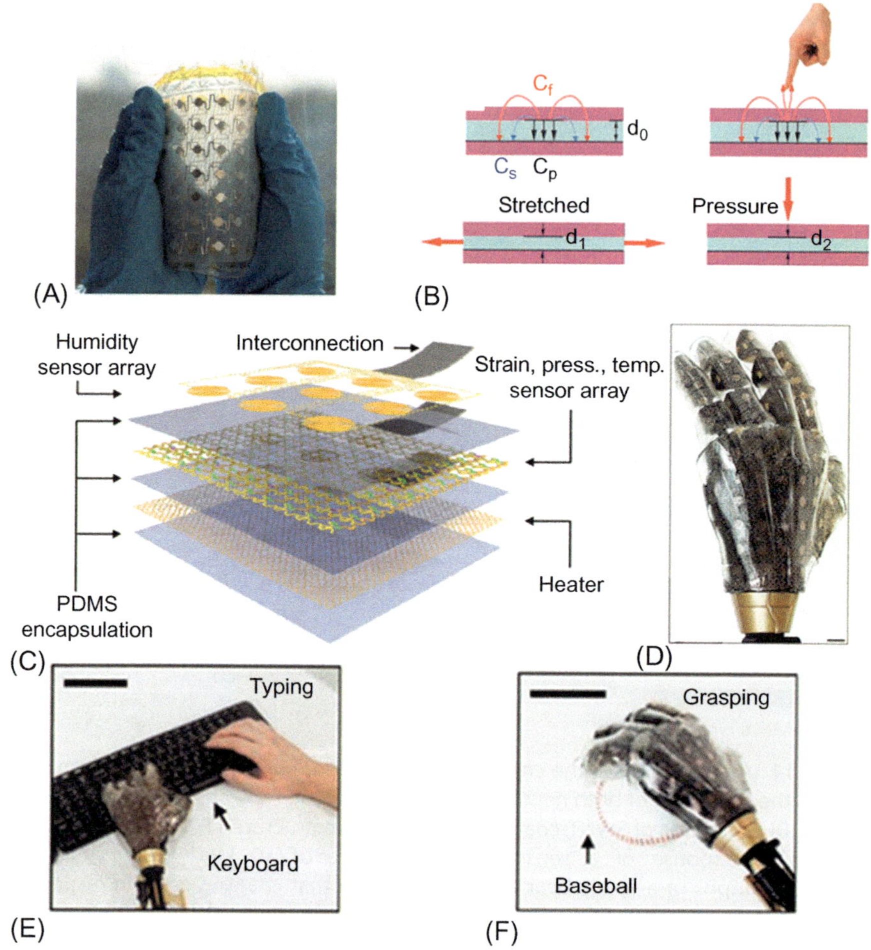

Fig. 8.15 (A) Photograph of a wearable piezopotential e-skin based on a pressure sensor array and (B) sensing modes of the sensor array.[148] (C) Schematic of the laminated system of six stacked sensor layers used to construct an e-skin, and photographs of (D) an artificial hand attached with the multifunctional e-skin and (E–F) its practical applications. *Panels (A) and (B) Copyright 2015, Wiley-VCH. Panels (C–F) reproduced with permission from Kim, J.; Lee, M.; Shim, H. J.; Ghaffari, R.; Cho, H. R.; Son, D.; Jung, Y. H.; Soh, M.; Choi, C.; Jung, S. Stretchable Silicon Nanoribbon Electronics for Skin Prosthesis. Nat. Commun.* ***2014****, 5, 5747. Copyright 2014, Macmillan Publishers Ltd.*

to inside of the Ecoflex layer. The response time (~100 ms) appears to suffer. At the same time, the sensitivity of the e-skin may be limited; only the detection of large strains from arm bending, rather than minute strains (e.g., pulse measurements), was demonstrated.

Apart from the sensing of pressure, sensing of temperature or humility is also important for e-skin. Through the incorporation of heat sensors, multifunctional e-skin has been produced. For instance, Kim and co-workers developed a multifunctional e-skin that is a laminated system of several sensors for temperature and strain and supporting electronics (Fig. 8.15C).[117] The serpentine interconnections give the device stretchability, while linear nanoribbon electronics generate sensitivity. This e-skin can conformably adhere to the complex geometry of the human hand through van der Waals forces, because of the micrometer-scale thickness (Fig. 8.15D). The e-skin also has strong human-machine interfacial ability; it can lead an artificial hand to do many complex operations, e.g., keyboard tapping (Fig. 8.15E), holding a cup, etc.

Another e-skin with multi-function and ultrathin thickness, which renders the e-skin seem like a tattoo was reported by the Rogers group.[149] The fabrication of such multifunctional e-skins typically involves printing[132] or etching procedures[150] for producing precise patterned circuits. Thus, efficiently, conveniently, and precisely producing ultrathin sensors is a research priority for the development of e-skin. Additionally, more innovative and functional e-skin could be created in the future. This can be achieved by combining material science, which could create more flexible and stretchable structures for sensors, with electronic techniques, which could invent more precise data transfer methods.

8.4.4 Stretchable/flexible FETs

Stretchable/flexible FETs have extensive applications in wearable electronics. These can be used as a fundamental component to control or process signals in wearable electronic devices such as displays, sensor arrays, and thin-film circuits.

As described in Section 8.2.2, semiconductive SWCNTs with a high mobility of $\sim$40 cm^2 V^{-1} s^{-1} and an on/off ratio of $\sim 10^5$ are ideal candidates for use as a channel material. A CNT-based FET was fabricated using poly (9,9-fluorene) (PFO)-wrapped semiconductive SWCNTs as the channel layer, a flexible ion-gel as the dielectric layer, and metal films as electrodes.[151] This FET can maintain its performance until 50% strain (Fig. 8.16A). This 50% stretchability is realized using a wavy CNT channel layer, which is produced by a prestretching-release-buckling process (Fig. 8.16B).

Apart from the wavy configuration, serpentine interconnection was utilized within an array (17×15) of a semiconducting SWNT-based

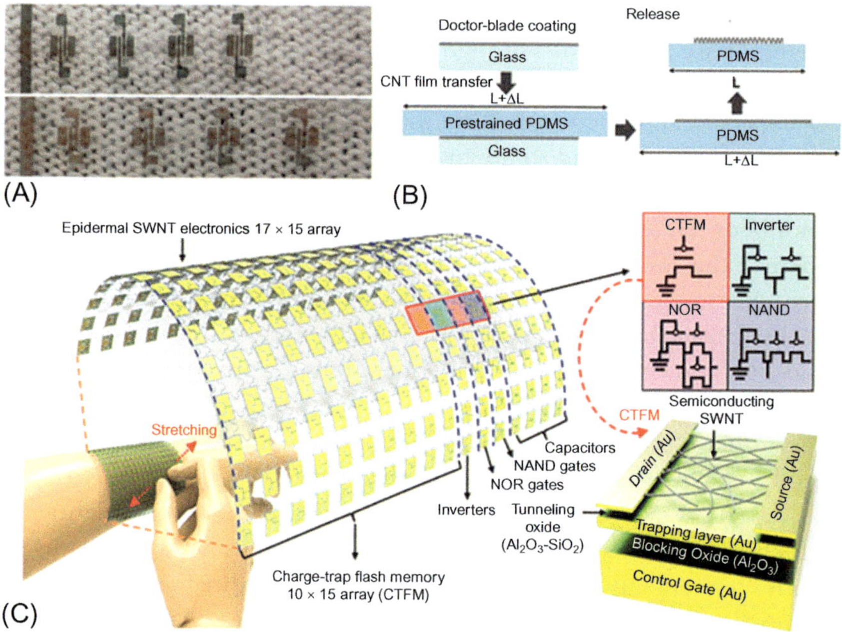

Fig. 8.16 (A) A CNT FET under normal condition (*top*) and 50% of tensile strain (*bottom*) and (B) prestrain-release-buckling process to produce a wavy structure of the CNT channel. (C) Schematic of the semiconductive SWNT-based wearable electronic device, which is composed of a memory unit, capacitors, and logic circuits. Simple circuitry schematics (CTFM, inverter, and NAND/NOR gates) are shown in the *top right* colored sections. The *bottom right* frame shows the cross-sectional layer of the CTFM.[152] *Panels (A) and (B) reproduced with permission from Xu, F.; Wu, M.-Y.; Safron, N. S.; Roy, S. S.; Jacobberger, R. M.; Bindl, D. J.; Seo, J.-H.; Chang, T.-H.; Ma, Z.; Arnold, M. S. Highly Stretchable Carbon Nanotube Transistors With Ion Gel Gate Dielectrics. Nano Lett.* ***2014****, 14, 682–686. Copyright 2014, American Chemical Society. (C) Copyright 2015, American Chemical Society.*

stretchable electronic device, consisting of logic gates (inverters and HAND/NOR gates), charge-trap floating-gate memory units (CTFMs), and capacitors.[152] The structure of the device is shown in Fig. 8.16C. All electrodes, charge-trap layers, and serpentine interconnections are formed using photolithography and lift-off techniques. The SWNT networks are deposited by dip coating and patterned by photolithography to fabricate serpentine interconnections and channels. The thin film-based active islands (<3 μm), serpentine interconnection designs, and neutral mechanical plane layout enable the device to endure a ~20% applied strain while minimizing deformations/strains in the channels and active regions.

In addition to CNTs, single-layer graphene is another promising material for semiconductive channels, because of its flexibility, high carrier mobility

($10{,}000\ cm^2\ V^{-1}\ s^{-1}$), and zero bandgap structure.[153] In one example, graphene is applied as a channel material, and a graphene/AgNW hybrid material is applied as a source/drain electrode in a FET.[154] The continuous graphene film bridges source/drain electrodes and channel and reduces contact resistance between the channel and electrodes to 6–9 kΩ μm. As a result, electron mobility of the FET was $2925 \pm 78\ cm^2\ V^{-1}\ S^{-1}$ for a channel length of 70 μm, which is similar to or slightly better than graphene FETs using Cr/Au as source/drain electrodes. When organizing several FETs into an array structure, FET arrays can display negligible change of resistance after stretching to 20%.

Due to their high conductivity and flexibility, AgNWs can be used as electrodes in stretchable FETs. In one such FET, a AgNW-polyurethane acrylate composite, an SWCNT network, and PU-co-polyethylene glycol (PU-co-PEG) served as stretchable transparent electrodes, channel, and dielectric layer, respectively.[155] The source/drain electrodes were fabricated by spray-printing AgNW solution through a shadow mask and wet transferring the AgNW films onto a PU substrate. AgNWs were encapsulated within the elastic PU substrate, so these could slide across each other and maintain connection following deformation of the substrate. Therefore, the devices can be stretched with up to 50% strain and subjected to 500 cycles of repeated stretching to 20% strain without significant loss in electrical properties.

8.4.5 Stretchable supercapacitors

Stretchable supercapacitors have attracted increased attention for advanced energy storage because of their high capability, long cycle life, fast charging rate, and easy fabrication. The basic structure of a stretchable supercapacitor includes two stretchable hybrid capacitors, an electrolyte, and an ion-permeable separator.

Each hybrid capacitor consists of an electrical double-layer capacitor and a pseudocapacitor. The electrical layer capacitor accommodates electrostatic charge. The pseudocapacitor provides enhanced energy density through fast and reversible redox reactions. For a stretchable supercapacitor, the materials for electrical layer mainly are stretchable, porous, conductive materials, e.g., CVD graphene foam.[156] The materials used in pseudocapacitors include conductive polymers, e.g., polyaniline, polypyrrole, and poly[3,4-ethylenedioxythiophene], or transition metal oxides, e.g., MnO_2, NiO, RuO_2, V_2O_5, and TiO_2.[157]

8.4.5.1 Film structure supercapacitor

To endow supercapacitors with stretchability, the strategy of prestrain-release-buckling has been applied. Using this technique, the resistance of the electrode can remain fairly unchanged before applied tensile strain reaches the prestrain.[64] For instance, CVD graphene films were fabricated as transparent wavy electrodes on PDMS.[158] The supercapacitor employed two pieces of transparent, wavy graphene/PDMS films that were electrically separated by a H_2SO_4-PVA gel electrolyte. The H_2SO_4-PVA gel served as both separator and electrolyte, prevented short-circuits, and ensured the integrity of the supercapacitor under stretching. This supercapacitor displayed capacitance of 13.9, 15.7, 16.7, and 17.3 F g^{-1} at tensile strains of 0%, 20%, 30%, and 40%, respectively. Thus, this prestrain-release-buckling strategy provides an efficient and effective path to fabricate wavy electrodes for stretchable supercapacitors.

MnO_2/CNT/nylon fiber supercapacitors, whose structure is shown in Fig. 8.17A_1, was reported.[162] Nylon fibers were used as a core and were coiled with CNT ribbons to form the active electrode (Fig. 8.17A_2). MnO_2 was electrochemically deposited onto the coiled CNT surface to create pseudocapacitance. The highest values of the length- and area-normalized specific capacitances were 5.4 mF cm^{-1} and 40.9 mF cm^{-2}, respectively. The supercapacitor is capable of being stretched to 150% strain, only exhibiting a change of capacitance at 10% strain. This change is caused by the elimination of inter-coil contact, which forces charge carriers to loop along the coils, rather than being directly transported between adjacent contacting coils.

However, most electrocapacitative materials are not intrinsically stretchable, so the stretchability of these supercapacitors is determined by the prepared structure and substrate. As a result, the stretchability of these stretchable supercapacitors is lower than 100%.[64, 163] Also, low stretchability of PVA-based acidic electrolytes ($<$100%) limits the stretchability of conventional supercapacitors. Thus, approaches for realizing high stretchability of supercapacitors will be a research focus in the future. In general, high stretchability of supercapacitors might be realized by fabricating electrode materials into high stretchable configurations and replacing rigid electrolytes with stretchable electrolytes.

8.4.5.2 Fiber structure supercapacitors

Fiber structured electrodes are also applied in stretchable supercapacitors. Under stretching, the fiber structure can be straightened, thus protecting the microstructure and materials within the device. Among various fiber electrodes, such as PANI-graphene fiber,[164] ZnO nanowire-coated poly(methyl methacrylate) fibers,[165] and reduced GO-coated Au fibers,[166] CNT fibers

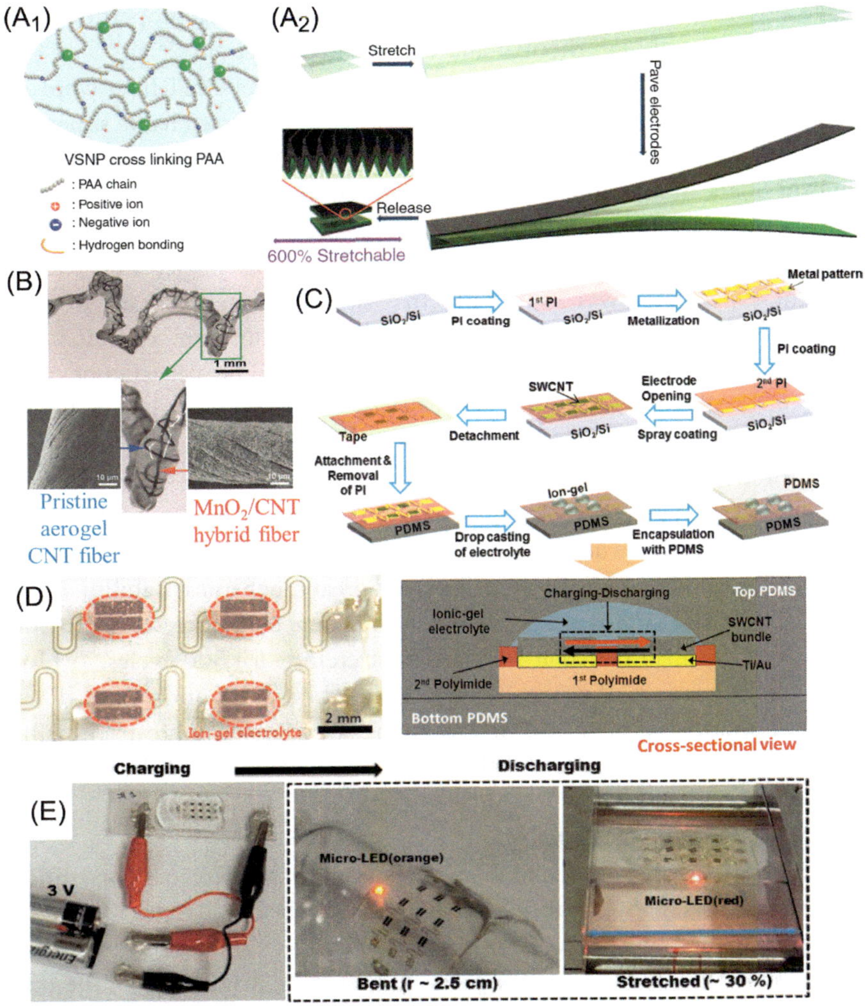

Fig. 8.17 Schematic of (A_1) a VSNP-PAA polyelectrolyte and (A_2) the fabrication of a self-healable and highly stretchable supercapacitor, including (*top*) prestrain of VSNP-PAA polyelectrolyte and (*bottom*) buckling of PPy/CNT paper.[159] (B) Photograph of an asymmetric supercapacitor (*right*); inset shows the SEM micrograph of a MnO_2/CNT hybrid fiber and (*left*) pristine aerogel CNT fiber.[160] (C) Schematic fabrication of a stretchable MSC array on a PDMS substrate and related cross-sectional view. (D) Top view of the MSC array and (E) charging state at 3 V and discharging state to light μ-LEDs under a bent and 30% stretched state.[161] *Panels (A_1) and (A_2) Copyright 2015, Nature Publishing Group. Panel (B) Copyright 2015, American Chemical Society. Panels (C–E) Copyright 2013, American Chemical Society.*

have advantages such as light weight, excellent mechanical properties, chemical inertness, and high thermal and electrical conductivity.[124a, 167]

The MnO_2/CNT hybrid fiber and pristine aerogel CNT fiber were used as positive and negative electrodes, respectively, in a stretchable supercapacitor.[160] A supercapacitor was fabricated by intertwining MnO_2/CNT hybrid fibers and pristine aerogel CNT fibers and separating the two fibers with the KOH-PVA gel electrolyte (Fig. 8.17B). A prestrain-releasing-buckling strategy was applied to form the flexuous shape. The resultant specific capacitance was 157.53 $\mu F\ cm^{-1}$ at a scan rate of 50 $mV\ s^{-1}$ with corresponding areal specific capacitance and volumetric capacitance of 16.87 $mF\ cm^{-2}$ and 21.21 $F\ cm^{-3}$, respectively. The capacitance values were generally constant under 20 cycles of stretching-releasing at different tensile strains (0%, 40%, 70%, and 100%).

Moreover, two PANI/CNT/Ecoflex fibers were used as electrodes and separated by PVA gel electrolytes in an elastic supercapacitor. Due to the highly stretchable structure, where CNTs were coiled around an elastic Ecoflex core, this supercapacitor displayed great mechanical stability. Its specific capacitance ($111.6 \pm 2.3\ Fg^{-1}$) was unchanged after stretching to 400% strain and was maintained at 74.9% after stretching for 500 cycles.[130] However, the procedure of prestrain also increased the fabrication time. Therefore, prestrain-free methods for producing stretchable supercapacitors may generate research interest, and these will be discussed in the next section.

8.4.5.3 Micro-supercapacitors

In addition to the wavy and fiber configuration supercapacitors, micro-supercapacitors (MSCs) can also be endowed with stretchability. With their planar structure and micrometer thickness, these could exhibit better flexibility than their buckle counterparts. In particular, patterning of the interspace between two planar electrodes in a supercapacitor on the micrometer scale can increase its capacitance by reducing the path length of ions in the electrolyte.[168, 169]

To realize stretchability, elastomeric substrates and serpentine interconnections were utilized in MSCs.[161] Fig. 8.17C schematically shows the fabrication process of a stretchable solid-state MSC array on a PDMS substrate. The process included the following steps: spin-coating polyimide film on a SiO_2/Si substrate, deposition of patterned electrodes and interconnections of Ti/Au, spray-coating of SWCNT electrodes, spin-coating polyimide, transferring the entire device to a PDMS substrate, drop-casting ion-gel electrolyte, and final encapsulation within a PDMS film.[161] The final

MSC array is shown in Fig. 8.17D in which each MSC is connected by serpentine interconnections. These serpentine interconnections can promote the stretchability of the whole device, because under stretching, these can change their shape to absorb strain and thus avoid any significant effect on MSC islands.[109] Each supercapacitor consists of 3×3 MSC arrays and exhibits less specific electrode capacitances of 55.3 and 34.2 Fg^{-1} at scan rates of 0.5 and 4 V/s, respectively. Under tensile strain <30%, no significant degradation was reported. After 60 s of charging at 3 V, the supercapacitor array was able to light LEDs under bending and stretching (Fig. 8.17E).

8.5 Conclusions and perspectives

This review highlighted the primary nanomaterial possessing for desirable chemical and physical properties of stretchable electronics, as well as common structural configurations used to realize flexibility/stretchability. The applications of flexible/stretchable electronics have been achieved by integrating electrically conductive nanomaterials with stretchable elastomers in various unique stretchable configurations.

The stretchable configurations, e.g., wavy or serpentine configurations, allow conductors to work under high tensile strains. The sensitive piezoresistivity of sensors dramatically enhances their applications from human health monitoring to remote machine control. The use of transistors with microstructural components enables the matrix to reduce signal crosstalk between pixels and promotes rapid addressing and low power consumption. The described supercapacitors could provide energy to light LEDs as a proof of concept, and thus could have the ability to power other electronics.

Breakthroughs in flexible/stretchable electronics require the support of multiple academic disciplines, including nanomaterial synthesis, advanced microfabrication techniques, fundamental mechanistic research, etc. For example, advancements in large-scale, low-cost synthesis of high-quality nanomaterials would promote the progress of flexible/stretchable electronics. The large-scale synthesis of single-crystalline graphene and flexible silver nanowires would have the potential to replace traditional silicon and indium tin oxide-based traditional electronics, respectively. At the same time, advanced microfabrication techniques, such as direct writing, could provide a precise and fast fabrication of structural materials with complex 3D configurations with low material consumption.

However, certain challenges remain for practical applications. For example, new materials and novel transduction mechanisms should be further investigated to realize stretchability. Although supercapacitors have been reported in recent studies, energy storage capacity should be further increased to satisfy the demands of various applications. Different types of artificial smart e-skins are urgently required for emerging fields in healthcare, including health monitoring, prosthesis techniques, and clinical medicine. Moreover, the production of devices with low power consumption or the ability to self-power remains a topic worthy of in-depth study.

In conclusion, although challenges and limitations exist, impressive advances have been made in the field of flexible/stretchable electronics over the past five years. In the future, new concepts for structural design, innovative patterning approaches, and other advancements in the field will hopefully allow these novel electronic devices to find practical applications in daily life, such as in smart homes and wearable healthcare.

References

1. Boland, J. J. Flexible Electronics: Within Touch of Artificial Skin. *Nat. Mater.* **2010**, *9*, 790–792.
2. Lee, J.; Lee, P.; Lee, H. B.; Hong, S.; Lee, I.; Yeo, J.; Lee, S. S.; Kim, T. S.; Lee, D.; Ko, S. H. Room-Temperature Nanosoldering of a Very Long Metal Nanowire Network by Conducting-Polymer-Assisted Joining for a Flexible Touch-Panel Application. *Adv. Funct. Mater.* **2013**, *23*, 4171–4176.
3. Lu, C.-C.; Lin, Y.-C.; Yeh, C.-H.; Huang, J.-C.; Chiu, P.-W. High Mobility Flexible Graphene Field-Effect Transistors With Self-Healing Gate Dielectrics. *ACS Nano* **2012**, *6*, 4469–4474.
4. Choi, C.; Lee, J. A.; Choi, A. Y.; Kim, Y. T.; Lepró, X.; Lima, M. D.; Baughman, R. H.; Kim, S. J. Flexible Supercapacitor Made of Carbon Nanotube Yarn With Internal Pores. *Adv. Mater.* **2014**, *26*(13), 2059–2065.
5. (a)Bolotin, K. I.; Sikes, K.; Jiang, Z.; Klima, M.; Fudenberg, G.; Hone, J.; Kim, P.; Stormer, H. Ultrahigh Electron Mobility in Suspended Graphene. *Solid State Commun.* **2008**, *146*, 351–355; (b)Morozov, S.; Novoselov, K.; Katsnelson, M.; Schedin, F.; Elias, D.; Jaszczak, J.; Geim, A. Giant Intrinsic Carrier Mobilities in Graphene and Its Bilayer. *Phys. Rev. Lett.* **2008**, *100*, 016602.
6. Li, X.; Zhu, Y.; Cai, W.; Borysiak, M.; Han, B.; Chen, D.; Piner, R. D.; Colombo, L.; Ruoff, R. S. Transfer of Large-Area Graphene Films for High-Performance Transparent Conductive Electrodes. *Nano Lett.* **2009**, *9*, 4359–4363.
7. Hu, L.; Kim, H. S.; Lee, J.-Y.; Peumans, P.; Cui, Y. Scalable Coating and Properties of Transparent, Flexible, Silver Nanowire Electrodes. *ACS Nano* **2010**, *4*, 2955–2963.
8. Leach, A. M.; McDowell, M.; Gall, K. Deformation of Top-Down and Bottom-Up Silver Nanowires. *Adv. Funct. Mater.* **2007**, *17*, 43–53.
9. Lipomi, D. J.; Vosgueritchian, M.; Tee, B. C.; Hellstrom, S. L.; Lee, J. A.; Fox, C. H.; Bao, Z. Skin-Like Pressure and Strain Sensors Based on Transparent Elastic Films of Carbon Nanotubes. *Nat. Nanotechnol.* **2011**, *6*, 788–792.

10. Lu, N.; Lu, C.; Yang, S.; Rogers, J. Highly Sensitive Skin-Mountable Strain Gauges Based Entirely on Elastomers. *Adv. Funct. Mater.* **2012**, *22*, 4044–4050.
11. (a)Choi, W. M.; Song, J.; Khang, D.-Y.; Jiang, H.; Huang, Y. Y.; Rogers, J. A. Biaxially Stretchable "Wavy" Silicon Nanomembranes. *Nano Lett.* **2007**, *7*, 1655–1663; (b)Xu, F.; Wang, X.; Zhu, Y.; Zhu, Y. Wavy Ribbons of Carbon Nanotubes for Stretchable Conductors. *Adv. Funct. Mater.* **2012**, *22*, 1279–1283.
12. Lin, L.; Liu, S.; Fu, S.; Zhang, S.; Deng, H.; Fu, Q. Fabrication of Highly Stretchable Conductors via Morphological Control of Carbon Nanotube Network. *Small* **2013**, *9*, 3620–3629.
13. Shin, M. K.; Oh, J.; Lima, M.; Kozlov, M. E.; Kim, S. J.; Baughman, R. H. Elastomeric Conductive Composites Based on Carbon Nanotube Forests. *Adv. Mater.* **2010**, *22*, 2663–2667.
14. Chen, Z.; Ren, W.; Gao, L.; Liu, B.; Pei, S.; Cheng, H.-M. Three-Dimensional Flexible and Conductive Interconnected Graphene Networks Grown by Chemical Vapour Deposition. *Nat. Mater.* **2011**, *10*, 424–428.
15. Cheng, Y.; Wang, R.; Sun, J.; Gao, L. Highly Conductive and Ultrastretchable Electric Circuits from Covered Yarns and Silver Nanowires. *ACS Nano* **2015**, *9*, 3887–3895.
16. Xu, S.; Zhang, Y.; Cho, J.; Lee, J.; Huang, X.; Jia, L.; Fan, J. A.; Su, Y.; Su, J.; Zhang, H. Stretchable Batteries With Self-Similar Serpentine Interconnects and Integrated Wireless Recharging Systems. *Nat. Commun.* **2013**, *4*, 1543.
17. Yang, P. K.; Lin, L.; Yi, F.; Li, X.; Pradel, K. C.; Zi, Y.; Wu, C. I.; He, H., Jr.; Zhang, Y.; Wang, Z. L. A Flexible, Stretchable and Shape-Adaptive Approach for Versatile Energy Conversion and Self-Powered Biomedical Monitoring. *Adv. Mater.* **2015**, *27*, 3817–3824.
18. Fan, J. A.; Yeo, W.-H.; Su, Y.; Hattori, Y.; Lee, W.; Jung, S.-Y.; Zhang, Y.; Liu, Z.; Cheng, H.; Falgout, L. Fractal Design Concepts for Stretchable Electronics. *Nat. Commun.* **2014**, *5*, 3266.
19. He, W.; Li, G.; Zhang, S.; Wei, Y.; Wang, J.; Li, Q.; Zhang, X. Polypyrrole/Silver Coaxial Nanowire Aero-Sponges for Temperature-Independent Stress Sensing and Stress-Triggered Joule Heating. *ACS Nano* **2015**, *9*, 4244–4251.
20. Kim, K. S.; Zhao, Y.; Jang, H.; Lee, S. Y.; Kim, J. M.; Kim, K. S.; Ahn, J.-H.; Kim, P.; Choi, J.-Y.; Hong, B. H. Large-Scale Pattern Growth of Graphene Films for Stretchable Transparent Electrodes. *Nature* **2009**, *457*, 706–710.
21. Lee, J.; Lee, P.; Lee, H.; Lee, D.; Lee, S. S.; Ko, S. H. Very Long Ag Nanowire Synthesis and Its Application in a Highly Transparent, Conductive and Flexible Metal Electrode Touch Panel. *Nanoscale* **2012**, *4*, 6408–6414.
22. Logothetidis, S. Flexible Organic Electronic Devices: Materials, Process and Applications. *Mater. Sci. Eng. B* **2008**, *152*, 96–104.
23. Yan, F.; Zhang, M.; Li, J. Solution-Gated Graphene Transistors for Chemical and Biological Sensors. *Adv. Healthc. Mater.* **2014**, *3*, 313–331.
24. Park, S. I.; Ahn, J. H.; Feng, X.; Wang, S.; Huang, Y.; Rogers, J. A. Theoretical and Experimental Studies of Bending of Inorganic Electronic Materials on Plastic Substrates. *Adv. Funct. Mater.* **2008**, *18*, 2673–2684.
25. Thongrattanasiri, S.; Koppens, F. H.; de Abajo, F. J. G. Complete Optical Absorption in Periodically Patterned Graphene. *Phys. Rev. Lett.* **2012**, *108047401*.
26. Stankovich, S.; Dikin, D. A.; Dommett, G. H.; Kohlhaas, K. M.; Zimney, E. J.; Stach, E. A.; Piner, R. D.; Nguyen, S. T.; Ruoff, R. S. Graphene-Based Composite Materials. *Nature* **2006**, *442*, 282–286.
27. Lee, C.; Wei, X.; Kysar, J. W.; Hone, J. Measurement of the Elastic Properties and Intrinsic Strength of Monolayer Graphene. *Science* **2008**, *321*, 385–388.
28. Avouris, P.; Dimitrakopoulos, C. Graphene: Synthesis and Applications. *Mater. Today* **2012**, *15*, 86–97.

29. Hernandez, Y.; Nicolosi, V.; Lotya, M.; Blighe, F. M.; Sun, Z.; De, S.; McGovern, I.; Holland, B.; Byrne, M.; Gun'Ko, Y. K. High-Yield Production of Graphene by Liquid-Phase Exfoliation of Graphite. *Nat. Nanotechnol.* **2008**, *3*, 563–568.
30. Du, W.; Jiang, X.; Zhu, L. From Graphite to Graphene: Direct Liquid-Phase Exfoliation of Graphite to Produce Single- and Few-Layered Pristine Graphene. *J. Mater. Chem. A* **2013**, *1*, 10592–10606.
31. Haar, S.; Ciesielski, A.; Clough, J.; Yang, H.; Mazzaro, R.; Richard, F.; Conti, S.; Merstorf, N.; Cecchini, M.; Morandi, V. A Supramolecular Strategy to Leverage the Liquid-Phase Exfoliation of Graphene in the Presence of Surfactants: Unraveling the Role of the Length of Fatty Acids. *Small* **2015**, *11*, 1691–1702.
32. Matsumoto, M.; Saito, Y.; Park, C.; Fukushima, T.; Aida, T. Ultrahigh-Throughput Exfoliation of Graphite Into Pristine 'Single-Layer' Graphene Using Microwaves and Molecularly Engineered Ionic Liquids. *Nat. Chem.* **2015**, *7*, 730–736.
33. Paton, K. R.; Varrla, E.; Backes, C.; Smith, R. J.; Khan, U.; O'Neill, A.; Boland, C.; Lotya, M.; Istrate, O. M.; King, P. Scalable Production of Large Quantities of Defect-Free Few-Layer Graphene by Shear Exfoliation in Liquids. *Nat. Mater.* **2014**, *13*, 624–630.
34. Zhang, P.; Zeng, J.; Zhai, S.; Xian, Y.; Yang, D.; Li, Q. Thermal Properties of Graphene Filled Polymer Composite Thermal Interface Materials. *Macromol. Mater. Eng.* **2017**, *302*, 1700068.
35. Zaman, I.; Kuan, H. C.; Meng, Q.; Michelmore, A.; Kawashima, N.; Pitt, T.; Zhang, L.; Gouda, S.; Luong, L.; Ma, J. A Facile Approach to Chemically Modified Graphene and Its Polymer Nanocomposites. *Adv. Funct. Mater.* **2012**, *22*, 2735–2743.
36. Li, X.; Magnuson, C. W.; Venugopal, A.; Tromp, R. M.; Hannon, J. B.; Vogel, E. M.; Colombo, L.; Ruoff, R. S. Large-Area Graphene Single Crystals Grown by Low-Pressure Chemical Vapor Deposition of Methane on Copper. *J. Am. Chem. Soc.* **2011**, *133*, 2816–2819.
37. Ramón, M. E.; Gupta, A.; Corbet, C.; Ferrer, D. A.; Movva, H. C.; Carpenter, G.; Colombo, L.; Bourianoff, G.; Doczy, M.; Akinwande, D. CMOS-Compatible Synthesis of Large-Area, High-Mobility Graphene by Chemical Vapor Deposition of Acetylene on Cobalt Thin Films. *ACS Nano* **2011**, *5*, 7198–7204.
38. Guermoune, A.; Chari, T.; Popescu, F.; Sabri, S. S.; Guillemette, J.; Skulason, H. S.; Szkopek, T.; Siaj, M. Chemical Vapor Deposition Synthesis of Graphene on Copper With Methanol, Ethanol, and Propanol Precursors. *Carbon* **2011**, *49*, 4204–4210.
39. Patera, L. L.; Africh, C.; Weatherup, R. S.; Blume, R.; Bhardwaj, S.; Castellarin-Cudia, C.; Knop-Gericke, A.; Schloegl, R.; Comelli, G.; Hofmann, S. In Situ Observations of the Atomistic Mechanisms of Ni Catalyzed Low Temperature Graphene Growth. *ACS Nano* **2013**, 7, 7901–7912.
40. Wu, W.; Jauregui, L. A.; Su, Z.; Liu, Z.; Bao, J.; Chen, Y. P.; Yu, Q. Growth of Single Crystal Graphene Arrays by Locally Controlling Nucleation on Polycrystalline Cu Using Chemical Vapor Deposition. *Adv. Mater.* **2011**, *23*, 4898–4903.
41. Tang, S.; Wang, H.; Wang, H. S.; Sun, Q.; Zhang, X.; Cong, C.; Xie, H.; Liu, X.; Zhou, X.; Huang, F. Silane-Catalysed Fast Growth of Large Single-Crystalline Graphene on Hexagonal Boron Nitride. *Nat. Commun.* **2015**, *6*, 6499.
42. Bae, S.; Kim, H.; Lee, Y.; Xu, X.; Park, J.-S.; Zheng, Y.; Balakrishnan, J.; Lei, T.; Kim, H. R.; Song, Y. I. Roll-to-Roll Production of 30-inch Graphene Films for Transparent Electrodes. *Nat. Nanotechnol.* **2010**, *5*, 574.
43. Mattevi, C.; Kim, H.; Chhowalla, M. A Review of Chemical Vapour Deposition of Graphene on Copper. *J. Mater. Chem.* **2011**, *21*, 3324–3334.
44. (a)Ma, T.; Ren, W.; Liu, Z.; Huang, L.; Ma, L.-P.; Ma, X.; Zhang, Z.; Peng, L.-M.; Cheng, H.-M. Repeated Growth–Etching–Regrowth for Large-Area Defect-Free Single-Crystal Graphene by Chemical Vapor Deposition. *ACS Nano* **2014**, *8*,

12806–12813; (b)Zhou, H.; Yu, W. J.; Liu, L.; Cheng, R.; Chen, Y.; Huang, X.; Liu, Y.; Wang, Y.; Huang, Y.; Duan, X. Chemical Vapour Deposition Growth of Large Single Crystals of Monolayer and Bilayer Graphene. *Nat. Commun.* **2013**, *4*, 2096.

45. Gan, L.; Luo, Z. Turning Off Hydrogen to Realize Seeded Growth of Subcentimeter Single-Crystal Graphene Grains on Copper. *ACS Nano* **2013**, *7*, 9480–9488.
46. Zhang, Y.; Gomez, L.; Ishikawa, F. N.; Madaria, A.; Ryu, K.; Wang, C.; Badmaev, A.; Zhou, C. Comparison of Graphene Growth on Single-Crystalline and Polycrystalline Ni by Chemical Vapor Deposition. *J. Phys. Chem. Lett.* **2010**, *1*, 3101–3107.
47. Ogawa, Y.; Hu, B.; Orofeo, C. M.; Tsuji, M.; Ikeda, K.-I.; Mizuno, S.; Hibino, H.; Ago, H. Domain Structure and Boundary in Single-Layer Graphene Grown on Cu (111) and Cu (100) Films. *J. Phys. Chem. Lett.* **2012**, *3*, 219–226.
48. Lee, J.-H.; Lee, E. K.; Joo, W.-J.; Jang, Y.; Kim, B.-S.; Lim, J. Y.; Choi, S.-H.; Ahn, S. J.; Ahn, J. R.; Park, M.-H. Wafer-Scale Growth of Single-Crystal Monolayer Graphene on Reusable Hydrogen-Terminated Germanium. *Science* **2014**, *344*, 286–289.
49. Dai, B.; Fu, L.; Zou, Z.; Wang, M.; Xu, H.; Wang, S.; Liu, Z. Rational Design of a Binary Metal Alloy for Chemical Vapour Deposition Growth of Uniform Single-Layer Graphene. *Nat. Commun.* **2011**, *2*, 522.
50. Geng, D.; Wu, B.; Guo, Y.; Huang, L.; Xue, Y.; Chen, J.; Yu, G.; Jiang, L.; Hu, W.; Liu, Y. Uniform Hexagonal Graphene Flakes and Films Grown on Liquid Copper Surface. *Proc. Natl. Acad. Sci.* **2012**, *109*, 7992–7996.
51. (a)Gomez De Arco, L.; Zhang, Y.; Schlenker, C. W.; Ryu, K.; Thompson, M. E.; Zhou, C. Continuous, Highly Flexible, and Transparent Graphene Films by Chemical Vapor Deposition for Organic Photovoltaics. *ACS Nano* **2010**, *4*, 2865–2873; (b)Suk, J. W.; Kitt, A.; Magnuson, C. W.; Hao, Y.; Ahmed, S.; An, J.; Swan, A. K.; Goldberg, B. B.; Ruoff, R. S. Transfer of CVD-grown monolayer graphene Onto Arbitrary Substrates. *ACS Nano* **2011**, *5*, 6916–6924.
52. Gao, L.; Ren, W.; Xu, H.; Jin, L.; Wang, Z.; Ma, T.; Ma, L.-P.; Zhang, Z.; Fu, Q.; Peng, L.-M. Repeated Growth and Bubbling Transfer of Graphene With Millimetre-Size Single-Crystal Grains Using Platinum. *Nat. Commun.* **2012**, *3*, 699.
53. Gao, L.; Ni, G.-X.; Liu, Y.; Liu, B.; Neto, A. H. C.; Loh, K. P. Face-to-Face Transfer of Wafer-Scale Graphene Films. *Nature* **2014**, *505*, 190–194.
54. Cao, X.; Shi, Y.; Shi, W.; Lu, G.; Huang, X.; Yan, Q.; Zhang, Q.; Zhang, H. Preparation of Novel 3D Graphene Networks for Supercapacitor Applications. *Small* **2011**, *7*, 3163–3168.
55. (a)Wang, Y.; Wang, L.; Yang, T.; Li, X.; Zang, X.; Zhu, M.; Wang, K.; Wu, D.; Zhu, H. Wearable and Highly Sensitive Graphene Strain Sensors for Human Motion Monitoring. *Adv. Funct. Mater.* **2014**, *24*, 4666–4670; (b)Jeong, Y. R.; Park, H.; Jin, S. W.; Hong, S. Y.; Lee, S. S.; Ha, J. S. Highly Stretchable and Sensitive Strain Sensors Using Fragmentized Graphene Foam. *Adv. Funct. Mater.* **2015**, *25*, 4228–4236.
56. Journet, C.; Maser, W.; Bernier, P.; Loiseau, A.; de La Chapelle, M. L.; Lefrant, D. L. S.; Deniard, P.; Lee, R.; Fischer, J. Large-Scale Production of Single-Walled Carbon Nanotubes by the Electric-Arc Technique. *Nature* **1997**, *388*, 756–758.
57. Morales, A. M.; Lieber, C. M. A Laser Ablation Method for the Synthesis of Crystalline Semiconductor Nanowires. *Science* **1998**, *279*, 208–211.
58. Chhowalla, M.; Teo, K.; Ducati, C.; Rupesinghe, N.; Amaratunga, G.; Ferrari, A.; Roy, D.; Robertson, J.; Milne, W. Growth Process Conditions of Vertically Aligned Carbon Nanotubes Using Plasma Enhanced Chemical Vapor Deposition. *J. Appl. Phys.* **2001**, *90*, 5308–5317.

59. Moore, K. E.; Tune, D. D.; Flavel, B. S. Double-Walled Carbon Nanotube Processing. *Adv. Mater.* **2015**, *27*, 3105–3137.
60. Araby, S.; Meng, Q.; Zhang, L.; Zaman, I.; Majewski, P.; Ma, J. Elastomeric Composites Based on Carbon Nanomaterials. *Nanotechnology* **2015**, *26*, 112001.
61. Tanaka, T.; Jin, H.; Miyata, Y.; Fujii, S.; Suga, H.; Naitoh, Y.; Minari, T.; Miyadera, T.; Tsukagoshi, K.; Kataura, H. Simple and Scalable Gel-Based Separation of Metallic and Semiconducting Carbon Nanotubes. *Nano Lett.* **2009**, *9*, 1497–1500.
62. Matsuda, Y.; Tahir-Kheli, J.; Goddard, W. A., III Definitive Band Gaps for Single-Wall Carbon Nanotubes. *J. Phys. Chem. Lett.* **2010**, *1*, 2946–2950.
63. De Volder, M. F.; Tawfick, S. H.; Baughman, R. H.; Hart, A. J. Carbon Nanotubes: Present and Future Commercial Applications. *Science* **2013**, *339*, 535–539.
64. Yu, C.; Masarapu, C.; Rong, J.; Wei, B.; Jiang, H. Stretchable Supercapacitors Based on Buckled Single-Walled Carbon-Nanotube Macrofilms. *Adv. Mater.* **2009**, *21*, 4793–4797.
65. Chen, T.; Peng, H.; Durstock, M.; Dai, L. High-Performance Transparent and Stretchable All-Solid Supercapacitors Based on Highly Aligned Carbon Nanotube Sheets. *Sci. Rep.* **2014**, *4*, 3612.
66. Zu, M.; Li, Q.; Wang, G.; Byun, J. H.; Chou, T. W. Carbon Nanotube Fiber Based Stretchable Conductor. *Adv. Funct. Mater.* **2013**, *23*, 789–793.
67. Yamada, T.; Hayamizu, Y.; Yamamoto, Y.; Yomogida, Y.; Izadi-Najafabadi, A.; Futaba, D. N.; Hata, K. A Stretchable Carbon Nanotube Strain Sensor for Human-Motion Detection. *Nat. Nanotechnol.* **2011**, *6*, 296–301.
68. Ryu, S.; Lee, P.; Chou, J. B.; Xu, R.; Zhao, R.; Hart, A. J.; Kim, S.-G. Extremely Elastic Wearable Carbon Nanotube Fiber Strain Sensor for Monitoring of Human Motion. *ACS Nano* **2015**, *9*, 5929–5936.
69. Song, W.; Jeon, C.; Kim, Y. S.; Kwon, Y. T.; Jung, D. S.; Jang, S. W.; Choi, W. C.; Park, J. S.; Saito, R.; Park, C.-Y. Synthesis of Bandgap-Controlled Semiconducting Single-Walled Carbon Nanotubes. *ACS Nano* **2010**, *4*, 1012–1018.
70. Park, S.; Vosguerichian, M.; Bao, Z. A Review of Fabrication and Applications of Carbon Nanotube Film-Based Flexible Electronics. *Nanoscale* **2013**, *5*, 1727–1752.
71. Moshammer, K.; Hennrich, F.; Kappes, M. M. Selective Suspension in Aqueous Sodium Dodecyl Sulfate According to Electronic Structure Type Allows Simple Separation of Metallic From Semiconducting Single-Walled Carbon Nanotubes. *Nano Res.* **2009**, *2*, 599–606.
72. Wang, W. Z.; Li, W. F.; Pan, X. Y.; Li, C. M.; Li, L. J.; Mu, Y. G.; Rogers, J. A.; Chan-Park, M. B. Degradable Conjugated Polymers: Synthesis and Applications in Enrichment of Semiconducting Single-Walled Carbon Nanotubes. *Adv. Funct. Mater.* **2011**, *21*, 1643–1651.
73. (a)Brady, G. J.; Joo, Y.; Roy, S. S.; Gopalan, P.; Arnold, M. S. High Performance Transistors via Aligned Polyfluorene-Sorted Carbon Nanotubes. *Appl. Phys. Lett.* **2014**, *104*, 083107; (b)Islam, A. E.; Rogers, J. A.; Alam, M. A. Recent Progress in Obtaining Semiconducting Single-Walled Carbon Nanotubes for Transistor Applications. *Adv. Mater.* **2015**, *27*, 7908–7937.
74. Liu, H.; Nishide, D.; Tanaka, T.; Kataura, H. Large-Scale Single-Chirality Separation of Single-Wall Carbon Nanotubes by Simple Gel Chromatography. *Nat. Commun.* **2011**, *2*, 309.
75. Wang, H.; Hsieh, B.; Jiménez-Osés, G.; Liu, P.; Tassone, C. J.; Diao, Y.; Lei, T.; Houk, K. N.; Bao, Z. Solvent Effects on Polymer Sorting of Carbon Nanotubes With Applications in Printed Electronics. *Small* **2015**, *11*, 126–133.
76. Derenskyi, V.; Gomulya, W.; Rios, J. M. S.; Fritsch, M.; Fröhlich, N.; Jung, S.; Allard, S.; Bisri, S. Z.; Gordiichuk, P.; Herrmann, A. Carbon Nanotube Network Ambipolar Field-Effect Transistors with 108 On/Off Ratio. *Adv. Mater.* **2014**, *26*, 5969–5975.

77. Jariwala, D.; Sangwan, V. K.; Lauhon, L. J.; Marks, T. J.; Hersam, M. C. Carbon Nanomaterials for Electronics, Optoelectronics, Photovoltaics, and Sensing. *Chem. Soc. Rev.* **2013**, *42*, 2824–2860.
78. Ye, S.; Rathmell, A. R.; Chen, Z.; Stewart, I. E.; Wiley, B. J. Metal Nanowire Networks: The Next Generation of Transparent Conductors. *Adv. Mater.* **2014**, *26*, 6670–6687.
79. (a)Kim, Y. H.; Lee, J.; Hofmann, S.; Gather, M. C.; Müller-Meskamp, L.; Leo, K. Achieving High Efficiency and Improved Stability in ITO-Free Transparent Organic Light-Emitting Diodes With Conductive Polymer Electrodes. *Adv. Funct. Mater.* **2013**, *23*, 3763–3769; (b)Han, T.-H.; Lee, Y.; Choi, M.-R.; Woo, S.-H.; Bae, S.-H.; Hong, B. H.; Ahn, J.-H.; Lee, T.-W. Extremely Efficient Flexible Organic Light-Emitting Diodes With Modified Graphene Anode. *Nat. Photonics* **2012**, *6*, 105–110.
80. Chen, Y.; Ding, X.; Steven Lin, S.-C.; Yang, S.; Huang, P.-H.; Nama, N.; Zhao, Y.; Nawaz, A. A.; Guo, F.; Wang, W. Tunable Nanowire Patterning Using Standing Surface Acoustic Waves. *ACS Nano* **2013**, *7*, 3306–3314.
81. Zhang, W.; Chen, P.; Gao, Q.; Zhang, Y.; Tang, Y. High-Concentration Preparation of Silver Nanowires: Restraining In Situ Nitric Acidic Etching by Steel-Assisted Polyol Method. *Chem. Mater.* **2008**, *20*, 1699–1704.
82. Schuette, W. M.; Buhro, W. E. Silver Chloride as a Heterogeneous Nucleant for the Growth of Silver Nanowires. *ACS Nano* **2013**, *7*, 3844–3853.
83. Gong, C.; Liang, J.; Hu, W.; Niu, X.; Ma, S.; Hahn, H. T.; Pei, Q. A Healable, Semitransparent Silver Nanowire-Polymer Composite Conductor. *Adv. Mater.* **2013**, *25*, 4186–4191.
84. Korte, K. E.; Skrabalak, S. E.; Xia, Y. Rapid Synthesis of Silver Nanowires Through a CuCl-or CuCl 2-Mediated Polyol Process. *J. Mater. Chem.* **2008**, *18*, 437–441.
85. Garnett, E. C.; Cai, W.; Cha, J. J.; Mahmood, F.; Connor, S. T.; Christoforo, M. G.; Cui, Y.; McGehee, M. D.; Brongersma, M. L. Self-Limited Plasmonic Welding of Silver Nanowire Junctions. *Nat. Mater.* **2012**, *11*, 241–249.
86. Hauger, T. C.; Al-Rafia, S. I.; Buriak, J. M. Rolling Silver Nanowire Electrodes: Simultaneously Addressing Adhesion, Roughness, and Conductivity. *ACS Appl. Mater. Interfaces* **2013**, *5*, 12663–12671.
87. Jurewicz, I.; Fahimi, A.; Lyons, P. E.; Smith, R. J.; Cann, M.; Large, M. L.; Tian, M.; Coleman, J. N.; Dalton, A. B. Insulator-Conductor Type Transitions in Graphene-Modified Silver Nanowire Networks: A Route to Inexpensive Transparent Conductors. *Adv. Funct. Mater.* **2014**, *24*, 7580–7587.
88. Jin, Y.; Li, L.; Cheng, Y.; Kong, L.; Pei, Q.; Xiao, F. Cohesively Enhanced Conductivity and Adhesion of Flexible Silver Nanowire Networks by Biocompatible Polymer Sol–Gel Transition. *Adv. Funct. Mater.* **2015**, *25*, 1581–1587.
89. Lu, H.; Zhang, D.; Ren, X.; Liu, J.; Choy, W. C. Selective Growth and Integration of Silver Nanoparticles on Silver Nanowires at Room Conditions for Transparent Nano-Network Electrode. *ACS Nano* **2014**, *8*, 10980–10987.
90. Liang, J.; Li, L.; Tong, K.; Ren, Z.; Hu, W.; Niu, X.; Chen, Y.; Pei, Q. Silver Nanowire Percolation Network Soldered With Graphene Oxide at Room Temperature and Its Application for Fully Stretchable Polymer Light-Emitting Diodes. *ACS Nano* **2014**, *8*, 1590–1600.
91. Lee, J.; Woo, J. Y.; Kim, J. T.; Lee, B. Y.; Han, C.-S. Synergistically Enhanced Stability of Highly Flexible Silver Nanowire/Carbon Nanotube Hybrid Transparent Electrodes by Plasmonic Welding. *ACS Appl. Mater. Interfaces* **2014**, *6*, 10974–10980.
92. Kim, J.-H.; Park, J.-W. Foldable Transparent Substrates with Embedded Electrodes for Flexible Electronics. *ACS Appl. Mater. Interfaces* **2015**, *7*, 18574–18580.
93. Lee, H.; Lee, K.; Park, J. T.; Kim, W. C.; Lee, H. Well-Ordered and High Density Coordination-Type Bonding to Strengthen Contact of Silver Nanowires on Highly Stretchable Polydimethylsiloxane. *Adv. Funct. Mater.* **2014**, *24*, 3276–3283.

94. Xu, Q.; Song, T.; Cui, W.; Liu, Y.; Xu, W.; Lee, S.-T.; Sun, B. Solution-Processed Highly Conductive PEDOT: PSS/AgNW/GO Transparent Film for Efficient Organic-Si Hybrid Solar Cells. *ACS Appl. Mater. Interfaces* **2015**, 7, 3272–3279.
95. (a)Hong, X.; Tan, C.; Chen, J.; Xu, Z.; Zhang, H. Synthesis, Properties and Applications of One- and Two-Dimensional Gold Nanostructures. *Nano Res.* **2015**, *8*, 40–55; (b)Lohse, S. E.; Murphy, C. J. The Quest for Shape Control: A History of Gold Nanorod Synthesis. *Chem. Mater.* **2013**, *25*, 1250–1261; (c)Drack, M.; Graz, I.; Sekitani, T.; Someya, T.; Kaltenbrunner, M.; Bauer, S. An Imperceptible Plastic Electronic Wrap. *Adv. Mater.* **2015**, *27*, 34–40.
96. Kimling, J.; Maier, M.; Okenve, B.; Kotaidis, V.; Ballot, H.; Plech, A. Turkevich Method for Gold Nanoparticle Synthesis Revisited. *J. Phys. Chem. B* **2006**, *110*, 15700–15707.
97. Schmid, G.; Corain, B. Nanoparticulated Gold: Syntheses, Structures, Electronics, and Reactivities. *Eur. J. Inorg. Chem.* **2003**, *2003*, 3081–3098.
98. Khan, Y.; Pavinatto, F. J.; Lin, M. C.; Liao, A.; Swisher, S. L.; Mann, K.; Subramanian, V.; Maharbiz, M. M.; Arias, A. C. Inkjet-Printed Flexible Gold Electrode Arrays for Bioelectronic Interfaces. *Adv. Funct. Mater.* **2016**, *26*, 1004–1013.
99. Kim, Y.; Zhu, J.; Yeom, B.; Di Prima, M.; Su, X.; Kim, J.-G.; Yoo, S. J.; Uher, C.; Kotov, N. A. Stretchable Nanoparticle Conductors With Self-Organized Conductive Pathways. *Nature* **2013**, *500*, 59–63.
100. Han, S. T.; Zhou, Y.; Xu, Z. X.; Huang, L. B.; Yang, X. B.; Roy, V. Microcontact Printing of Ultrahigh Density Gold Nanoparticle Monolayer for Flexible Flash Memories. *Adv. Mater.* **2012**, *24*, 3556–3561.
101. Gong, S.; Schwalb, W.; Wang, Y.; Chen, Y.; Tang, Y.; Si, J.; Shirinzadeh, B.; Cheng, W. A Wearable and Highly Sensitive Pressure Sensor With Ultrathin Gold Nanowires. *Nat. Commun.* **2014**, *5*, 3132.
102. Forrer, P.; Schlottig, F.; Siegenthaler, H.; Textor, M. Electrochemical Preparation and Surface Properties of Gold Nanowire Arrays Formed by the Template Technique. *J. Appl. Electrochem.* **2000**, *30*, 533–541.
103. Halder, A.; Ravishankar, N. Ultrafine Single-Crystalline Gold Nanowire Arrays by Oriented Attachment. *Adv. Mater.* **2007**, *19*, 1854–1858.
104. (a)Feng, H.; Yang, Y.; You, Y.; Li, G.; Guo, J.; Yu, T.; Shen, Z.; Wu, T.; Xing, B. Simple and Rapid Synthesis of Ultrathin Gold Nanowires, Their Self-Assembly and Application in Surface-Enhanced Raman Scattering. *Chem. Commun.* **2009**, *15*, 1984–1986; (b)Huo, Z.; Tsung, C.-K.; Huang, W.; Zhang, X.; Yang, P. Sub-Two Nanometer Single Crystal Au Nanowires. *Nano Lett.* **2008**, *8*, 2041–2044; (c)Lu, X.; Yavuz, M. S.; Tuan, H.-Y.; Korgel, B. A.; Xia, Y. Ultrathin Gold Nanowires Can Be Obtained by Reducing Polymeric Strands of Oleylamine – AuCl Complexes Formed via Aurophilic Interaction. J. Am. Chem. Soc. **2008**, *130*, 8900–8901.
105. Chen, Y.; Ouyang, Z.; Gu, M.; Cheng, W. Mechanically Strong, Optically Transparent, Giant Metal Superlattice Nanomembranes From Ultrathin Gold Nanowires. *Adv. Mater.* **2013**, *25*, 80–85.
106. Gong, S.; Lai, D. T.; Su, B.; Si, K. J.; Ma, Z.; Yap, L. W.; Guo, P.; Cheng, W. Highly Stretchy Black Gold E-Skin Nanopatches as Highly Sensitive Wearable Biomedical Sensors. *Adv. Electron. Mater.* **2015**, *1*, 1400063.
107. Yoo, J.; Jeong, S.; Kim, S.; Je, J. H. A Stretchable Nanowire UV–vis–NIR Photodetector With High Performance. *Adv. Mater.* **2015**, *27*, 1712–1717.
108. Khang, D.-Y.; Jiang, H.; Huang, Y.; Rogers, J. A. A Stretchable Form of Single-Crystal Silicon for High-Performance Electronics on Rubber Substrates. *Science* **2006**, *311*, 208–212.
109. Xie, K.; Wei, B. Materials and Structures for Stretchable Energy Storage and Conversion Devices. *Adv. Mater.* **2014**, *26*, 3592–3617.

110. Yu, Y.; Luo, S.; Sun, L.; Wu, Y.; Jiang, K.; Li, Q.; Wang, J.; Fan, S. Ultra-Stretchable Conductors Based on Buckled Super-Aligned Carbon Nanotube Films. *Nanoscale* **2015**, 7, 10178–10185.
111. Qi, D.; Liu, Z.; Liu, Y.; Leow, W. R.; Zhu, B.; Yang, H.; Yu, J.; Wang, W.; Wang, H.; Yin, S. Suspended Wavy Graphene Microribbons for Highly Stretchable Microsupercapacitors. *Adv. Mater.* **2015**, *27*, 5559–5566.
112. Weng, W.; Sun, Q.; Zhang, Y.; He, S.; Wu, Q.; Deng, J.; Fang, X.; Guan, G.; Ren, J.; Peng, H. A Gum-Like Lithium-Ion Battery Based on a Novel Arched Structure. *Adv. Mater.* **2015**, *27*, 1363–1369.
113. Xu, F.; Zhu, Y. Highly Conductive and Stretchable Silver Nanowire Conductors. *Adv. Mater.* **2012**, *24*, 5117–5122.
114. Cai, L.; Song, L.; Luan, P.; Zhang, Q.; Zhang, N.; Gao, Q.; Zhao, D.; Zhang, X.; Tu, M.; Yang, F. Super-Stretchable, Transparent Carbon Nanotube-Based Capacitive Strain Sensors for Human Motion Detection. *Sci. Rep.* **2013**, *3*, 3048.
115. Cheng, T.; Zhang, Y.; Lai, W. Y.; Huang, W. Stretchable Thin-Film Electrodes for Flexible Electronics With High Deformability and Stretchability. *Adv. Mater.* **2015**, *27*, 3349–3376.
116. Lu, N.; Yang, S. Mechanics for Stretchable Sensors. *Curr. Opinion Solid State Mater. Sci.* **2015**, *19*, 149–159.
117. Kim, J.; Lee, M.; Shim, H. J.; Ghaffari, R.; Cho, H. R.; Son, D.; Jung, Y. H.; Soh, M.; Choi, C.; Jung, S. Stretchable Silicon Nanoribbon Electronics for Skin Prosthesis. *Nat. Commun.* **2014**, *5*, 5747.
118. Widlund, T.; Yang, S.; Hsu, Y.-Y.; Lu, N. Stretchability and Compliance of Free-standing Serpentine-Shaped Ribbons. *Int. J. Solids Struct.* **2014**, *51*, 4026–4037.
119. Choi, S.; Park, J.; Hyun, W.; Kim, J.; Kim, J.; Lee, Y. B.; Song, C.; Hwang, H. J.; Kim, J. H.; Hyeon, T. Stretchable Heater Using Ligand-Exchanged Silver Nanowire Nanocomposite for Wearable Articular Thermotherapy. *ACS Nano* **2015**, *9*, 6626–6633.
120. Moreno, S.; Baniasadi, M.; Mohammed, S.; Mejia, I.; Chen, Y.; Quevedo-Lopez, M. A.; Kumar, N.; Dimitrijevich, S.; Minary-Jolandan, M. Biocompatible Collagen Films as Substrates for Flexible Implantable Electronics. *Adv. Electron. Mater.* **2015**, *1*, 1500154.
121. Xu, L.; Gutbrod, S. R.; Bonifas, A. P.; Su, Y.; Sulkin, M. S.; Lu, N.; Chung, H.-J.; Jang, K.-I.; Liu, Z.; Ying, M. 3D Multifunctional Integumentary Membranes for Spatiotemporal Cardiac Measurements and Stimulation Across the Entire Epicardium. *Nat. Commun.* **2014**, *5*, 3329.
122. Xu, B.; Akhtar, A.; Liu, Y.; Chen, H.; Yeo, W. H.; Park, S. I.; Boyce, B.; Kim, H.; Yu, J.; Lai, H. Y. An Epidermal Stimulation and Sensing Platform for Sensorimotor Prosthetic Control, Management of Lower Back Exertion, and Electrical Muscle Activation. *Adv. Mater.* **2016**, *28*, 4462–4471.
123. Zeng, W.; Shu, L.; Li, Q.; Chen, S.; Wang, F.; Tao, X. M. Fiber-Based Wearable Electronics: A Review of Materials, Fabrication, Devices, and Applications. *Adv. Mater.* **2014**, *26*, 5310–5336.
124. (a)Zhang, D.; Miao, M.; Niu, H.; Wei, Z. Core-Spun Carbon Nanotube Yarn Supercapacitors for Wearable Electronic Textiles. *ACS Nano* **2014**, *8*, 4571–4579; (b)Li, C.; Cui, Y.-L.; Tian, G.-L.; Shu, Y.; Wang, X.-F.; Tian, H.; Yang, Y.; Wei, F.; Ren, T.-L. Flexible CNT-Array Double Helices Strain Sensor With High Stretchability for Motion Capture. *Sci. Rep.* **2015**, *5*, 15554.
125. Zhang, Y.; Bai, W.; Cheng, X.; Ren, J.; Weng, W.; Chen, P.; Fang, X.; Zhang, Z.; Peng, H. Flexible and Stretchable Lithium-Ion Batteries and Supercapacitors Based on Electrically Conducting Carbon Nanotube Fiber Springs. *Angew. Chem.* **2014**, *126*, 14792–14796.

126. Pan, S.; Lin, H.; Deng, J.; Chen, P.; Chen, X.; Yang, Z.; Peng, H. Novel Wearable Energy Devices Based on Aligned Carbon Nanotube Fiber Textiles. *Adv. Energy Mater.* **2015**, *5*, 1401438.
127. Lee, S.; Shin, S.; Lee, S.; Seo, J.; Lee, J.; Son, S.; Cho, H. J.; Algadi, H.; Al-Sayari, S.; Kim, D. E. Ag Nanowire Reinforced Highly Stretchable Conductive Fibers for Wearable Electronics. *Adv. Funct. Mater.* **2015**, *25*, 3114–3121.
128. Ma, R.; Kang, B.; Cho, S.; Choi, M.; Baik, S. Extraordinarily High Conductivity of Stretchable Fibers of Polyurethane and Silver Nanoflowers. *ACS Nano* **2015**, *9*, 10876–10886.
129. Yang, Z.; Deng, J.; Chen, X.; Ren, J.; Peng, H. A Highly Stretchable, Fiber-Shaped Supercapacitor. *Angew. Chem. Int. Ed.* **2013**, *52*, 13453–13457.
130. Zhang, Z.; Deng, J.; Li, X.; Yang, Z.; He, S.; Chen, X.; Guan, G.; Ren, J.; Peng, H. Superelastic Supercapacitors With High Performances During Stretching. *Adv. Mater.* **2015**, *27*, 356–362.
131. Liu, Z.; Fang, S.; Moura, F.; Ding, J.; Jiang, N.; Di, J.; Zhang, M.; Lepró, X.; Galvão, D.; Haines, C. Hierarchically Buckled Sheath-Core Fibers for Superelastic Electronics, Sensors, and Muscles. *Science* **2015**, *349*, 400–404.
132. Tybrandt, K.; Vörös, J. Fast and Efficient Fabrication of Intrinsically Stretchable Multilayer Circuit Boards by Wax Pattern Assisted Filtration. *Small* **2016**, *12*, 180–184.
133. Sekitani, T.; Nakajima, H.; Maeda, H.; Fukushima, T.; Aida, T.; Hata, K.; Someya, T. Stretchable Active-Matrix Organic Light-Emitting Diode Display Using Printable Elastic Conductors. *Nat. Mater.* **2009**, *8*, 494–499.
134. Patel, S.; Park, H.; Bonato, P.; Chan, L.; Rodgers, M. A Review of Wearable Sensors and Systems With Application in Rehabilitation. *J. Neuroeng. Rehabil.* **2012**, *9*, 1.
135. Roh, E.; Hwang, B.-U.; Kim, D.; Kim, B.-Y.; Lee, N.-E. Stretchable, Transparent, Ultrasensitive, and Patchable Strain Sensor for Human–Machine Interfaces Comprising a Nanohybrid of Carbon Nanotubes and Conductive Elastomers. *ACS Nano* **2015**, *9*, 6252–6261.
136. Lee, J.; Kwon, H.; Seo, J.; Shin, S.; Koo, J. H.; Pang, C.; Son, S.; Kim, J. H.; Jang, Y. H.; Kim, D. E. Conductive Fiber-Based Ultrasensitive Textile Pressure Sensor for Wearable Electronics. *Adv. Mater.* **2015**, *27*, 2433–2439.
137. Pang, C.; Lee, G.-Y.; Kim, T.-I.; Kim, S. M.; Kim, H. N.; Ahn, S.-H.; Suh, K.-Y. A Flexible and Highly Sensitive Strain-Gauge Sensor Using Reversible Interlocking of Nanofibres. *Nat. Mater.* **2012**, *11*, 795–801.
138. Sun, Q.; Seung, W.; Kim, B. J.; Seo, S.; Kim, S. W.; Cho, J. H. Active Matrix Electronic Skin Strain Sensor Based on Piezopotential-Powered Graphene Transistors. *Adv. Mater.* **2015**, *27*, 3411–3417.
139. Kim, K. K.; Hong, S.; Cho, H. M.; Lee, J.; Suh, Y. D.; Ham, J.; Ko, S. H. Highly Sensitive and Stretchable Multidimensional Strain Sensor With Prestrained Anisotropic Metal Nanowire Percolation Networks. *Nano Lett.* **2015**, *15*, 5240–5247.
140. Gong, S.; Lai, D. T.; Wang, Y.; Yap, L. W.; Si, K. J.; Shi, Q.; Jason, N. N.; Sridhar, T.; Uddin, H.; Cheng, W. Tattoolike Polyaniline Microparticle-Doped Gold Nanowire Patches as Highly Durable Wearable Sensors. *ACS Appl. Mater. Interfaces* **2015**, *7*, 19700–19708.
141. Shi, G.; Zhao, Z.; Pai, J. H.; Lee, I.; Zhang, L.; Stevenson, C.; Ishara, K.; Zhang, R.; Zhu, H.; Ma, J. Highly Sensitive, Wearable, Durable Strain Sensors and Stretchable Conductors Using Graphene/Silicon Rubber Composites. *Adv. Funct. Mater.* **2016**, *26*, 7614–7625.
142. Lim, S.; Son, D.; Kim, J.; Lee, Y. B.; Song, J. K.; Choi, S.; Lee, D. J.; Kim, J. H.; Lee, M.; Hyeon, T. Transparent and Stretchable Interactive Human Machine Interface Based on Patterned Graphene Heterostructures. *Adv. Funct. Mater.* **2015**, *25*, 375–383.

143. Son, D.; Lee, J.; Qiao, S.; Ghaffari, R.; Kim, J.; Lee, J. E.; Song, C.; Kim, S. J.; Lee, D. J.; Jun, S. W. Multifunctional Wearable Devices for Diagnosis and Therapy of Movement Disorders. *Nat. Nanotechnol.* **2014**, *9*, 397–404.
144. Hammock, M. L.; Chortos, A.; Tee, B. C. K.; Tok, J. B. H.; Bao, Z. 25th Anniversary Article: The Evolution of Electronic Skin (E-Skin): A Brief History, Design Considerations, and Recent Progress. *Adv. Mater.* **2013**, *25*, 5997–6038.
145. Schwartz, G.; Tee, B. C.-K.; Mei, J.; Appleton, A. L.; Kim, D. H.; Wang, H.; Bao, Z. Flexible polymer transistors with high Pressure Sensitivity for Application in Electronic Skin and Health Monitoring. *Nat. Commun.* **2013**, *4*, 1859.
146. Park, Y.-L.; Chen, B.-R.; Wood, R. J. Design and Fabrication of Soft Artificial Skin Using Embedded Microchannels and Liquid Conductors. *IEEE Sens. J.* **2012**, *12*, 2711–2718.
147. Chou, H.-H.; Nguyen, A.; Chortos, A.; To, J. W.; Lu, C.; Mei, J.; Kurosawa, T.; Bae, W.-G.; Tok, J. B.-H.; Bao, Z. A Chameleon-Inspired Stretchable Electronic Skin With Interactive Colour Changing Controlled by Tactile Sensing. *Nat. Commun.* **2015**, *6*, 8011.
148. Zhao, X.; Hua, Q.; Yu, R.; Zhang, Y.; Pan, C. Flexible, Stretchable and Wearable Multifunctional Sensor Array as Artificial Electronic Skin for Static and Dynamic Strain Mapping. *Adv. Electron. Mater.* **2015**, *1*, 1500142.
149. Jeong, J. W.; Yeo, W. H.; Akhtar, A.; Norton, J. J.; Kwack, Y. J.; Li, S.; Jung, S. Y.; Su, Y.; Lee, W.; Xia, J. Materials and Optimized Designs for Human-Machine Interfaces via Epidermal Electronics. *Adv. Mater.* **2013**, *25*, 6839–6846.
150. Harada, S.; Kanao, K.; Yamamoto, Y.; Arie, T.; Akita, S.; Takei, K. Fully Printed Flexible Fingerprint-Like Three-Axis Tactile and Slip Force and Temperature Sensors for Artificial Skin. *ACS Nano* **2014**, *8*, 12851–12857.
151. Xu, F.; Wu, M.-Y.; Safron, N. S.; Roy, S. S.; Jacobberger, R. M.; Bindl, D. J.; Seo, J.-H.; Chang, T.-H.; Ma, Z.; Arnold, M. S. Highly Stretchable Carbon Nanotube Transistors With Ion Gel Gate Dielectrics. *Nano Lett.* **2014**, *14*, 682–686.
152. Son, D.; Koo, J. H.; Song, J.-K.; Kim, J.; Lee, M.; Shim, H. J.; Park, M.; Lee, M.; Kim, J. H.; Kim, D.-H. Stretchable Carbon Nanotube Charge-Trap Floating-Gate Memory and Logic Devices for Wearable Electronics. *ACS Nano* **2015**, *9*, 5585–5593.
153. Zhan, B.; Li, C.; Yang, J.; Jenkins, G.; Huang, W.; Dong, X. Graphene Field-Effect Transistor and Its Application for Electronic Sensing. *Small* **2014**, *10*, 4042–4065.
154. Kim, J.; Lee, M. S.; Jeon, S.; Kim, M.; Kim, S.; Kim, K.; Bien, F.; Hong, S. Y.; Park, J. U. Highly Transparent and Stretchable Field-Effect Transistor Sensors Using Graphene–Nanowire Hybrid Nanostructures. *Adv. Mater.* **2015**, *27*, 3292–3297.
155. Liang, J.; Li, L.; Chen, D.; Hajagos, T.; Ren, Z.; Chou, S.-Y.; Hu, W.; Pei, Q. Intrinsically Stretchable and Transparent Thin-Film Transistors Based on Printable Silver Nanowires, Carbon Nanotubes and an Elastomeric Dielectric. *Nat. Commun.* **2015**, *6*, 7647.
156. Chen, L.; Liu, Y.; Zhao, Y.; Chen, N.; Qu, L. Graphene-Based Fibers for Supercapacitor Applications. *Nanotechnology* **2015**, *27032001*.
157. Lu, Q.; Chen, J. G.; Xiao, J. Q. Nanostructured Electrodes for High-Performance Pseudocapacitors. *Angew. Chem. Int. Ed.* **2013**, *52*, 1882–1889.
158. Xu, P.; Kang, J.; Choi, J.-B.; Suhr, J.; Yu, J.; Li, F.; Byun, J.-H.; Kim, B.-S.; Chou, T.-W. Laminated Ultrathin Chemical Vapor Deposition Graphene Films Based Stretchable and Transparent High-Rate Supercapacitor. *ACS Nano* **2014**, *8*, 9437–9445.
159. Huang, Y.; Zhong, M.; Huang, Y.; Zhu, M.; Pei, Z.; Wang, Z.; Xue, Q.; Xie, X.; Zhi, C. A Self-Healable and Highly Stretchable Supercapacitor Based on a Dual Crosslinked Polyelectrolyte. *Nat. Commun.* **2015**, *6*, 10310.

160. Xu, P.; Wei, B.; Cao, Z.; Zheng, J.; Gong, K.; Li, F.; Yu, J.; Li, Q.; Lu, W.; Byun, J.-H. Stretchable Wire-Shaped Asymmetric Supercapacitors Based on Pristine and MnO2 Coated Carbon Nanotube Fibers. *ACS Nano* **2015**, *9*, 6088–6096.
161. Kim, D.; Shin, G.; Kang, Y. J.; Kim, W.; Ha, J. S. Fabrication of a Stretchable Solid-State Micro-Supercapacitor Array. *ACS Nano* **2013**, 7, 7975–7982.
162. Choi, C.; Kim, S. H.; Sim, H. J.; Lee, J. A.; Choi, A. Y.; Kim, Y. T.; Lepró, X.; Spinks, G. M.; Baughman, R. H.; Kim, S. J. Stretchable, Weavable Coiled Carbon Nanotube/mno2/Polymer Fiber Solid-State Supercapacitors. Sci. Rep. **2015**, *5*, 9387.
163. (a)Li, X.; Gu, T.; Wei, B. Dynamic and Galvanic Stability of Stretchable Supercapacitors. *Nano Lett.* **2012**, *12*(12), 6366–6371; (b)Nam, I.; Kim, G.-P.; Park, S.; Han, J. W.; Yi, J. All-Solid-State, Origami-Type Foldable Supercapacitor Chips With Integrated Series Circuit Analogues. *Energy Environ. Sci.* **2014**, 7, 1095–1102.
164. Wu, Q.; Xu, Y.; Yao, Z.; Liu, A.; Shi, G. Supercapacitors Based on Flexible Graphene/Polyaniline Nanofiber Composite Films. *ACS Nano* **2010**, *4*, 1963–1970.
165. Bae, J.; Song, M. K.; Park, Y. J.; Kim, J. M.; Liu, M.; Wang, Z. L. Fiber Supercapacitors Made of Nanowire-Fiber Hybrid Structures for Wearable/Flexible Energy Storage. *Angew. Chem. Int. Ed.* **2011**, *50*, 1683–1687.
166. Li, Y.; Sheng, K.; Yuan, W.; Shi, G. A High-Performance Flexible Fibre-Shaped Electrochemical Capacitor Based on Electrochemically Reduced Graphene Oxide. *Chem. Commun.* **2013**, *49*, 291–293.
167. (a)Xu, P.; Gu, T.; Cao, Z.; Wei, B.; Yu, J.; Li, F.; Byun, J. H.; Lu, W.; Li, Q.; Chou, T. W. Carbon Nanotube Fiber Based Stretchable Wire-Shaped Supercapacitors. *Adv. Energy Mater.* **2014**, *4*, 1300759; (b)Kou, L.; Huang, T.; Zheng, B.; Han, Y.; Zhao, X.; Gopalsamy, K.; Sun, H.; Gao, C. Coaxial Wet-Spun Yarn Supercapacitors for High-Energy Density and Safe Wearable Electronics. *Nat. Commun.* **2014**, *5*, 3754; (c)Ma, Y.; Li, P.; Sedloff, J. W.; Zhang, X.; Zhang, H.; Liu, J. Conductive Graphene Fibers for Wire-Shaped Supercapacitors Strengthened by Unfunctionalized Few-Walled Carbon Nanotubes. *ACS Nano* **2015**, *9*, 1352–1359.
168. Hong, S. Y.; Yoon, J.; Jin, S. W.; Lim, Y.; Lee, S.-J.; Zi, G.; Ha, J. S. High-Density, Stretchable, All-Solid-State Microsupercapacitor Arrays. *ACS Nano* **2014**, *8*, 8844–8855.
169. Liu, W.; Lu, C.; Wang, X.; Tay, R. Y.; Tay, B. K. High-Performance Microsupercapacitors Based on Two-Dimensional Graphene/Manganese Dioxide/Silver Nanowire Ternary Hybrid Film. *ACS Nano* **2015**, *9*, 1528–1542.

CHAPTER 9

2D materials for bio-photonic applications

Hui Ying Hoh[a,b], Bannur Nanjunda Shivananju[c,d], Chang-Ming Li[c,e], Qiaoliang Bao[d]

[a]Key Laboratory of Optoelectronic Devices and Systems of Ministry of Education and Guangdong Province, College of Optoelectronic Engineering, Shenzhen University, Shenzhen, China
[b]Institute of Microscale Optoelectronics, Shenzhen University, Shenzhen, China
[c]Institute of Advanced Cross-field Science, College of Life Science, Qingdao University, Qingdao, China
[d]Department of Materials Science and Engineering, ARC Centre of Excellence in Future Low-Energy Electronics Technologies (FLEET), Monash University, Clayton, VIC, Australia
[e]Institute of Material Science and Devices, Suzhou University of Science and Technology, Suzhou, China

9.1 Introduction

Owning to its unique structure and exceptional properties, two-dimensional (2D) material graphene has demonstrated immerse potential in photonics and optoelectronics, especially in lasing and energy harvesting applications.[1, 2] Similarly, other 2D materials such as transition metal dichalcogenides, topological insulators, boron nitride, and black phosphorus (BP) have attracted tremendous research efforts. With high surface-to-volume ratios, 2D materials would be sensitive to modify their surfaces or adsorbates. Coupled with biocompatibility and excellent optical properties, 2D materials are increasingly being considered for biomedical applications. In 2015, it was noted that graphene-based 2D materials are the most extensively studied, followed by silicate clays and layered double hydroxides.[3] There were few reports on transition metal oxides and chalcogenides or BP; since then, however, reports on these materials have been emerging steadily. The 2015 review also finds that most of the biomedical applications of 2D nanomaterials are utilized in the areas of biosensors and drug delivery, making up almost 90% of the publications.[3] We recognize that 2D materials are also potential candidates for bio-photonic applications. Therefore, in this chapter, we consider the different classes of 2D materials: graphene [usually in the form of graphene oxide (GO)], transition metal chalcogenides, and BP, and discuss their potential uses of bio-photonics. We also review the current research progress of these 2D materials in bio-related applications.

2D Materials for Photonic and Optoelectronic Applications
https://doi.org/10.1016/B978-0-08-102637-3.00009-7

9.2 Biocompatibility and cytotoxicity of 2D materials

When selecting materials used for biomedical applications, the first factor to consider would be biocompatibility. Biocompatibility refers to the behavior of a material in a biological system; a biocompatible material does not cause toxic effects or immune response while performing its desired function. Another important consideration for drug delivery and therapeutic applications is cytotoxicity. Cytotoxicity of a material indicates the degree to which extent that material is harmful to cells. Most of the 2D materials exhibited low cytotoxicity, although it must be noted that variations in size, thickness, shape, concentration, and surface functional groups should be taken into consideration.

9.2.1 Graphene

The biological and environmental effects of carbon-based nanomaterials have been studied for decades since the proliferation of uses involving carbon black, fullerenes, and carbon nanotubes (CNTs). In the case of graphene, it was found that graphene demonstrates different cytotoxic effects from single-walled CNTs, despite the two being similar in chemical composition and crystal structure.[4] Therefore, for graphitic materials, the shape plays a critical role in the way it interacts with biological systems. The cytotoxicity of graphene arises from direct interactions between the nanosheets and the cell membrane, causing physical damage to the cell membrane.[5] However, when proteins are present at a sufficient concentration, the nanosheets can conjugate (coronas) with proteins, mitigating the cytotoxicity effect. Most toxicity studies of graphene refer to the oxidized form: GO, which is slightly different from graphene in terms of structure (containing more sp^3 carbon) and functional groups (–O, –OH). GO also displayed mutagenic potential, and thus may induce germline mutation, posing a hazard to future generations.[6] It is believed that that graphene poses less risk compared to GO, as shown in both in vitro and in vivo studies on mammalian cells and amphibian models.[7] For an in-depth analysis on the biocompatibility and toxicology investigations of graphene and GO, the reviews by Volkov et al.[8] and Seabra et al.[9] are excellent resources.

9.2.2 Transitional metal chalcogenides

For the transitional metal chalcogenides (TMDs), due to the large varieties of 2D TMDs, we will not discuss all of them, but will focus on the sulfites, mainly MoS_2, which appear to have lower toxicity than selenides and tellurides. Readers interested in a broad summary may wish to refer to the work by Nath et al., which provided an excellent summary of toxicity studies of 2D materials and fibrous materials, focusing on the transition metal oxides and chalcogenides, with a brief mention of h-BN.[10]

The most commonly studied member of the TMDs is MoS_2, followed by WS_2. Both have displayed biocompatibility and low toxicity[11]; in vitro studies with human lung carcinoma epithelial cells also showed that MoS_2 and WS_2 exhibited lower cytotoxicity than GO and halogenated graphene.[12] However, in a study analyzing the pulmonary hazard potential of MoS_2 suspensions, Wang et al. reported that a suspension consisting of aggregated MoS_2 flakes invokes strong proinflammatory responses in vitro and in mouse lungs.[13] On the other hand, MoS_2 suspensions exfoliated by lithium ions and surfactant do not exhibit similar effect. Nevertheless, the authors also note that there is no evidence of pulmonary fibrosis in response to all forms of MoS_2 in a sub-chronic study, suggesting low cytotoxicity of MoS_2 suspensions. Therefore, the method of production plays a significant role in defining the properties of the material.

CVD-grown MoS_2 is shown to be biocompatible and nontoxic.[14] First, cultures of PC12 pheochromocytoma neuronal origin cells and human endothelial HUVEC cells were grown on substrates with none, high-density polyethylene (HDPE), zinc diethyldithiocarbamate (ZDEC), and MoS_2 continuous film. HDPE is a standard reference material for biocompatibility testing while ZDEC is a positive (cytotoxic) control. As depicted in Fig. 9.1A and B, PC12 and HUVEC cells cultured on HDPE and MoS_2 films did not show any changes in viability, with a greater proportion of live cells (AnnexinV−/7-AAD−). On the other hand, on the ZDEC film, much higher percentage of dying cells was observed (AnnexinV+/7-AAD+: late apoptotic, AnnexinV+/7-AAD−: early apoptotic, AnnexinV−/7-AAD+: necrotic cell). Cell culture studies on as-grown monolayer MoS_2, isolated flakes, and MoS_2 dissolved in phosphate buffered saline (PBS) solution yielded similar results.

In addition, the long-term biocompatibility and cytotoxicity of MoS_2 was also demonstrated in an animal model. MoS_2 layers on PET substrates were implanted subcutaneously into BALB/c mice and the lymphocyte distributions in the peripheral blood of the mice were monitored for 4 weeks. An increase in the amount of immune cells (CD4+T, CD8+T, B, NK, neutrophil, and monocytes cells) will indicate immunologic or inflammatory response toward the MoS_2 implant. As seen in Fig. 9.1C, the percentage of immune cells found in the peripheral blood did not

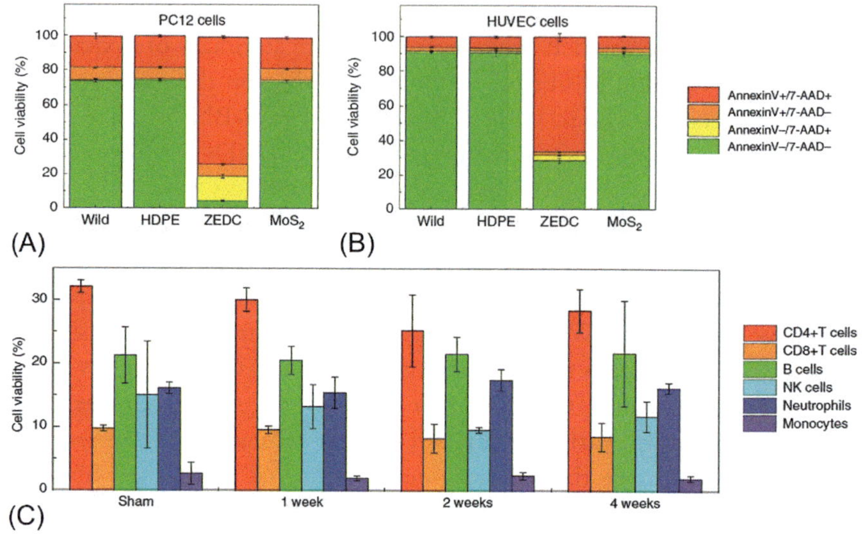

Fig. 9.1 (A) and (B) Cell viability of PC12 and HUVEC cultures with none, HDPE, ZDEC, and MoS_2 continuous film. The condition of the cells is assessed as AnnexinV+/7-AAD+: late apoptotic, AnnexinV+/7-AAD−: early apoptotic, AnnexinV−/7-AAD+: necrotic cell, and AnnexinV−/7-AAD−: live cells. The populations in each condition is averaged and plotted with standard error. (C) Immunoprofiling of cell populations in peripheral blood after implantation of MoS_2 in mice for 4 weeks. The individual subsets (CD4+T, CD8+T, B, NK, neutrophils, and monocytes) are presented as a percentage of immune cells found in the peripheral blood. Error bars of the experimental data represent standard error of mean.[14] *Reproduced under the Creative Commons Attribution 4.0 International License.*

change significantly throughout the 4 weeks, and the distributions of these cells is similar to those in the sham control group. The authors had also established that at physiological pH and temperature, a polycrystalline MoS_2 monolayer with grain size ~200 nm will take 2–3 months to dissolve completely in PBS solutions. Thus, a MoS_2-based device in vivo is expected to be broken down eventually without causing much side effects. Moreover, the dissolution rate (and breakdown time) could be tailored by adjusting the grain size and density of defects in the MoS_2 layer. Hence, MoS_2 would be suitable for many bio-photonic applications such as implantable sensors and drug delivery.

9.2.3 Black phosphorus

Black phosphorus (BP) exhibits low cytotoxicity, that is, lower toxicity compared to graphene but higher than MoS_2. The cytotoxicity of few-layer

BP flakes to human lung carcinoma epithelial cells BP is investigated alongside that of graphene and MoS_2.[15] It was found that at concentrations below 50 ppm, the cytotoxicity of BP is dependent on the concentration level. In a follow-up study, the size effect was examined.[16] Three samples of BP suspensions were studied, as summarized below:

BP-1, with flakes having average thickness of 92 nm and lateral size of 884 nm;

BP-2, with flakes having average thickness of 27 nm and lateral size of 426 nm; and

BP-3, with flakes having average thickness of 17 nm and lateral size of 209 nm.

In this study, the authors found their results are contrary to popular belief that thinner flakes are more cytotoxic. BP-1 with the largest lateral size and thickness exhibited the highest cytotoxicity, while BP-2 and BP-3 displayed moderate toxicity. Two mechanisms for cytotoxicity were proposed: generation of intracellular reactive oxygen species (ROS) and disruption of cell membrane integrity. It was found that BP-1 causes a significant larger disruption to the cell membrane integrity, possibly damage. Therefore, while BP is viewed to be of low toxicity, the dimension of the flakes should be properly considered, such that it is suitable for the type of cells the flakes may contact.

To improve the biocompatibility of BP and reduce inflammatory responses, the flakes may be chemically modified with appropriate ligands. For example, in vitro experiments showed that BP flakes modified with titanium sulfonate ligands displayed lower cytotoxicity compared to bare BP flakes.[17] In vivo studies also showed that the modified BP flakes do not trigger immune responses.

While toxicity studies have shown encouraging results so far, we must keep in mind about the complexity of physiological and biological systems. Nanomaterials such as 2D materials, when introduced to a biological system, will interact with the biomolecules present, subsequently forming a complex known as corona. Put simply, it may be viewed as a "surface modification," altering the properties of the material. When exposed to plasma proteins, BP nanomaterials (nanosheets and quantum dots) form corona complexes which are enriched in immune proteins, as shown in Fig. 9.2.[18] Moreover, the uptake efficiency of BP-corona complexes by macrophages is higher, although the cytotoxicity of BP was significantly reduced after formation of the complexes. The size of BP nanomaterials also significantly affects corona formation, and thus must be carefully considered. Therefore, current

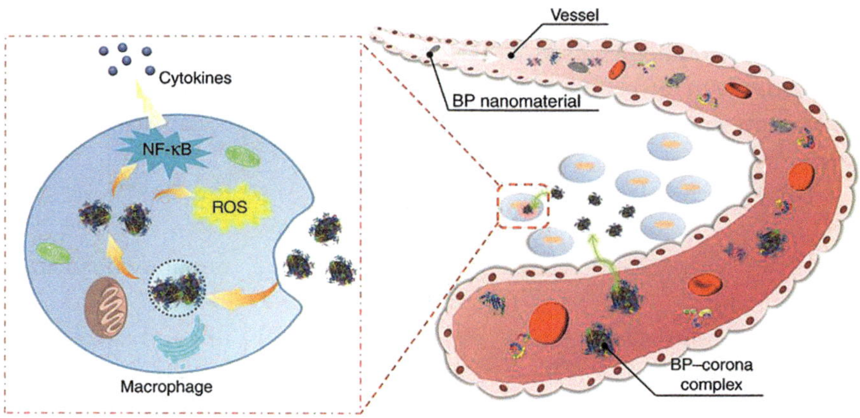

Fig. 9.2 Formation of BP-corona complexes in blood and their immunoregulation on macrophages, as depicted by the nuclear factor (NF)-κB pathway and reactive oxygen species (ROS).[18] *Reproduced under the Creative Commons Attribution 4.0 International License.*

understanding on 2D materials in biological systems is still limited and more efforts should be carried out. We also recognize that a thorough discussion is beyond the scope of this chapter and we implore readers to refer to other works.[19–21]

9.3 Biosensors

Biosensors are devices which detect the presence and/or concentration of biological analytes, such as enzymes, antigens, antibodies, proteins, or DNA. A biosensor is made up of three parts: the probe (which detects the analyte of interest, the transducer component (which converts the biorecognition event into a measurable signal), and the electronic system for readable output (often combined with the transducer). One well-known example of a biosensor is a pregnancy test kit, which utilizes the specific antigen-antibody bonding. Another example is a glucose monitoring device, a type of electrochemical biosensor. Since 2D materials possess a high surface-to-volume ratio and are ultrasensitive to minute changes on their surfaces with lower limit of detection (LOD), they are excellent candidates for biosensing applications.

There are different types of biosensors depending on their biological recognition element (probe). For instance, an aptamer is an oligonucleotide or a peptide molecule which binds to a specific target molecule.

Examples of aptamers include synthetic antibodies, single-stranded DNA, or RNA molecules. A biosensor which exploits such specific interactions is known as an aptasensor. Recent works on aptasensors based on 2D materials are promising. For example, BP was employed for the electrochemical detection of myoglobin (Mb), a cardiovascular disease biomarker. Few-layered BP nanosheets were functionalized with poly-L-lysine (PLL), after which PLL-BP electrodes were fabricated to detect anti- Mb aptamers. Myoglobin is a small, oxygen-binding protein found in heart and skeletal muscles, and can be used as an indicator for heart attacks. The detection limit of the PLL-BP aptasensor is 0.52 ng/mL, outperforming a reduced GO-based aptasensor, with a detection limit of 1.3 ng/mL.[22] A gold nanoparticle-BP hybrid has also demonstrated encouraging results in the detection of carcinoembryonic antigen (CEA) with a wide detection range (1 pg mL^{-1} to -10 μg mL^{-1}), high sensitivity (0.20 pg mL^{-1}), and good selectivity.[23]

Biosensors can also be classified according to the transducer component. For example, in electrochemical sensors the detection will be performed electrochemically; field-effect transistor (FET) sensors use a FET component as the transducer. In this chapter, we focus on the optical biosensors which exploit the optical properties of the 2D materials. In general, such optical biosensors are simpler to operate and provide quick feedback, rendering them suited for continuous monitoring such as studying events in real time. Readers interested in other types of biosensors are encouraged to refer to the reviews by Ping et al.,[24] Kalantar-zadeh and Ou,[25] Zhu et al.[26] and Peña-Bahamonde et al.[27]

9.3.1 Optical biosensors

In this section, we will look at optical biosensors that do not require a probe or label. In these sensors, the detection is usually through changes in refractive index measured through surface plasmon resonance. It is well known that for 2D materials, a nanosheet is extremely sensitive to changes on its surface, and there have been many reports highlighting the potential of 2D materials in optical biosensors.[28–30] For instance, a graphene-based optical biosensor demonstrated flow sensing of a single cell with high resolution of 1.7×10^{-8} and excellent sensitivity of 4.3×10^{7} mV/RIU (Fig. 9.3A).[31] This sensor is able to differentiate the smaller normal lymphocyte cells and larger cancerous cells (Jurkat), up to the single-cell level, as seen in the variation in peak voltage (Fig. 9.3B–D). Moreover, the dynamic range is

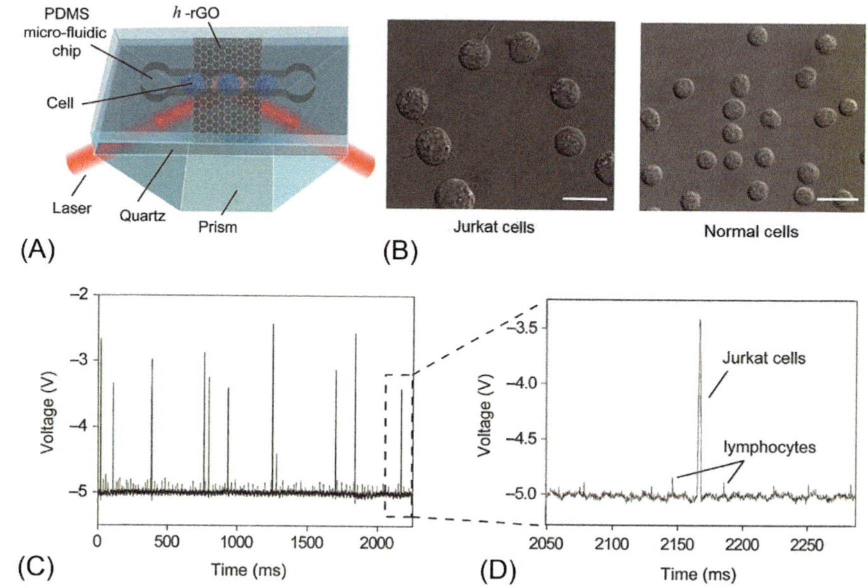

Fig. 9.3 (A) Schematic representation of the graphene-based (*h*-rGO) optical biosensor, depicting the cells moving in the micro-fluidic channel across the *h*-rGO detection window. (B) Microscopic images of Jurkat (*left*) and normal cells (*right*). The scale bar is 15 μm. (C) Discrete time-dependent changes in voltage corresponding to normal lymphocytes and Jurkat cells as they move across the detection window. (D) Zoom-in image of (C) a for certain positions in which the voltage signals are clearly depicted. The high and low signals represent Jurkat cells and normal cells, respectively. *Adapted with permission from Xing, F.; Meng, G.-X.; Zhang, Q.; Pan, L.-T.; Wang, P.; Liu, Z.-B.; Jiang, W.-S.; Chen, Y.; Tian, J.-G., Ultrasensitive Flow Sensing of a Single Cell Using Graphene-Based Optical Sensors. Nano Lett.* ***2014****, 14(6), 3563–3569. Copyright (2014) American Chemical Society.*

extensive and can be conveniently tuned by adjusting the incidence angle and incident power.

Recently, MoS_2 has been used in a photo-electrochemical sensor for the detection of miRNAs.[32] The photoanode is made up of gold nanoparticles (AuNP) and MoS_2 nanosheets heterojunction on indium tin oxide (ITO). The AuNPs improves the light harvesting efficiency of MoS_2 in the visible range, and the MoS_2-AuNPs heterojunction acts as the photoresponsive material. Photocurrent measurements showed that the MoS_2-AuNPs/ITO sensor exhibited excellent sensitivity for miRNA detection, with the ability to distinguish complementary and single-base mismatched miRNA. The authors also evaluated the performance of this sensor in human serum and found little deviation (3.3%–7.9%), therefore, it will be well-suited for biological samples.

9.3.2 Fluorescent biosensors

As the name implies, a fluorescent biosensor relies on change in fluorescence properties for detection. It can be assembled by integrating a biological receptor, a fluorophore whose emission properties is sensitive to change in the local environment and the electronic system. One advantage of fluorescent sensors is that it does not require additional reagents like mediators used in electrochemical sensors. In addition, they are more versatile and can be used for solid samples and tissue or cells.

A fluorescent sensing platform for DNA detection was developed with BP nanoparticles acting as nanofluorophores.[33] The detection assay was able to detect DNA concentrations with low limits of detection (LOD) and quantification (LOQ) values as low as 5.9 pM and 19.7 pM, respectively, and demonstrated a large linear dynamic detection range from <4 to 4000 pM with good linearity ($r=0.91$). It is also able to distinguish complementary DNA and DNA with three mismatched base pairs.

9.3.3 Förster (or fluorescence) resonance energy transfer biosensors

Förster (or fluorescence) resonance energy transfer (FRET) refers to the energy transfer from a donor fluorophore to an acceptor molecule following photoexcitation. It is widely used as a spectroscopic technique for biological applications, such as life-cell imaging, structural elucidation of biological molecules (e.g., protein folding), and analysis of their interactions (e.g., between an antigen with an antibody). Recently, FRET has also found applications in materials chemistry and was used to study the formation of nanoparticles.[34]

The efficiency of the FRET process depends on a few factors (Fig. 9.4). First, spectral overlap between the emission of the donor and the absorbance of the acceptor is required for FRET to occur. Secondly, the distance between the donor and acceptor should be in the range of 1–10 nm. The efficiency of the FRET process is inversely proportional to the sixth power of the distance between the donor and acceptor, and thus FRET is very sensitive to change in distance. In view of the interesting electronic properties of graphene, it has been proposed that graphene could be a highly efficient quencher of fluorophores, useful for distances beyond 10 nm. Indeed, when

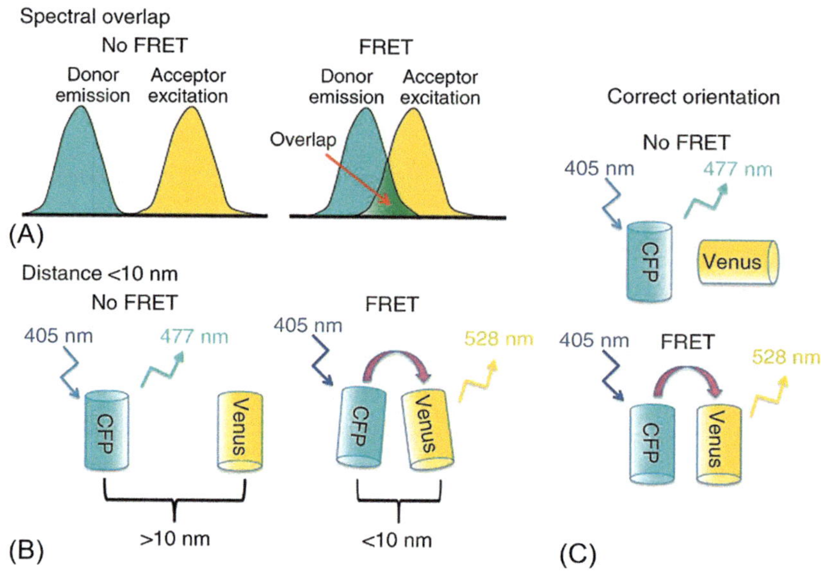

Fig. 9.4 Factors affecting the efficiency of the FRET process: (A) spectral overlap, (B) distance, and (C) orientation. The donor and acceptor molecules are represented by CFP (*blue*) and Venus (*yellow*), respectively.[35]

GO was used as an acceptor in a DNA-length-dependent fluorescence quenching study, quenching is observed up to 23.8 nm.[36] Although the mechanism of the energy transfer to GO is not well understood, this study demonstrates the excellent quenching efficiency of graphene and photonic materials. Lastly, FRET is also dependent on the spectral characteristics of the chromophores and their relative orientation.

9.3.3.1 Graphene-based FRET devices

A novel graphene-based "FRET on fiber" device utilizing refractive index changes for sensing biochemical molecules was developed.[37] A partially reduced graphene oxide (prGO) film was deposited on a fiber-optic modal interferometer. As mentioned previously, 2D materials, being ultrathin, are extremely sensitive to the change on their surfaces. The "FRET on fiber" device exploits this as well as the fluorescence quenching property. When a dye such as Rhodamine 6G (Rh6G) is adsorbed on the prGO surface, prGO quenches the fluorescence (Fig. 9.5A). When target analyte molecules are introduced, binding competition occurs and Rh6G

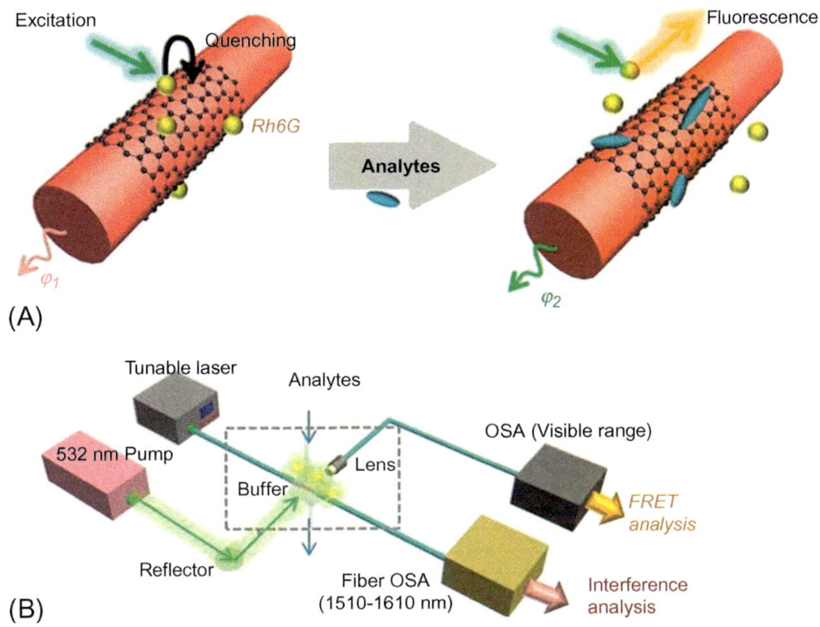

Fig. 9.5 (A) Partially reduced GO (prGO) is coated around an optical fiber, with Rh6G molecules adsorbed on the prGO surface. (B) The detection system contains two channels, one for fluorescence detection, and the other is for all-fiber interference-phase measurement.[37] *Adapted under Creative Commons Attribution 4.0 International License.*

molecules are released, thus restoration of fluorescent intensity could be observed by FRET analysis. At the same time, the effective refractive index of the prGO-coated area changes according to the analyte concentration, which is observed via interference analysis (Fig. 9.5B). The prGO film can be tailored to sense for different molecules. As a proof of concept, the authors used the prGO-coated fiber-optic modal interferometer to detect Cd^{2+}, dopamine (DA), and single-strand (ss) DNA (Fig. 9.6A–C). The corresponding fluorescence and optical spectra are shown in Figs. 9.6D–F and G–I, respectively. The interferometer demonstrated high sensitivities, with detection limits of 1.2 nM, 1.3 μM, and 1 pM for Cd^{2+}, DA, and ssDNA, respectively. Moreover, the "FRET on fiber" device is compact and offers a much quicker response time compared to common FRET methods (seconds compared to minutes). Therefore, it will be an excellent candidate for real-time sensors in the biochemical and environmental fields.

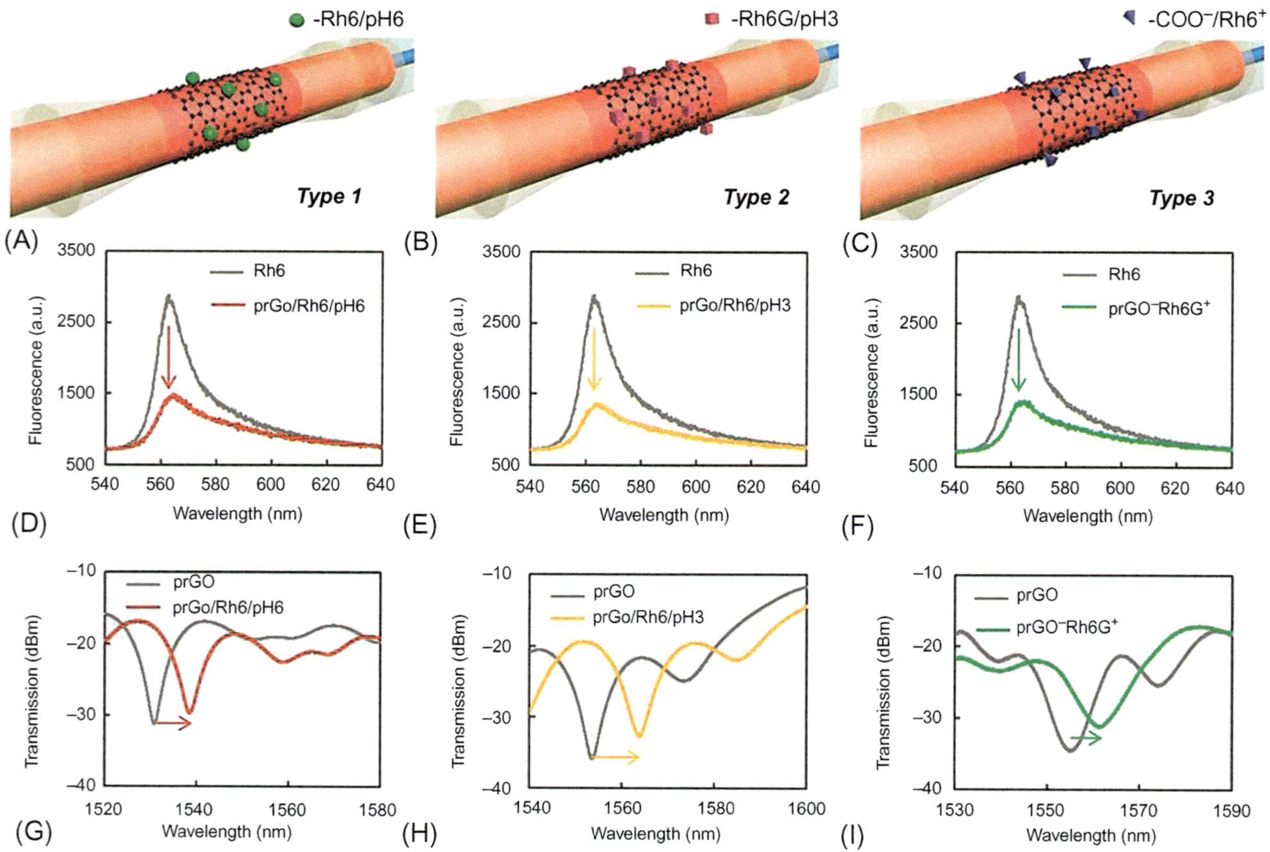

Fig. 9.6 (A–C) Modified prGO such that the coated interferometer is *Type* 1, 2, and 3 to detect Cd^{2+}, DA, and ssDNA, respectively. (D–F) Fluorescence quenching by the prGO film of the (D) *Type 1* (*red* curve), (E) *Type 2* (*yellow* curve), and (F) *Type 3* (*green* curve). The gray curves depicted the fluorescence of the Rh6G. (G–I) Optical spectra of the (G) *Type 1* with dip location of 1538.2 nm (*red* curve), (H) *Type 2* with dip location of 1563.7 nm (*yellow* curve), (I) *Type 3* with dip location of 1561.1 nm (*green* curve). The gray curves show the spectra before adding Rh6G.[37] *Reproduced under Creative Commons Attribution 4.0 International License.*

9.3.3.2 MoS_2-based FRET devices

TMDs have also found applications in FRET biosensors. Lim's group fabricated a FRET aptasensor using MoS_2 nanosheets for the detection of a malarial biomarker, *Plasmodium*lactate dehydrogenase (pLDH).[38] Malaria is a serious infectious disease and early diagnosis is critical for appropriate treatment so as to reduce sickness and death. The traditional method of diagnosis is microscopy based, which is slow and tedious, and thus there is a need to develop alternatives. One way is to identify and detect malarial biomarkers in blood samples. The pLDH protein is a highly expressed malarial biomarker produced by all *Plasmodium* species at both sexual and asexual parasitic stages.[39, 40] Moreover, the pLDH concentrations correspond directly to the parasitaemia levels.[41] The MoS_2-based FRET aptasensor showed good specificity and selectivity, being able to discriminate the target pLDH protein from individual random proteins, such as insulin, albumin, globulin, and fibrinogen, as well as distinguishing the target in a heterogeneous mixture of proteins.[38] In addition, the aptasensor displayed exceptional sensitivity, with a limit of detection estimated to be 550 pM, far exceeding the requirement for clinical applications. The detection is also quick; the time required was merely 10 min.

With these encouraging results, the authors followed up with another work which improves the sensing and specificity of the MoS_2-based FRET aptasensor using a surface blocking strategy.[42] Nonspecific adsorption of proteins onto the fluorescence quencher results in the displacement of targeted aptamers from the quencher surface. To reduce interference from nontargeted proteins, a surface blocking agent, bovine serum albumin (BSA), can be applied to the quenching material. In unblocked samples, the pLDH fluorescence recovery was about twice as high as nonspecific protein recovery, compared to at least 4 times as high in blocked samples. Therefore, surface functionalization is a simple and effective way to improve the performance of such sensors.

The same group continued to develop the MoS_2-based FRET aptasensor on paper for malaria diagnosis (Fig. 9.7).[43] The working principle remains similar, but adapting paper sensors will allow rapid, portable, and low-cost diagnosis; this is vital to the successful management of malaria. In this work, the authors aim to investigate the effects of the choice of paper. They found that only printer paper worked as a FRET aptasensor, demonstrating significant fluorescence recovery in 30 min. Conversely, other types of paper such as the commonly used chromatography paper prove to be unsuitable. This was attributed to lower wettability and smaller pore size

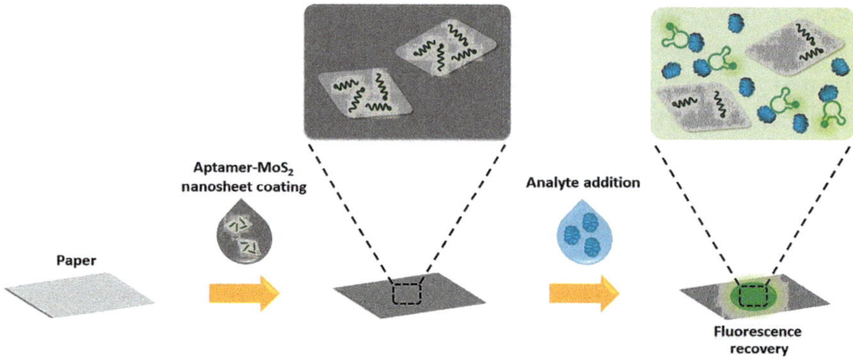

Fig. 9.7 MoS_2-based FRET aptasensor on paper. To form a test strip, the paper is first dipped in a solution containing aptamers and MoS_2 nanosheets and then fully dried. As in previous studies, the MoS_2 nanosheets act as fluorescence quenchers here. When the test strip comes into contact with analytes containing the target pLDH molecules, the fluorescently labeled aptamers will be released from the quencher. The fluorescence increase can then be detected.[43] *Reproduced under Creative Commons Attribution 4.0 International License.*

of the printer paper, thus preventing the aptamers from being displaced or washed out of the test strip. Having successfully demonstrated the sensing capability of the MoS_2-based aptasensor, future studies should focus on improving the sensitivity and specificity of the sensor. It may also be worthwhile to explore printable sensors, with the aptamer and MoS_2 solution printed on paper. In this way, the production of the test strips could be automated and cost effective.

9.4 Cancer diagnosis and treatment: Imaging, drug delivery, photothermal/photodynamic therapy

Cancer is one of the top leading cause of death, killing about 9 million annually.[44] On top of that, being a chronic disease, it affects an increasing number of patients every year. This is a concerning public health issue. Cancer cells grow and reproduce uncontrollably, creating tumors in a specific part of the body and spreading to other areas, where they attach to and destroy surrounding healthy tissues and organs and ultimately cause death. The type of cancer is usually named after the origin: the tissue or organ where the cancerous cells start to form, for example: skin cancer, lung cancer, and brain cancer. Typically, treatment plans involve a combination of surgery, chemotherapy, radiation therapy, and targeted therapy. In recent years, new methods such as photodynamic therapy (PDT) and photothermal therapy (PTT) have been developed. These methods are less invasive and less harmful side effects than traditional treatments, and are thus gaining attention. From the last few years, the potential of nanomaterials for cancer treatment has been widely studied. Lately, 2D materials, being ultrathin with extremely high surface-to-volume ratio and possessing excellent optical properties, are increasingly being considered for imaging and combinative therapy.

9.4.1 Graphene

The ultra-broadband optical absorbance of graphene including in the near-infrared (NIR) wavelength may be exploited for in vivo photothermal therapy (PTT). PTT is a treatment that employs a photothermal therapeutic agent which absorbs electromagnetic radiation (usually IR) to generate heat for thermal ablation of cancer cells.

In 2010, Liu's group demonstrated the photothermal ablation of tumors in a mouse model by using biocompatible graphene.[45] Unlike previous works, the photothermal therapeutic agent was not injected directly into the tumor; instead it was intravenously administrated. Systemic administration offers the advantage of reaching all lesions or abnormal cells. GO sheets functionalized with polyethylene glycol (PEG) are highly soluble and stable in physiological solutions, and high tumor accumulation of the GO-PEG was observed in several tumor models. In vivo phototherapy was then carried out with low-power near-IR irradiation and no tumor regrowth was observed in the treated mice when the study ended after 40 days. The mice also remained alive with no abnormalities to their major organs. In contrast, tumors in mice which had only GO-PEG injections or only irradiation grew rapidly, and these mice died in about 16 days.

The group also considered the potential of GO-PEG in PDT.[46] PDT differs from PTT in that an additional photosensitizer or photosensitizing agent is used in conjunction with a specific wavelength of light. After exposure to this light, the excited photosensitizer molecule transfers the energy to surrounding oxygen molecules, generating reactive oxygen species (ROS) such as singlet oxygen and free radicals to kill nearby cells. In cancer treatments, the photosensitizers chosen are retained in cancer cells for a longer period than normal cells. In Liu's work, Chlorin e6 (Ce6), a photosensitizer molecule, is loaded on GO-PEG via supramolecular π–π stacking. The

GO-PEG-Ce6 complex showed a significantly improved cancer cell photodynamic destruction effect on human nasopharyngeal epidermal carcinoma KB cells under irradiation with a 660 nm laser, compared to free Ce6. The GO-PEG aids the delivery of Ce6 into the cancer cells, possibly through endocytosis. In addition, the photothermal effect of graphene was employed to further enhance the delivery of Ce6 molecules, subsequently improving the efficiency of the photodynamic treatment (Fig. 9.8). After the KB cells were exposed to a near-IR source (808 nm), the cellular uptake of GO-PEG-Ce6 was significantly increased compared to cells incubated in the dark. The preirradiated KB cells showed further reduced cell viabilities after exposure to the 660 nm laser, verifying that the photothermal effect of graphene was able to promote the delivery of photosensitizing agents, subsequently enhancing the efficiency of PDT.

Later, Nurunnabi et al. made a significant progress by utilizing graphene in optical imaging, photothermal, and photodynamic therapies, in one system simultaneously[47] (Fig. 9.9). Carboxylated graphene nanodots (cGdots) possessing photoluminescent properties were designed and synthesized. The cGdots consists of mostly monolayer graphene sheets with carboxyl acid and hydroxyl

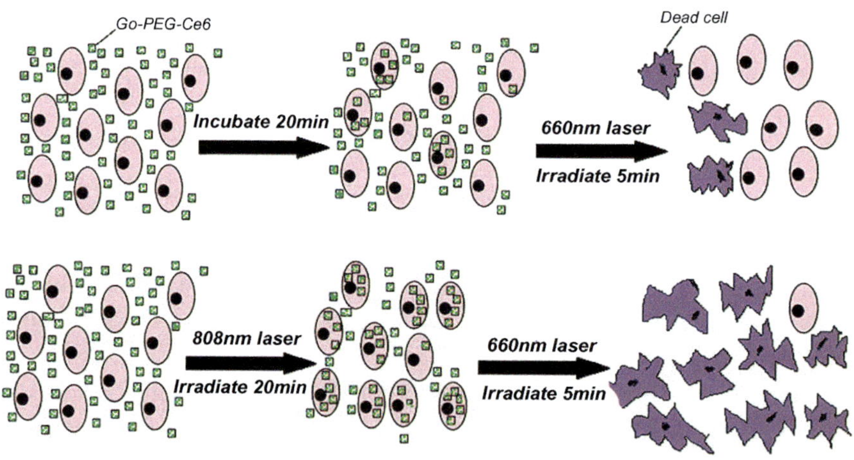

Fig. 9.8 Photothermally enhanced photodynamic therapy using GO-PEG-Ce6. *Top*: KB cells were incubated with GO-PEG-Ce6 for 20 min in the dark and then irradiated by the 660-nm laser (50 mW/cm^2, 5 min, 15 J/cm^2). *Bottom*: GO-PEG-Ce6 incubated KB cells were first exposed to a 808-nm laser (0.3 W/cm^2, 20 min, 360 J/cm^2) first before irradiation by the 660-nm laser. *Reprinted with permission from Tian, B.; Wang, C.; Zhang, S.; Feng, L.; Liu, Z., Photothermally Enhanced Photodynamic Therapy Delivered by Nano-Graphene Oxide. ACS Nano* ***2011****, 5(9) 7000–7009. Copyright (2011) American Chemical Society.*

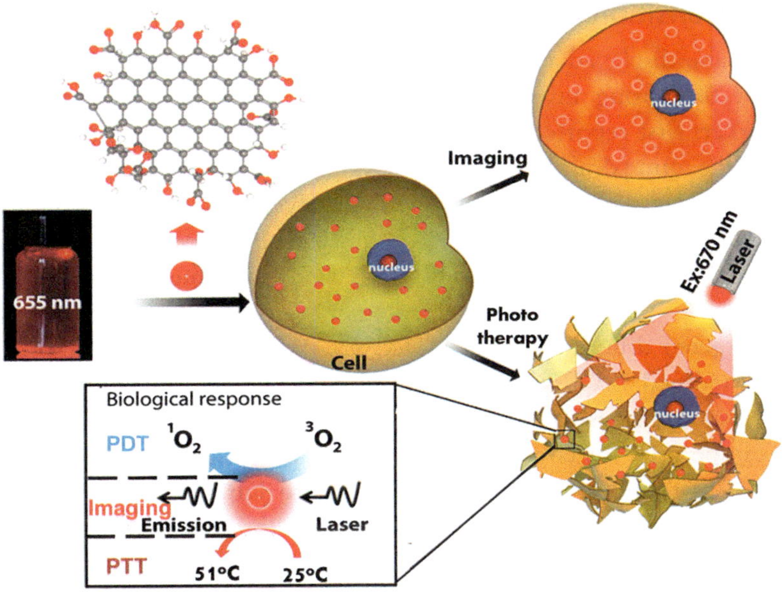

Fig. 9.9 Schematic representation depicting structure of cGdots and application in optical imaging, photodynamic, and photothermal therapy. The cGdots display photoluminescence when excited with near IR light (655 nm). *Reprinted with permission from Nurunnabi, M.; Khatun, Z.; Reeck, G. R.; Lee, D. Y.; Lee, Y.-k., Photoluminescent Graphene Nanoparticles for Cancer Phototherapy and Imaging. ACS Appl. Mater. Interfaces* ***2014****, 6(15) 12413–12421. Copyright (2014) American Chemical Society.*

functional groups on the edges. However, they are referred to nanodots as they are small in size, with diameters ranging from 3 to 9 nm. The cellular uptake of cGdots can be easily monitored in real time using confocal laser scanning microscopy. When irradiated with a 670-nm laser source, the cGdots with a high level of unsaturated bonds absorbed the energy and transferred it to surrounding oxygen molecules and unsaturated bonds, thus generating ROS. The temperature also increased to 50°C after 30 min of laser irradiation. In vitro studies showed that for MDA-MB231 breast cancer cells, cellular damage was only observed in cells treated with cGdots and laser irradiation; 72% of the cGdots with laser-treated cells were in the apoptosis state compared to 0.5% of the control cells. The cGdots were also administered intravenously to tumor-bearing mice and it was found that phototherapy reduced the tumor size by 70% compared to control group. Although this indicates that phototherapy with cGdots is only useful for the inhibition of tumor growth, the ease of this method, coupled with few harmful effects, still hold great promise in cancer treatments.

9.4.2 Transition metal chalcogenides

TMDs also displayed a great potential in cancer treatments. Similar to graphene, the strong near IR absorbance of MoS_2 is exploited. Liu and coworkers first reported the use of MoS_2 nanosheets as a multifunctional agent, executing drug delivery for chemotherapy and PTT.[48] MoS_2 was first functionalized with lipoic acid modified polyethylene glycol (LA-PEG) to improve their biocompatibility and stability in physiological conditions. The drug loading capacity of MoS_2-PEG was found to be higher than PEG functionalized GO, with the highest loading of ~239% for doxorubicin (DOX). The targeted drug delivery of DOX was then performed using folic acid (FA) modified LA-PEG to obtain MoS_2-PEG-FA. The folate receptor (FR) is over-expressed in several types of cancer cells and the FA-FR binding can be used for targeted drug delivery. As expected, MoS_2-PEG-FA/DOX was found to be much more toxic to KB cells than MoS_2-PEG/DOX or free DOX. More importantly, when the cells were irradiated with an 808 nm laser, most of the MoS_2-PEG-FA/DOX treated cells died (95%). Laser irradiation produced no noticeable effect on free DOX-treated cells, while the MoS_2-PEG/DOX treated cells only showed reduced viabilities (~33%). In vivo and animal studies based on MoS2-PEG-FA/DOX showed encouraging results. The efficacy of combining chemotherapy and PTT were evaluated in tumor-carrying mice. The mice were injected with the test solutions (intratumoral and intravenous) and subjected to 808 nm NIR radiation for 20 min (Fig. 9.10A and E). As seen in Fig. 9.10B and C, apparent temperature increases were noticed for tumors injected with MoS_2-PEG or MoS_2-PEG/DOX, reaching up to 45°C. In contrast, tumors injected with PBS or free DOX exhibited reduced changes and temperatures remained below 39°C. Similar observations were made for mice that underwent intravenous injections (Fig. 9.10F and G). The overall efficiency of the combination therapy can be seen in Fig. 9.10D and H. In mice injected with PBS, free DOX, and MoS_2-PEG, the tumors grew quickly. In mice injected with MoS_2-PEG/DOX but not subjected to NIR, the tumor growth was only slightly slower. Clearly, combination therapy (chemotherapy and PTT) using MoS_2-PEG/DOX offers the best outcome in hindering tumor growths.

Other studies also reported similar results. By combining chemotherapy and PTT, MoS_2 nanosheets functionalized with chitosan and loaded with DOX exhibited remarkable cell-killing ability towards KB cells and pancreatic cancer cells (Panc-1).[49] In vivo investigations also found that the combined treatment successfully inhibit tumor growth for pancreatic cancer.

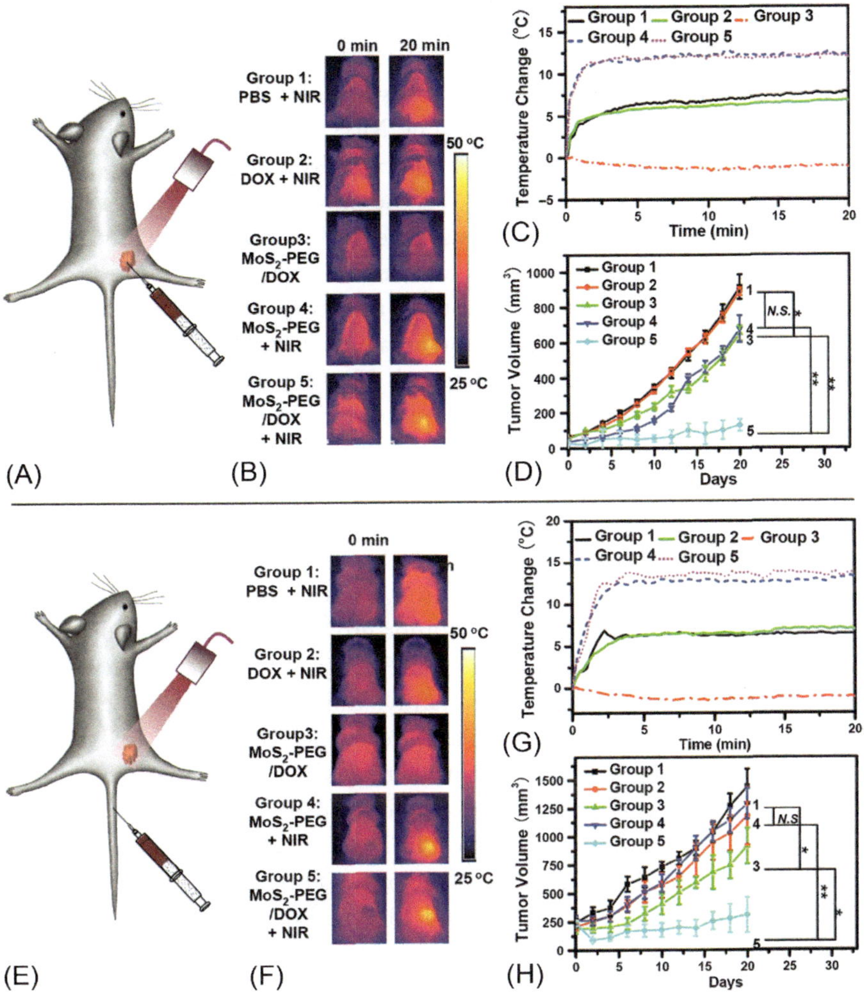

Fig. 9.10 Evaluation of MoS_2-PEG/DOX for in vivo combination therapy (chemotherapy and PTT). Group 1: PBS+NIR; Group 2: DOX+NIR; Group 3: MoS_2-PEG/DOX without NIR irradiation; Group 4: MoS_2-PEG+NIR; Group 5: MoS_2-PEG/DOX+NIR. (A) Scheme of combination therapy based on intratumorly injected MoS_2-PEG/DOX. (B) IR thermal images of tumor-bearing mice recorded by an IR camera. (C) Temperature change of tumors monitored by the IR thermal camera in different groups during NIR irradiation as indicated in (A). (D) Tumor volume growth curves of different groups of mice after various treatments (five mice for each group). Error bars are based on standard errors of the mean (SEM). *P* values: $^{**}P<0.01$, $^{*}P<0.05$, N. S. (no significant difference) $P>0.05$, are calculated by ANOVA with Tukey's post-test. (E) Scheme of combination therapy based on intravenously injected MoS_2-PEG/DOX. (F) IR thermal images of tumor-bearing mice recorded by an IR camera. (G) Temperature change of tumors monitored by the IR thermal camera in different groups during NIR irradiation as indicated in (A). (H) Tumor volume growth curves of different groups of mice after various treatments (five mice for each group). *Reproduced with permission from Liu, T.; Wang, C.; Gu, X.; Gong, H.; Cheng, L.; Shi, X.; Feng, L.; Sun, B.; Liu, Z., Drug Delivery With PEGylated MoS2 Nano-Sheets for Combined Photothermal and Chemotherapy of Cancer. Adv. Mater.* ***2014****, 26(21), 3433–3440, John Wiley and Sons.*

More importantly, tumor regrowth was not observed again in the treated group until the end of the study. The MoS_2 nanosheets can also be used as a contrast agent in computed tomography (CT) imaging, providing important diagnostics. To improve the imaging capability of MoS_2, a Bi_2S_3/MoS_2-PEG composite was synthesized to harness the advantages of Bi_2S_3 and MoS_2.[50] Bi_2S_3 exhibits excellent X-ray attenuation in CT imaging[51] and dose-enhancing capability in radiotherapy (RT).[52] The in vitro X-ray attenuation (HU value) and imaging properties of the composite was studied with iobitridol as a control. The Bi_2S_3/MoS_2-PEG composite outperformed iobitridol, providing significant higher HU value and CT image brightness than iobitridol under the same radiodense element concentrations. The absorbance of near IR radiation by MoS_2 is also exploited for in vivo photoacoustic (PA) imaging. The efficacy of the Bi_2S_3/MoS_2-PEG composite in PTT and RT is also evaluated (Fig. 9.11). After successive exposure to NIR and X-ray irradiation, tumor-carrying mice treated with the composite displayed a slower tumor growth compared to mice which undergo only PTT or RT (Fig. 9.11A). Mice treated with MoS_2-PEG did not fare as well. More importantly, most of the mice injected with the Bi_2S_3/MoS_2-PEG composite and treated with PTT and RT remained healthy for 28 days posttreatment, with no tumor recurrence observed (Fig. 9.11B and D). Hematoxylin and eosin staining (H&E staining) allows a better picture at the cellular level. In Fig. 9.11C, it can be seen that the mice in the control group displayed no visible cell apoptosis. In mice injected with Bi_2S_3/MoS_2-PEG and MoS_2-PEG and treated with X-ray, the tumors were partially destroyed, although when Bi_2S_3/MoS_2-PEG was used, a more pronounced effect can be observed (more vacuoles, condensed nuclei, and changed cell shapes). Indisputably, the anticancer effect is most noticeable in mice injected with Bi_2S_3/MoS_2-PEG and MoS_2-PEG and treated with both X-ray and NIR radiation, highlighting the potential of combinative treatments towards cancer.

Despite the exciting results, the same group found that a considerable amount of MoS_2 nanosheets were found in the bloodstream and later accumulated in liver, spleen, and lung tissues, prompting safety concerns.[53] They therefore proposed using a different polymer, poly(lactic-co-glycolic acid) (PLGA) which offer several advantages.[54] PLGA is biocompatible and used extensively in various biomedical applications.[55, 56] It also dissolves easily in N-methylpyrrolidone (NMP) to form an oleosol. Since MoS_2 nanosheets and DOX can also be uniformly dispersed or dissolved into NMP, NMP can be used to prepare PLGA/MoS_2/DOX (PMD) oleosol. Being

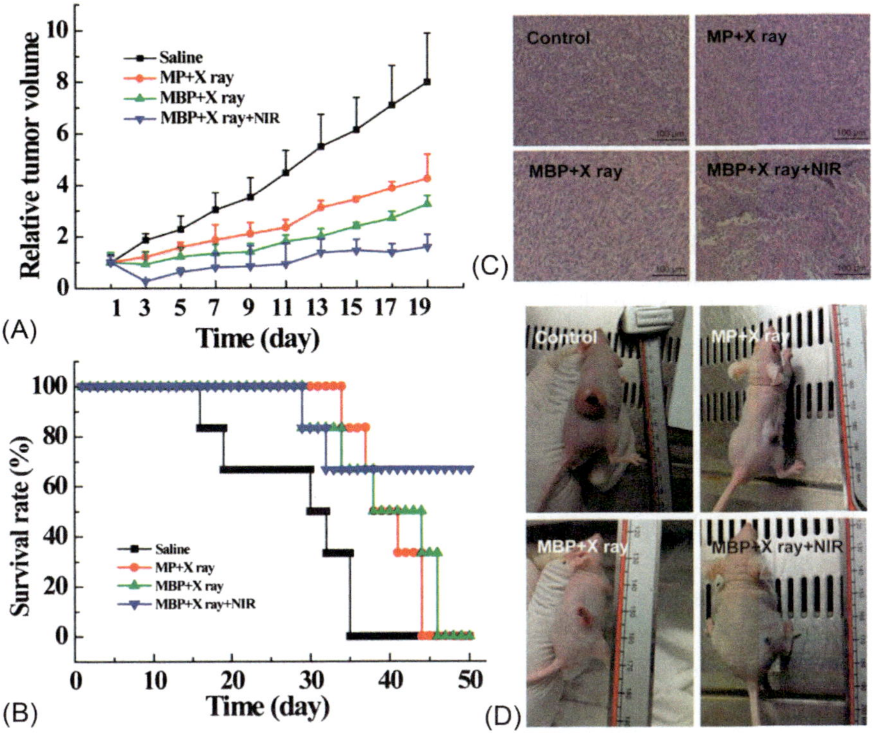

Fig. 9.11 Evaluating Bi_2S_3/MoS_2-PEG for in vivo combination therapy (PTT and RT). MP: MoS_2-PEG; MBP: Bi_2S_3/MoS_2-PEG (A) Tumour growth profiles of tumours after various treatments as depicted. (B) Survival rates of tumour-bearing mice as a function of time following treatments. (C) H&E staining of tumor sections after different treatments (D) Photos of tumor-bearing mice on day 28. *Reproduced with permission from Wang, S.; Li, X.; Chen, Y.; Cai, X.; Yao, H.; Gao, W.; Zheng, Y.; An, X.; Shi, J.; Chen, H., A Facile One-Pot Synthesis of a Two-Dimensional MoS2/Bi2S3 Composite Theranostic Nanosystem for Multi-Modality Tumor Imaging and Therapy. Adv. Mater.* ***2015****, 27(17), 2775–2782, John Wiley and Sons.*

hydrophobic, the PLGA polymer chains will undergo a liquid-solid-phase transformation readily upon contact with water, encapsulating the MoS_2 and DOX within the formed solid, then subsequently forming a PMD implant within a tumor. In this way, MoS_2 and DOX can be introduced without entering the bloodstream. The DOX is later released in a controlled manner by either reducing the pH to around 5.4 or NIR radiation. The efficacy of combining chemotherapy and PTT was evaluated in animal studies. Similar to previous investigations, both therapies are required for the best outcome. More remarkably, following PMD implant and NIR radiation, the tumor shrank and almost disappeared after 7 days. No tumor recurrence was observed in the next 2 months. This approach is also versatile and may be applied to deliver other drugs.

9.4.3 Black phosphorus

BP has also found applications for combinative therapies. Chen et al. reported exceptional high-loading capacity of DOX on BP sheets, up to 95.0% in weight.[57] This far surpasses the reported amounts for other 2D materials such as MoS_2.[48, 57] Since BP also exhibits absorbance across a broad spectrum, the group went on to develop a drug delivery system for synergistic photodynamic/photothermal/chemotherapy based on BP nanosheets. Similar to the MoS_2-based delivery system, the DOX release is accelerated in acidic tumor microenvironments or upon NIR radiation. PDT and PTT were performed successively with a 660 nm and 808 nm lasers, respectively. With the tri-therapy, a 95.5% tumor growth inhibition was accomplished; this is significantly higher than other treatment options (Table 9.1). Histological analysis corroborated this: after the tri-therapy, most of tumor tissue cells were killed and became necrotic, whereas cell viability in the other groups remains partially or mostly unchanged.

Other groups also reported success with BP nanosheets. For example, Tao et al. developed a PEGylated BP delivery platform for various theranostic agents, such as DOX for chemotherapy and cyanine7 (Cy7) for in vivo NIR imaging[58] (Fig. 9.12). The BP-PEG platform can also be functionalized with folic acid (FA) for targeted delivery, as in the case of MoS_2-PEG system, or fluorescein isothiocyanate (FITC) for imaging. To allow visualization of tumors with magnetic resonance imaging (MRI), Yang et al. fabricated a composite by assembling gold (Au) and iron oxide (Fe_3O_4) nanoparticles on BP sheets.[59] In addition, the Au nanoparticles absorb red or NIR light through surface plasmon resonance, while Fe_3O_4 nanoparticles are

Table 9.1 Tumor growth inhibition rates achieved by different treatments.

Treatment	Tumor growth inhibition (%)
BP-DOX with PDT and PTT	95.5
BP-DOX with PTT	84.8
BP-DOX with PDT	56.2
BP-DOX	47.4
Free DOX	43.6

The tumor volumes were measured by a caliper for 14 days and normalized against their initial sizes (0 day).

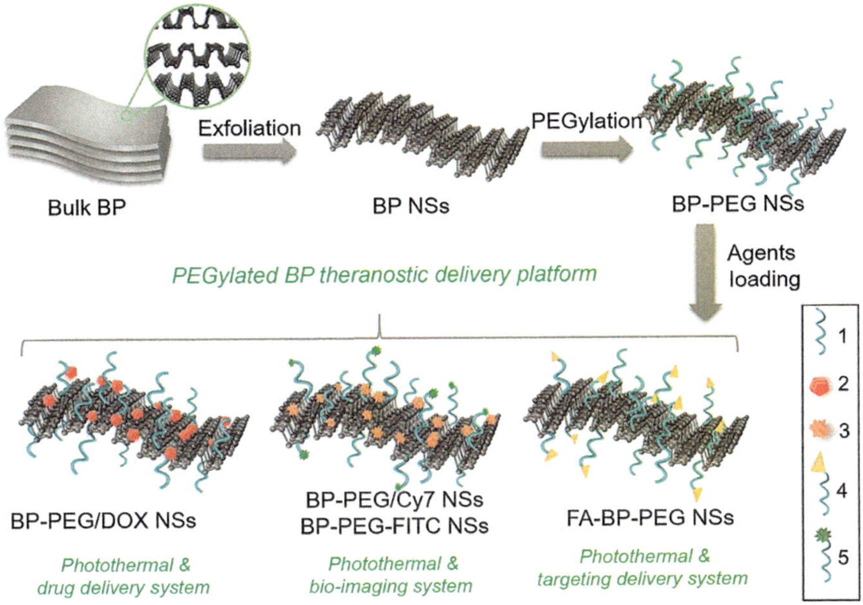

Fig. 9.12 Schematic representation of the PEGylated BP theranostic delivery platform. (1) PEG-NH_2 (surface modification), (2) DOX, (3) Cy7-NH_2, (4) FA-PEG-NH_2, and (5) FITC-PEG-NH_2. *Reproduced with permission from Tao, W.; Zhu, X.; Yu, X.; Zeng, X.; Xiao, Q.; Zhang, X.; Ji, X.; Wang, X.; Shi, J.; Zhang, H.; Mei, L., Black Phosphorus Nanosheets as a Robust Delivery Platform for Cancer Theranostics. Adv. Mater.* ***2017****, 29(1), 1603276, John Wiley and Sons.*

photoabsorbers. Therefore, the BPs@Au@Fe_3O_4 composite exhibited an enhanced efficiency in PTT. In another work, Chlorin e6 (Ce6), a common photosensitizer (see also Section 9.4.1) was loaded on PEGylated BP sheets.[60] By combining PTT and PDT using BP-PEG/Ce6, excellent ROS generation performance was observed in vitro. Animal studies also verified that the tumor growth inhibition is best with the combined approach.

The search for better cancer treatments is a never-ending one. Although a thorough discussion is beyond our scope, we hope that by highlighting these outstanding works, readers would be able to appreciate the role of 2D materials can play. One interesting point to note is that in all cases mentioned above, approaches which combines two or more therapies produce significantly better outcomes. This is in line with the direction of oncology is heading. Lastly, readers interested in further reading are encouraged to refer to the works by Chen et al.,[61] Gong et al.,[62] Fan et al.,[63] Shivananju et al.[64, 65] and Murugan et al.[66]

9.5 Conclusion

In this chapter, we have discussed how 2D materials can find potential use in bio-medical applications. In particular, we have brought to attention two applications which harness the exceptional optical properties of 2D materials: FRET biosensors and cancer diagnosis and treatment. Considering the numerous types of 2D materials, many of them have not been studied as extensively as the materials mentioned in this chapter. However, we will likely see more of them soon. For example, MXene nanosheets were utilized for combined photothermal, photodynamic, and chemotherapy recently.[67] Despite exciting results, one aspect that should be carefully investigated is the mutagenic potential of 2D materials. Even though the materials may not show acute effects, undesirable mutations may harm the health of future generation (of cells). This is especially critical for in vivo applications such as drug delivery and PTT.

References

1. Zhen, Z.; Zhu, H. 1—Structure and Properties of Graphene. In *Graphene, Academic Press;* **2018**, pp 1–12.
2. Bao, Q.; Loh, K. P. Graphene Photonics, Plasmonics, and Broadband Optoelectronic Devices. *ACS Nano* **2012**, *6*(5), 3677–3694.
3. Chimene, D.; Alge, D. L.; Gaharwar, A. K. Two-Dimensional Nanomaterials for Biomedical Applications: Emerging Trends and Future Prospects. *Adv. Mater.* **2015**, *27*(45), 7261–7284.
4. Zhang, Y.; Ali, S. F.; Dervishi, E.; Xu, Y.; Li, Z.; Casciano, D.; Biris, A. S. Cytotoxicity Effects of Graphene and Single-Wall Carbon Nanotubes in Neural Phaeochromocytoma-Derived PC12 Cells. *ACS Nano* **2010**, *4*(6), 3181–3186.
5. Hu, W.; Peng, C.; Lv, M.; Li, X.; Zhang, Y.; Chen, N.; Fan, C.; Huang, Q. Protein Corona-Mediated Mitigation of Cytotoxicity of Graphene Oxide. *ACS Nano* **2011**, *5*(5), 3693–3700.
6. Liu, Y.; Luo, Y.; Wu, J.; Wang, Y.; Yang, X.; Yang, R.; Wang, B.; Yang, J.; Zhang, N. Graphene Oxide Can Induce In Vitro and In Vivo Mutagenesis. *Sci. Rep.* **2013**, *3*, 3469.
7. Muzi, L.; Mouchet, F.; Cadarsi, S.; Janowska, I.; Russier, J.; Ménard-Moyon, C.; Risuleo, G.; Soula, B.; Galibert, A.-M.; Flahaut, E.; Pinelli, E.; Gauthier, L.; Bianco, A. Examining the Impact of Multi-Layer Graphene Using Cellular and Amphibian Models. *2D Mater.* **2016**, *3*(2), 025009.
8. Volkov, Y.; McIntyre, J.; Prina-Mello, A. Graphene Toxicity as a Double-Edged Sword of Risks and Exploitable Opportunities: A Critical Analysis of the Most Recent Trends and Developments. *2D Mater.* **2017**, *4*(2), 022001.
9. Seabra, A. B.; Paula, A. J.; de Lima, R.; Alves, O. L.; Durán, N. Nanotoxicity of Graphene and Graphene Oxide. *Chem. Res. Toxicol.* **2014**, *27*(2), 159–168.
10. Nath, N. C. D.; Debnath, T.; Nurunnabi, M.; Kim, E.-K. In Vitro Toxicity of 2D Materials. In *Biomedical Applications of Graphene and 2D Nanomaterials;* Nurunnabi, M.; McCarthy, J., Eds. Elsevier, **2019**, pp 165–186.
11. Appel, J. H.; Li, D. O.; Podlevsky, J. D.; Debnath, A.; Green, A. A.; Wang, Q. H.; Chae, J. Low Cytotoxicity and Genotoxicity of Two-Dimensional MoS_2 and WS_2. *ACS Biomater. Sci. Eng.* **2016**, *2*(3), 361–367.

12. Teo, W. Z.; Chng, E. L. K.; Sofer, Z.; Pumera, M. Cytotoxicity of Exfoliated Transition-Metal Dichalcogenides (MoS_2, WS_2, and WSe_2) Is Lower Than That of Graphene and Its Analogues. *Chem. Eur. J.* **2014**, *20*(31), 9627–9632.
13. Wang, X.; Mansukhani, N. D.; Guiney, L. M.; Ji, Z.; Chang, C. H.; Wang, M.; Liao, Y.-P.; Song, T.-B.; Sun, B.; Li, R.; Xia, T.; Hersam, M. C.; Nel, A. E. Differences in the Toxicological Potential of 2D Versus Aggregated Molybdenum Disulfide in the Lung. *Small* **2015**, *11*(38), 5079–5087.
14. Chen, X.; Park, Y. J.; Kang, M.; Seung-Kyun, K.; Koo, J.; Shinde, S. M.; Shin, J.; Jeon, S.; Park, G.; Yan, Y.; MacEwan, M. R.; Ray, W. Z.; Kyung-Mi, L.; Rogers, J. A.; Jong-Hyun, A. CVD-Grown Monolayer MoS_2 in Bioabsorbable Electronics and Biosensors. *Nat. Commun.* **2018**, *9*, 1–12.
15. Latiff, N. M.; Teo, W. Z.; Sofer, Z.; Fisher, A. C.; Pumera, M. The Cytotoxicity of Layered Black Phosphorus. *Chem. Eur. J.* **2015**, *21*(40), 13991–13995.
16. Zhang, X.; Zhang, Z.; Zhang, S.; Li, D.; Ma, W.; Ma, C.; Wu, F.; Zhao, Q.; Yan, Q.; Xing, B. Size Effect on the Cytotoxicity of Layered Black Phosphorus and Underlying Mechanisms. *Small* **2017**, *13*(32), 1701210.
17. Qu, G.; Liu, W.; Zhao, Y.; Gao, J.; Xia, T.; Shi, J.; Hu, L.; Zhou, W.; Gao, J.; Wang, H.; Luo, Q.; Zhou, Q.; Liu, S.; Yu, X.-F.; Jiang, G. Improved Biocompatibility of Black Phosphorus Nanosheets by Chemical Modification. *Angew. Chem. Int. Ed.* **2017**, *56*(46), 14488–14493.
18. Mo, J.; Xie, Q.; Wei, W.; Zhao, J. Revealing the Immune Perturbation of Black Phosphorus Nanomaterials to Macrophages by Understanding the Protein Corona. *Nat. Commun.* **2018**, *9*(1), 2480.
19. Palmieri, V.; Perini, G.; De Spirito, M.; Papi, M. Graphene Oxide Touches Blood: In Vivo Interactions of Bio-Coronated 2D Materials. *Nanoscale Horiz.* **2019**, *4*(2), 273–290.
20. Neagu, M.; Piperigkou, Z.; Karamanou, K.; Engin, A. B.; Docea, A. O.; Constantin, C.; Negrei, C.; Nikitovic, D.; Tsatsakis, A. Protein Bio-Corona: Critical Issue in Immune Nanotoxicology. *Arch. Toxicol.* **2017**, *91*(3), 1031–1048.
21. Monopoli, M. P.; Åberg, C.; Salvati, A.; Dawson, K. A. Biomolecular Coronas Provide the Biological Identity of Nanosized Materials. *Nat. Nanotechnol.* **2012**, 7, 779.
22. Kumar, V.; Brent, J. R.; Shorie, M.; Kaur, H.; Chadha, G.; Thomas, A. G.; Lewis, E. A.; Rooney, A. P.; Nguyen, L.; Zhong, X. L.; Burke, M. G.; Haigh, S. J.; Walton, A.; McNaughter, P. D.; Tedstone, A. A.; Savjani, N.; Muryn, C. A.; O'Brien, P.; Ganguli, A. K.; Lewis, D. J.; Sabherwal, P. Nanostructured Aptamer-Functionalized Black Phosphorus Sensing Platform for Label-Free Detection of Myoglobin, a Cardiovascular Disease Biomarker. *ACS Appl. Mater. Interfaces* **2016**, *8*(35), 22860–22868.
23. Peng, J.; Lai, Y.; Chen, Y.; Xu, J.; Sun, L.; Weng, J. Sensitive Detection of Carcinoembryonic Antigen Using Stability-Limited Few-Layer Black Phosphorus as an Electron Donor and a Reservoir. *Small* **2017**, *13*(15), 1603589.
24. Ping, J.; Zhou, Y.; Wu, Y.; Papper, V.; Boujday, S.; Marks, R. S.; Steele, T. W. J. Recent Advances in Aptasensors Based on Graphene and Graphene-like Nanomaterials. *Biosens. Bioelectron.* **2015**, *64*, 373–385.
25. Kalantar-zadeh, K.; Ou, J. Z. Biosensors Based on Two-Dimensional MoS_2. *ACS Sens.* **2016**, *1*(1), 5–16.
26. Zhu, C.; Du, D.; Lin, Y. Graphene-Like 2D Nanomaterial-Based Biointerfaces for Biosensing Applications. *Biosens. Bioelectron.* **2017**, *89*, 43–55.
27. Peña-Bahamonde, J.; Nguyen, H. N.; Fanourakis, S. K.; Rodrigues, D. F. Recent Advances in Graphene-Based Biosensor Technology With Applications in Life Sciences. *J. Nanobiotechnol.* **2018**, *16*(1).
28. Wu, L.; Chu, H. S.; Koh, W. S.; Li, E. P. Highly Sensitive Graphene Biosensors Based on Surface Plasmon Resonance. *Opt. Express* **2010**, *18*(14), 14395–14400.

29. Zeng, S.; Hu, S.; Xia, J.; Anderson, T.; Dinh, X.-Q.; Meng, X.-M.; Coquet, P.; Yong, K.-T. Graphene–MoS_2 Hybrid Nanostructures Enhanced Surface Plasmon Resonance Biosensors. *Sensors Actuators B Chem.* **2015**, *207*, 801–810.
30. Ouyang, Q.; Zeng, S.; Jiang, L.; Hong, L.; Xu, G.; Dinh, X.-Q.; Qian, J.; He, S.; Qu, J.; Coquet, P.; Yong, K.-T. Sensitivity Enhancement of Transition Metal Dichalcogenides/Silicon Nanostructure-Based Surface Plasmon Resonance Biosensor. *Sci. Rep.* **2016**, *6*, 28190.
31. Xing, F.; Meng, G.-X.; Zhang, Q.; Pan, L.-T.; Wang, P.; Liu, Z.-B.; Jiang, W.-S.; Chen, Y.; Tian, J.-G. Ultrasensitive Flow Sensing of a Single Cell Using Graphene-Based Optical Sensors. *Nano Lett.* **2014**, *14*(6), 3563–3569.
32. Fu, N.; Hu, Y.; Shi, S.; Ren, S.; Liu, W.; Su, S.; Zhao, B.; Weng, L.; Wang, L. Au Nanoparticles on Two-Dimensional MoS_2 Nanosheets as a Photoanode for Efficient Photoelectrochemical miRNA Detection. *Analyst* **2018**, *143*(7), 1705–1712.
33. Yew, Y. T.; Sofer, Z.; Mayorga-Martinez, C. C.; Pumera, M. Black Phosphorus Nanoparticles as a Novel Fluorescent Sensing Platform for Nucleic Acid Detection. *Mater. Chem. Front.* **2017**, *1*(6), 1130–1136.
34. Sanchez-Gaytan, B. L.; Fay, F.; Hak, S.; Alaarg, A.; Fayad, Z. A.; Pérez-Medina, C.; Mulder, W. J. M.; Zhao, Y. Real-Time Monitoring of Nanoparticle Formation by FRET Imaging. *Angew. Chem. Int. Ed.* **2017**, *56*(11), 2923–2926.
35. Broussard, J. A.; Rappaz, B.; Webb, D. J.; Brown, C. M. Fluorescence Resonance Energy Transfer Microscopy as Demonstrated by Measuring the Activation of the Serine/Threonine Kinase Akt. *Nat. Protoc.* **2013**, *8*, 265.
36. Huang, P.-J. J.; Liu, J. DNA-Length-Dependent Fluorescence Signaling on Graphene Oxide Surface. *Small* **2012**, *8*(7), 977–983.
37. Yao, B. C.; Wu, Y.; Yu, C. B.; He, J. R.; Rao, Y. J.; Gong, Y.; Fu, F.; Chen, Y. F.; Li, Y. R. Partially Reduced Graphene Oxide Based FRET on Fiber-Optic Interferometer for Biochemical Detection. *Sci. Rep.* **2016**, *6*, 23706.
38. Kenry, ; Geldert, A.; Zhang, X.; Zhang, H.; Lim, C. T. Highly Sensitive and Selective Aptamer-Based Fluorescence Detection of a Malarial Biomarker Using Single-Layer MoS_2 Nanosheets. *ACS Sens.* **2016**, *1*(11), 1315–1321.
39. McMorrow, M. L.; Aidoo, M.; Kachur, S. P. Malaria Rapid Diagnostic Tests in Elimination Settings—Can They Find the Last Parasite? *Clin. Microbiol. Infect.* **2011**, *17*(11), 1624–1631.
40. Anthony, M. Rapid Diagnostic Tests for Malaria Parasites. *Clin. Microbiol. Rev.* **2002**, *15*(1), 66.
41. Iqbal, J.; Sher, A.; Rab, A. Plasmodium Falciparum Histidine-Rich Protein 2-Based Immunocapture Diagnostic Assay for Malaria: Cross-Reactivity With Rheumatoid Factors. *J. Clin. Microbiol.* **2000**, *38*(3), 1184–1186.
42. Geldert, A.; Kenry, ; Zhang, X.; Zhang, H.; Lim, C. T. Enhancing the Sensing Specificity of a MoS_2 Nanosheet-Based FRET Aptasensor Using a Surface Blocking Strategy. *Analyst* **2017**, *142*(14), 2570–2577.
43. Geldert, A.; Kenry, ; Lim, C. T. Paper-Based MoS_2 Nanosheet-Mediated FRET Aptasensor for Rapid Malaria Diagnosis. *Sci. Rep.* **2017**, 7(1), 17510.
44. Global Health Data Exchange. http://ghdx.healthdata.org/.
45. Yang, K.; Zhang, S.; Zhang, G.; Sun, X.; Lee, S.-T.; Liu, Z. Graphene in Mice: Ultrahigh In Vivo Tumor Uptake and Efficient Photothermal Therapy. *Nano Lett.* **2010**, *10*(9), 3318–3323.
46. Tian, B.; Wang, C.; Zhang, S.; Feng, L.; Liu, Z. Photothermally Enhanced Photodynamic Therapy Delivered by Nano-Graphene Oxide. *ACS Nano* **2011**, *5*(9), 7000–7009.

47. Nurunnabi, M.; Khatun, Z.; Reeck, G. R.; Lee, D. Y.; Lee, Y.-k. Photoluminescent Graphene Nanoparticles for Cancer Phototherapy and Imaging. *ACS Appl. Mater. Interfaces* **2014**, *6*(15), 12413–12421.
48. Liu, T.; Wang, C.; Gu, X.; Gong, H.; Cheng, L.; Shi, X.; Feng, L.; Sun, B.; Liu, Z. Drug Delivery With PEGylated MoS_2 Nano-Sheets for Combined Photothermal and Chemotherapy of Cancer. *Adv. Mater.* **2014**, *26*(21), 3433–3440.
49. Yin, W.; Yan, L.; Yu, J.; Tian, G.; Zhou, L.; Zheng, X.; Zhang, X.; Yong, Y.; Li, J.; Gu, Z.; Zhao, Y. High-Throughput Synthesis of Single-Layer MoS_2 Nanosheets as a Near-Infrared Photothermal-Triggered Drug Delivery for Effective Cancer Therapy. *ACS Nano* **2014**, *8*(7), 6922–6933.
50. Wang, S.; Li, X.; Chen, Y.; Cai, X.; Yao, H.; Gao, W.; Zheng, Y.; An, X.; Shi, J.; Chen, H. A Facile One-Pot Synthesis of a Two-Dimensional MoS_2/Bi_2S_3 Composite Theranostic Nanosystem for Multi-Modality Tumor Imaging and Therapy. *Adv. Mater.* **2015**, *27*(17), 2775–2782.
51. Rabin, O.; Manuel Perez, J.; Grimm, J.; Wojtkiewicz, G.; Weissleder, R. An X-Ray Computed Tomography Imaging Agent Based on Long-Circulating Bismuth Sulphide Nanoparticles. *Nat. Mater.* **2006**, *5*(2), 118–122.
52. Ma, M.; Huang, Y.; Chen, H.; Jia, X.; Wang, S.; Wang, Z.; Shi, J. Bi_2S_3-Embedded Mesoporous Silica Nanoparticles for Efficient Drug Delivery and Interstitial Radiotherapy Sensitization. *Biomaterials* **2015**, *37*, 447–455.
53. Wang, S.; Li, K.; Chen, Y.; Chen, H.; Ma, M.; Feng, J.; Zhao, Q.; Shi, J. Biocompatible PEGylated MoS_2 Nanosheets: Controllable Bottom-Up Synthesis and Highly Efficient Photothermal Regression of Tumor. *Biomaterials* **2015**, *39*, 206–217.
54. Wang, S.; Chen, Y.; Li, X.; Gao, W.; Zhang, L.; Liu, J.; Zheng, Y.; Chen, H.; Shi, J. Injectable 2D MoS_2-Integrated Drug Delivering Implant for Highly Efficient NIR-Triggered Synergistic Tumor Hyperthermia. *Adv. Mater.* **2015**, *27*(44), 7117–7122.
55. Makadia, H. K.; Siegel, S. J. Poly Lactic-co-Glycolic Acid (PLGA) as Biodegradable Controlled Drug Delivery Carrier. *Polymers* **2011**, *3*(3).
56. Park, J. K.; Shim, J.-H.; Kang, K. S.; Yeom, J.; Jung, H. S.; Kim, J. Y.; Lee, K. H.; Kim, T.-H.; Kim, S.-Y.; Cho, D.-W.; Hahn, S. K. Solid Free-Form Fabrication of Tissue-Engineering Scaffolds with a Poly(lactic-co-glycolic acid) Grafted Hyaluronic Acid Conjugate Encapsulating an Intact Bone Morphogenetic Protein–2/Poly(ethylene glycol) Complex. *Adv. Funct. Mater.* **2011**, *21*(15), 2906–2912.
57. Chen, W.; Ouyang, J.; Liu, H.; Chen, M.; Zeng, K.; Sheng, J.; Liu, Z.; Han, Y.; Wang, L.; Li, J.; Deng, L.; Liu, Y.-N.; Guo, S. Black Phosphorus Nanosheet-Based Drug Delivery System for Synergistic Photodynamic/Photothermal/Chemotherapy of Cancer. *Adv. Mater.* **2017**, *29*(5), 1603864.
58. Tao, W.; Zhu, X.; Yu, X.; Zeng, X.; Xiao, Q.; Zhang, X.; Ji, X.; Wang, X.; Shi, J.; Zhang, H.; Mei, L. Black Phosphorus Nanosheets as a Robust Delivery Platform for Cancer Theranostics. *Adv. Mater.* **2017**, *29*(1), 1603276.
59. Yang, D.; Yang, G.; Yang, P.; Lv, R.; Gai, S.; Li, C.; He, F.; Lin, J. Assembly of Au Plasmonic Photothermal Agent and Iron Oxide Nanoparticles on Ultrathin Black Phosphorus for Targeted Photothermal and Photodynamic Cancer Therapy. *Adv. Funct. Mater.* **2017**, *27*(18), 1700371.
60. Yang, X.; Wang, D.; Shi, Y.; Zou, J.; Zhao, Q.; Zhang, Q.; Huang, W.; Shao, J.; Xie, X.; Dong, X. Black Phosphorus Nanosheets Immobilizing Ce6 for Imaging-Guided Photothermal/Photodynamic Cancer Therapy. *ACS Appl. Mater. Interfaces* **2018**, *10*(15), 12431–12440.
61. Chen, Y.; Wang, L.; Shi, J. Two-Dimensional Non-Carbonaceous Materials-Enabled Efficient Photothermal Cancer Therapy. *Nano Today* **2016**, *11*(3), 292–308.

62. Gong, L.; Yan, L.; Zhou, R.; Xie, J.; Wu, W.; Gu, Z. Two-Dimensional Transition Metal Dichalcogenide Nanomaterials for Combination Cancer Therapy. *J. Mater. Chem. B* **2017**, *5*(10), 1873–1895.
63. Fan, T.; Zhou, Y.; Qiu, M.; Zhang, H. Black Phosphorus: A Novel Nanoplatform With Potential in the Field of Bio-Photonic Nanomedicine. *J. Innov. Opt. Health Sci.* **2018**, *11*(6).
64. Shivananju, B. N.; Yu, W.; Liu, Y.; Zhang, Y.; Lin, B.; Li, S.; Bao, Q. The Roadmap of Graphene-Based Optical Biochemical Sensors. *Adv. Funct. Mater.* **2017**, *27*(19), 1603918.
65. Shivananju, B. N.; Hoh, H. Y.; Yu, W.; Bao, Q. Chapter 10—Optical Biochemical Sensors Based on 2D Materials. In *Fundamentals and Sensing Applications of 2D Materials,* Hywel, M.; Rout, C. S.; Late, D. J., Eds. Woodhead Publishing, **2019**, pp 379–406.
66. Murugan, C.; Sharma, V.; Murugan, R. K.; Malaimegu, G.; Sundaramurthy, A. Two-Dimensional Cancer Theranostic Nanomaterials: Synthesis, Surface Functionalization and Applications in Photothermal Therapy. *J. Control. Release* **2019**, *299*, 1–20.
67. Liu, G.; Zou, J.; Tang, Q.; Yang, X.; Zhang, Y.; Zhang, Q.; Huang, W.; Chen, P.; Shao, J.; Dong, X. Surface Modified Ti_3C_2 MXene Nanosheets for Tumor Targeting Photothermal/Photodynamic/Chemo Synergistic Therapy. *ACS Appl. Mater. Interfaces* **2017**, *9*(46), 40077–40086.

CHAPTER 10

Valleytronics in 2D semiconductors

John Schaibley
Department of Physics, University of Arizona, Tucson, AZ, United States

10.1 Introduction

In crystalline systems, the electronic band structure determines the relationship between the crystal momentum and energy of an electron. In intrinsic semiconductors, the Fermi energy lies between two bands, where typically the highest valence band below the Fermi energy and the lowest conduction band above the Fermi energy determine the electronic and optical physics. These extrema are referred to as crystal momentum space "valleys," which are stable points in the band structure that electrons or holes can occupy. Therefore, in addition to charge and spin, electrons in a crystalline solid also have a valley degree of freedom that determines the electron's position in crystal momentum space. The field of valleytronics seeks to use the electron's valley degree of freedom to encode and process information, analogous to charge in electronics or spin in spintronics (see Fig. 10.1). One motivation of valleytronics research (and spintronics) is an application to lower-energy information storage and processing devices. For example, a pure valley (or spin) current offers the possibility of carrying valley (or spin) information without the ohmic heating associated with a charge current. However, unlike charge that is easily manipulated by electric fields, it is more difficult to manipulate particles based on the valley degree of freedom. In this chapter, the progress in valleytronics in 2D semiconducting transition metal dichalcogenides (TMDs) with an emphasis on experimental results is reviewed.

10.2 Valley physics and contrasting optical selection rule of monolayer TMDs

Even though a variety of bulk semiconductors (e.g., Si, AlAs) have degenerate valleys that could potentially be used for valleytronics, it is difficult to dynamically control the valley degree of freedom in these materials. This is

2D Materials for Photonic and Optoelectronic Applications
https://doi.org/10.1016/B978-0-08-102637-3.00010-3

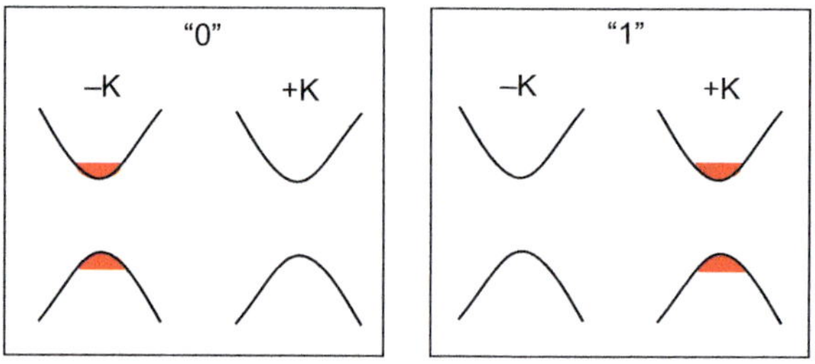

Fig. 10.1 Valley-based information. Information can potentially be stored in "valley polarizations" of 2D semiconductors by polarizing carriers into −K and +K momentum space valleys. Here, a −K valley polarization of electrons and holes represents "0," and a +K valley polarization represents "1."

not the case in 2D material systems, which can exhibit valleys with contrasting physical properties. In particular, the interest in 2D material valleytronics is largely motivated by the graphene-like band structure consisting of two degenerate valleys at the +K and −K (also referred to as K and K′) points in crystal momentum space. In fact, a variety of 2D "honeycomb" crystals share this basic band structure, including graphene, hexagonal boron nitride (hBN), and TMDs. It should be noted that intrinsic graphene is inversion-symmetric, which requires that the +K an −K valleys are indistinguishable; however, it has been pointed out by Xiao and colleagues[1] that, when this inversion symmetry is broken (typically by substrate or externally applied fields), the valleys become distinguishable, giving rise to a host of valley-dependent physical effects. Monolayer TMDs lack inversion symmetry and, therefore, naturally possess valley-contrasting physical effects.[2] In this section, one of the most important valleytronic effects is reviewed: valley-dependent optical selection rule of monolayer TMDs.

10.2.1 Optical properties of TMDs

Although this physics is general to several 2D crystals, I focus on the specific example of monolayer group VI TMD semiconductors with the chemical formula MX_2, where M=Mo, W, and X=S, Se, or Te. The crystal structure is shown in Fig. 10.2A and B. It should be noted that TMD semiconductors received considerable experimental investigations decades ago,[3–6] and were rediscovered in the context of 2D van der Waals materials following graphene.[7–9] One exciting feature of monolayer TMDs is that they exhibit strongly layer-dependent optical properties that were first revealed through photoluminescence (PL) measurements.[8, 9] In PL spectroscopy, focused light (usually from a laser) excites electrons and holes in a material.

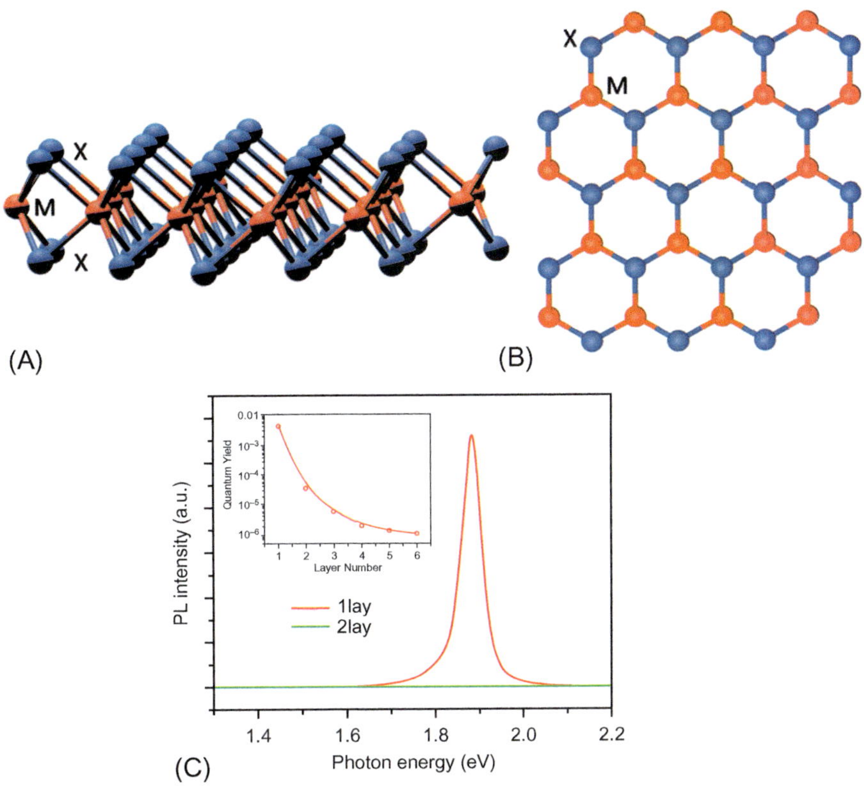

Fig. 10.2 (A, B) Side and top view of an MX_2 monolayer, respectively. (C) Layer-dependent PL from MoS_2. *Adapted from Mak, K. F.; Lee, C.; Hone, J.; Shan, J.; Heinz, T. F. Atomically Thin MoS2: A New Direct-Gap Semiconductor. Phys. Rev. Lett.* ***2010****, 105, 136805.*

These electrons and holes undergo relaxation processes and then recombine with each other emitting photons. Typically, the intensity of the PL is recorded as a function of emission energy (or wavelength) to give a PL spectrum as shown in Fig. 10.2C for MoS_2. Remarkably, it was observed that unlike bulk MoS_2, which is an indirect bandgap semiconductor (and therefore a poor light emitter), monolayer MoS_2 is a direct bandgap semiconductor efficiently emitting PL.[8, 9] This effect is a consequence of interlayer coupling that modifies the band structure (lowering the gamma point valence band energy). This discovery led to a tremendous number of optical investigations of TMDs.[10]

In 2012, Di Xiao and colleagues made a significant discovery that monolayer TMDs host coupled spin-valley physics and a valley-dependent optical selection rule.[2] Arguably the most important feature of TMDs for optically driven valleytronics is the valley-dependent optical polarization selection rule. It is important because it provides a simple way to write and read valley polarization by addressing specific valleys. This selection rule requires that normal-incidence circularly polarized light of a given helicity only couples to optical transitions in one valley. That is, $\sigma\pm$ (right or left circularly) polarized light only couples to the $\pm K$ valley (see Fig. 10.3A) for optical frequencies near the optical band edge. This rule is a consequence of the valley structure and broken inversion symmetry. In fact, this optical selection rule was first derived for graphene with broken inversion symmetry[11] and later applied to TMD monolayers.[2] In addition in monolayer TMDs, the strong spin-orbit coupling results in a locking effect between the spin and valley degrees of freedom. This spin-valley locking effect means that whenever a valley polarization is realized in a monolayer TMD, there is an accompanying spin polarization. The valley-dependent optical selection rule was first experimentally demonstrated using low-temperature polarization-dependent PL. In these experiments, excitons were excited by a circularly polarized laser, and PL emitted by the TMD was analyzed using polarization optics (typically an achromatic quarter-wave plate and polarizer) before it was detected. By rotating the polarization optics, one can measure the co-polarized *PL(co)* and cross-polarized *PL(cross)*. The valley polarization, η, provides a measure of the contrast between co- and cross-polarized components, and it is defined as the difference divided by the sum, $\eta = \frac{PL(co) - PL(cross)}{PL(co) + PL(cross)}$.

The first measurements were carried out on monolayer MoS_2,[12–14] which showed large (50%–100%) values of valley polarization for near-

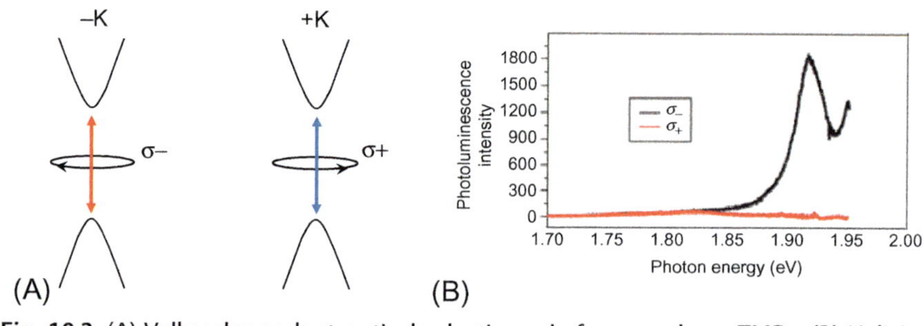

Fig. 10.3 (A) Valley-dependent optical selection rule for monolayer TMDs. (B) Helicity-dependent PL where the sample is pumped with σ_- polarized light. *Adapted from Mak, K. F.; He, K. L.; Shan, J.; Heinz, T. F. Control of Valley Polarization in Monolayer MoS_2 by Optical Helicity. Nat. Nanotechnol.* ***2012****, 7, 494-498.*

resonance excitation (see Fig. 10.3B). The next year, measurements on monolayer WSe_2 revealed that when the system was pumped with linearly polarized light, a coherent superposition of valley excitons could be generated and read out optically.[15]

The optical properties of monolayer TMDs are dominated by excitons (Coulomb-bound electron-hole pairs) with large ~500 meV binding energies.[16] This large binding energy is a consequence of the reduced screening of Coulomb interaction in 2D materials. This is in contrast to III-V semiconductors, which have exciton binding energies on the order of 5–10 meV. Another exciting feature of 2D materials is the ability to dope the 2D layer via electrostatic gating.[17] A typical configuration is depicted in Fig. 10.4A and B, where a voltage is applied to a 2D layer. This problem is usually modeled as a parallel plate capacitor, where the gate voltage is applied across the substrate (typically 285 nm SiO_2). Here, the voltage applied is limited by the dielectric breakdown of the dielectric layer. For high-quality SiO_2, the breakdown field is ~0.5 V/nm, which typically limits doping levels to $\sim 10^{13}$ cm^{-2}.

Early low-temperature measurements revealed multiple peaks in the low-temperature PL spectrum of MoS_2, which were attributed to charged excitons (trions) composed of two electrons (holes) bound to a single hole (electron).[18] Fig. 10.4C and D shows that the intensity of the neutral and charged exciton emission can be controlled by varying the charge density.[19] The binding energies of the trion can be determined by the energy difference between the trion and neutral exciton, typically on the order of

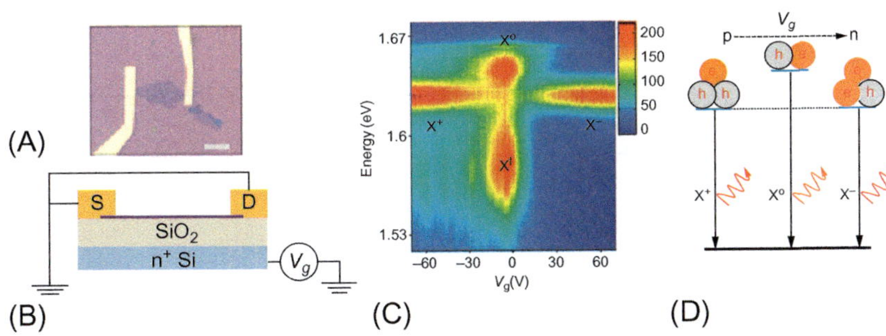

Fig. 10.4 Gate dependence of a monolayer TMD. (A, B) Optical image and cartoon depiction of a monolayer $MoSe_2$ field effect device. Scale bar is 5 μm. (C) Gate-dependent PL from $MoSe_2$ showing PL emission attributed to neutral (X^0) and charged ($X^{\pm}$) excitons. (D) Energy level diagram for neutral and charged excitons (trions) in monolayer $MoSe_2$. *Adapted from Ross, J. S.; et al. Electrical Control of Neutral and Charged Excitons in a Monolayer Semiconductor. Nat. Commun.* ***2013****, 4, 1474.*

10–20 meV. More recently, it has been pointed out that low-energy peaks in the PL spectra have alternative explanations that stem from intervalley Coulomb interactions and plasmonic effects.[20, 21]

Although early measurements on MoS_2 revealed near-unity valley polarization in helicity-dependent PL measurements, similar measurements on other TMD monolayers (with narrower exciton linewidths) reported lower degrees of valley polarization,[15] typically on the order of 30%–50%. This depolarization indicates an intrinsic intervalley scattering mechanism whereby an exciton in one valley can scatter to the other valley. Theoretical investigations attributed this effect to a Maialle-Sham type of electron-hole exchange interaction.[22] This intrinsic depolarization mechanism potentially limits the application of monolayer valley excitons to valleytronics and motivates new structures that can overcome this depolarization mechanism.

10.3 The valley Hall effect

Similar to the spin Hall effect, observed in III–V systems,[23] monolayer TMDs were predicted to host a valley Hall effect (VHE) that results in the accumulation of valley polarization on opposite edges during electron or hole transport (Fig. 10.5A). The VHE is a consequence of the non-zero

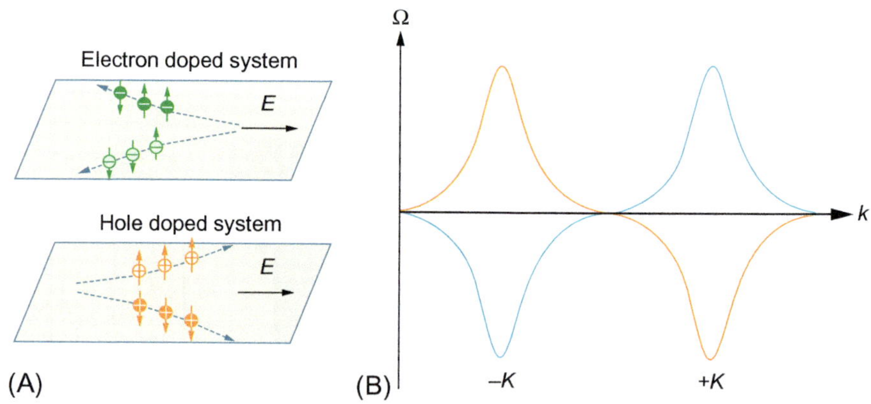

Fig. 10.5 (A) In monolayer TMDs, a VHE is predicted where spin-valley polarized carriers are predicted to flow to sample edges. (B) The VHE arises due to the large Berry curvature (Ω) at the −K and +K valleys. *Panel (A) adapted from Xiao, D., Liu, G.-B., Feng, W., Xu, X. & Yao, W. Coupled Spin and Valley Physics in Monolayers of MoS2 and Other Group-VI Dichalcogenides. Phys. Rev. Lett. 108, 196802 (2012) and (B) from Mak, K. F.; Xiao, D.; Shan, J. Light–Valley Interactions in 2D Semiconductors. Nat. Photonics **2018**, 12, 451.*

Berry curvature that is opposite for electrons and holes in opposite valleys (as shown in Fig. 10.5B). The Berry curvature is an effective field that results from the structure of the Bloch bands. In centrosymmetric crystals the Berry curvature must be zero, and in typical inversion asymmetric crystals (GaAs, for example), the effects are typically small since carriers are close to the gamma ($k=0$) point where the Berry curvature must go to zero due to time reversal symmetry.

The VHE is best understood in terms of the semiclassical transport equations of motion that simultaneously represent an electron wavepacket in real and crystal momentum space,[1, 24]

$$\dot{r} = \frac{1}{\hbar}\frac{\partial E_{n,k}}{\partial k} - \dot{k} \times \Omega_{n,k}, \quad \hbar\dot{k} = -eE - e\dot{r} \times B,$$

where $E_{n,k}(\Omega_{n,k})$ is the energy (Berry curvature) of the nth band, $k(r)$ is the crystal momentum (position) of the electron wavepacket, E is an external electric field, and B is an external magnetic field. By comparing the two equations, it is clear that the Berry curvature acts like a magnetic field in crystal momentum space. When these transport equations are solved, one finds that the carriers in opposite valleys accumulate on opposite edges of the sample, giving rise to a valley current in the transverse direction (Fig. 10.5A).

The VHE in TMDs was first demonstrated experimentally by Mak, McEuen, and colleagues using optical injection and electrical readout in a monolayer MoS_2 Hall bar structure (Fig. 10.6A and B).[25] Here, a circularly polarized pump laser (near the exciton resonance at 1.9 eV) injected valley-polarized electrons and holes. A voltage V_x provided an electric field, which resulted in hole and electron currents traveling in opposite directions—holes to the right and electrons to the left. The resulting valley Hall current in the transverse direction resulted in a preferential flow of holes toward the A contact and electrons toward the B contact, which yielded a transverse Hall voltage (V_H) between A and B (Fig. 10.6C). As expected the sign of Hall voltage flips with the helicity of the circularly polarized pump laser.

In later experiments, the VHE was mapped using spatially resolved Kerr rotation.[26] In these experiments, bilayer 2H-MoS_2 was studied, which allowed for the VHE to be controlled electronically. As discussed earlier, monolayer MoS_2 has broken inversion symmetry, which is a requirement to observe valley-contrasting physical effects. Bilayer MoS_2, on the other hand, is inversion-symmetric, so the VHE should be zero. However, it was shown earlier[27] that a sufficiently large out-of-plane electric field (provided by a vertical gate voltage) can be used to break the inversion symmetry

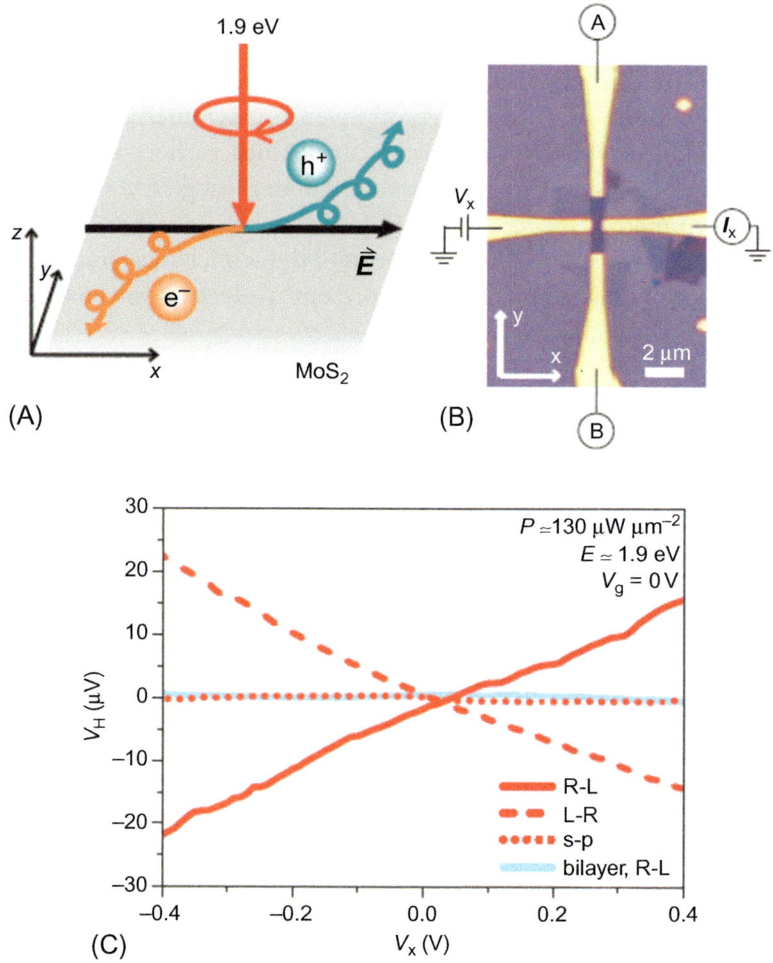

Fig. 10.6 (A) Circularly polarized light optically injects electrons and holes into monolayer MoS_2. The VHE causes electrons and holes to flow to opposite sample edges in the presence of in-plane electric field (E). (B) Optical microscope image of the MoS_2 Hall bar device. (C) Hall voltage (V_H) as a function of in-plane bias (V_x). The solid and dashed lines show that the Hall signal changes sign with pump helicity. *Adapted from Mak, K. F.; McGill, K. L.; Park, J.; McEuen, P. L. The Valley Hall Effect in MoS_2 Transistors. Science* ***2014****, 344, 1489-1492.*

of MoS_2. Therefore, by varying the gate voltage, the VHE magnitude and the resulting transverse valley current can be controlled. The VHE was observed by measuring the resulting polarizations on sample edges via Kerr rotation. In a Kerr rotation experiment, a linearly polarized laser was focused on the sample and the polarization of the back-reflected beam is measured. If

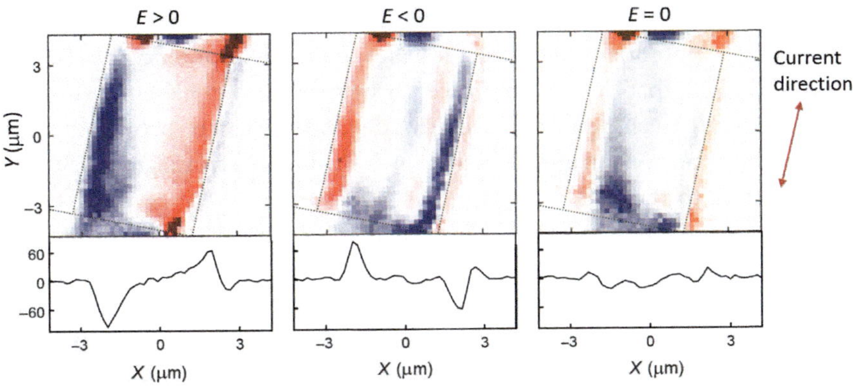

Fig. 10.7 Spatial Kerr rotation map of the VHE in bilayer MoS_2 showing spin-valley polarization accumulation on sample edges. The VHE can be controlled by applying an out-of-plane electric field that breaks the inversion symmetry of the bilayer. *Adapted from Lee, J.; Mak, K. F.; Shan, J. Electrical Control of the Valley Hall Effect in Bilayer MoS_2 Transistors. Nat. Nanotechnol.* ***2016****, 11, 421–425.*

the sample exhibited circular dichroism or birefringence, right and left circularly polarized light would reflect with different reflection coefficients, giving rise to a rotation of the reflected beam's polarization angle. Since TMDs (with broken inversion symmetry) exhibit a valley-dependent optical selection rule, a polarization of electrons in one valley will change that valley exciton's oscillator strength. Therefore, valley polarization results in a circular dichroism that can be read out by Kerr rotation. Fig. 10.7 shows a spatial map of the Kerr rotation signal, obtained by scanning the laser beam over the sample area. The data show clear Kerr rotation signals on the sample edges with opposite signs, consistent with opposite valley polarizations on the two edges. By varying the applied field ($E>0$, $E=0$, and $E<0$), the Kerr rotation signal flips sign, demonstrating the VHE can be controlled in bilayer MoS_2 by an external electric field.

10.4 Breaking the valley degeneracy

A pure valleytronic current requires the ability to control the flow of valley-polarized particles (such as electrons, holes, or excitons). We have already seen how the VHE can realize this effect by making use of the non-zero Berry curvature at the +K and −K valleys. However, this approach requires a longitudinal charge current to be flowing and, therefore, is less attractive for pure valleytronic devices, which seek to reduce or eliminate charge currents completely. This goal of manipulating valley-polarized particles without a charge current motivates the use of external fields to tune the valley energies, potentially allowing for valley-dependent potentials to be engineered. Recall that due to time reversal symmetry, the +K and −K valleys are degenerate in energy. However, this degeneracy can be broken by applying external optical, magnetic fields, or by using a ferromagnetic substrate that couples to the TMD layer.

One approach uses intense optical fields to break valley degeneracy. With polarized laser pulses it was shown that a valley-dependent energy shift on the order of 10 meV can be realized.[28, 29] In these experiments, polarization of the laser pulse allows the pulse to selectively couple to a single valley. Then the coupling between the optical field and the bands results in an optical Stark-like energy shift of the exciton energy while the pulse is on. This dynamical control of the valley energy is of particular interest to optical information processing.

Due to the analogy between spin and valley pseudo-spin degree of freedom, it is perhaps not surprising that an external magnetic field can be used to break the degeneracy of valleys through a valley Zeeman effect. The effect was first measured in monolayer TMDs by magneto-PL,[30–37] that is, by measuring the low-temperature polarization-dependent PL as a function of external magnetic field. One complication with this readout technique is that even though valley degeneracy may be broken by the external magnetic field, PL is dependent on the difference in energy between the conduction and valence band states, which to the lowest order does not shift with applied magnetic field. This can be seen in Fig. 10.8, which shows that +K (−K) valley conduction and valence bands are shifted up (down) together. It should be noted that there are actually three components to this energy splitting, since the spin (Δ_s), atomic orbital (Δ_a), and valley orbital (Δ_v) shifts all must be considered. The resulting valley-dependent energy shift in PL is, therefore, only a higher-order perturbation, which has to be carefully extracted from the data (Fig. 10.8B). To isolate the PL signal from a single valley, both the excitation laser and polarization detection axis are co-circularly polarized, again using the valley-dependent optical selection rule. The difference in PL center energy for the two different valleys was measured as a function of the external magnetic field (Fig. 10.8C), resulting in splitting on the order of 0.1–0.2 meV/T.[30–37] These studies showed that magnetic fields could be used to break valley degeneracy; however, the application of such large (~1 T) magnetic fields may not be practical for valleytronic devices. These results did motivate the investigation of magnetic materials that can be used to break valley degeneracy.

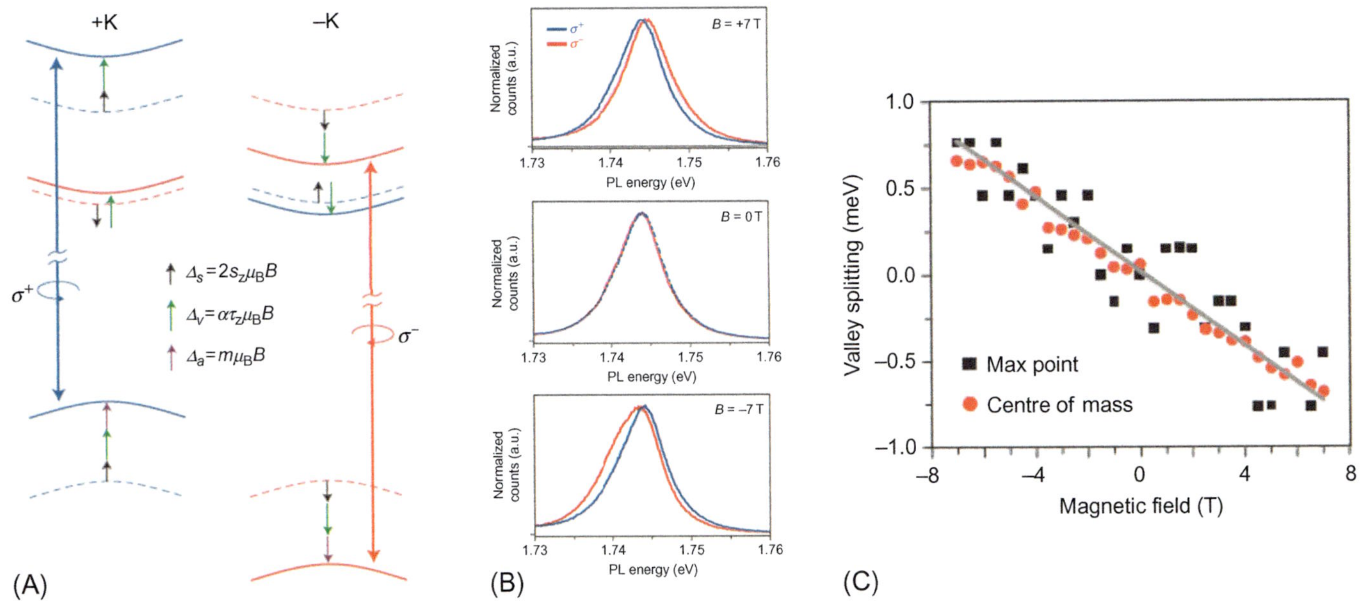

Fig. 10.8 (A) In the presence of an external magnetic field B, the +K and −K bands shift in energy. The spin (Δ_s), atomic orbital (Δ_a) and valley orbital (Δ_v) shifts must be considered. (B, C) Helicity-dependent PL for three different magnetic fields showing a small but observable energy shift of the WSe_2 exciton emission line. The energy difference is plotted in (C). *Adapted from Aivazian, G.; et al. Magnetic Control of Valley Pseudospin in Monolayer WSe_2. Nat. Phys.* ***2015****, 11, 148–152.*

In addition to the externally applied fields, it was shown that integrating TMDs with ferromagnetic substrates can break the valley degeneracy. This was first studied in WSe_2-CrI_3 heterostructures. CrI_3 is a 2D layered ferromagnetic insulator, and a Curie temperature of 61 K.[38] When monolayer WSe_2 was mechanically transferred (see Section 10.5) onto CrI_3, coupling between the layers resulted in a magnetic proximity effect that resulted in an effective magnetic field splitting of the WSe_2 layer equivalent to an externally applied magnetic field of 13 T[38, 39] (Fig. 10.9). This early result showed the promise of 2D semiconductor magnetic heterostructures. It should be noted that CrI_3 is a very unstable 2D material requiring the use of a high-purity oxygen-free glove box for van der Waals assembly. Future work in this 2D semiconductor valleytronics could be enhanced by the discovery and development of air-stable 2D ferromagnetic insulators with higher Curie temperatures and larger band gaps.

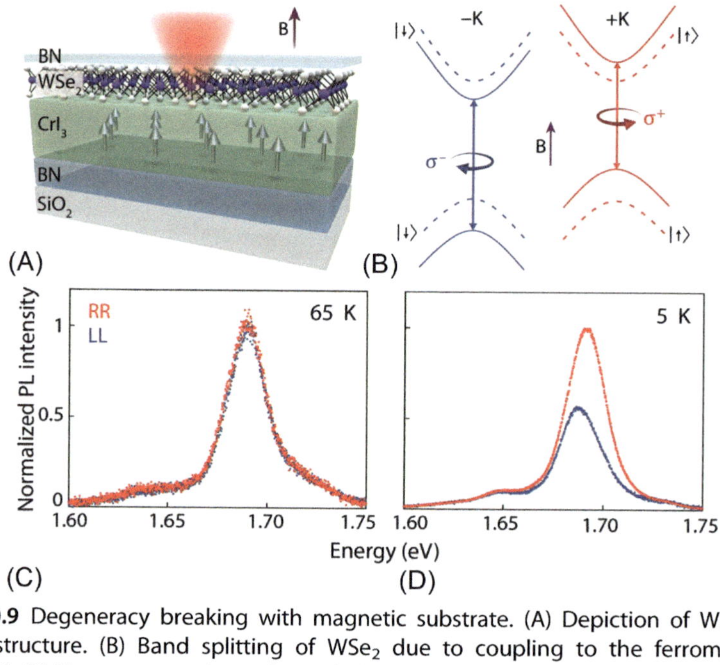

Fig. 10.9 Degeneracy breaking with magnetic substrate. (A) Depiction of WSe_2-CrI_3 heterostructure. (B) Band splitting of WSe_2 due to coupling to the ferromagnetic layer. (C, D) Temperature dependence helicity resolved PL showing that below the Curie temperature, WSe_2 PL becomes helicity-dependent due to the breaking valley degeneracy. *Adapted from Zhong, D.; et al. Van der Waals Engineering of Ferromagnetic Semiconductor Heterostructures for Spin and Valleytronics. Sci. Adv.* ***2017****, 3, e1603113.*

10.5 Valleytronics in 2D semiconductor heterostructures

One of the most exciting opportunities in the field of 2D layered materials is the ability to stack disparate layers together to realize van der Waals heterostructures. Currently, the most common approach is to first isolate individual monolayers either through exfoliation from "bulk flakes" using scotch tape or direct monolayer growth, typically using chemical vapor deposition. Through both methods it is fairly common to obtain a single crystal with areas of 10×10 μm or larger. Single monolayers are often first identified through optical contrast on 285-nm SiO_2/Si substrates whose thickness provides an advantageous contrast for many 2D materials.[40] Once single layers are identified they can be stacked together using a pick-and-place technique while imaging both monolayers under a microscope. The most common method currently used is based on a dry polymer stamping technique using a PDMS stamp on a glass slide to pick up the layers.[41] Once all the layers are picked up, these are transferred onto a target substrate, resulting in the desired device.

One of the first 2D heterostructures to be studied was graphene on hBN, where early work showed a huge improvement in the electronic performance of graphene due to the ultra-flat and low defect nature of the hBN substrate.[42] Later work showed that topological currents in graphene-hBN heterostructures provided some of the first evidence of the VHE in 2D material systems.[43] This capability to stack disparate 2D layers together has provided many new opportunities ranging from atomically thin PN junctions[44, 45] to superconducting states in bilayer heterostructures.[46] An important feature of these 2D heterostructures is that when layers are stacked together, the twist angle between layers can be controlled, allowing for further manipulation of these artificial materials. This is particularly important for assembled (homo) bilayers of the same material because a small twist angle results in a long-range periodic moiré structure.[46–48]

In 2D semiconductors, heterostructures play an important role in controlling valley polarizations. An early theoretical work by Kang et al.[49] showed that a number of combinations of 2D semiconductor bilayer heterostructures have resulted in type II band alignment (Fig. 10.10). In type II heterostructures, the lowest conduction band is in one layer, and the highest valence band is in the other layer (Fig. 10.10). Type II heterostructures are particularly important in electronics and optoelectronics because these have applications to devices such as diodes and light-emitting diodes. In the context of valleytronics, we will see how these type II heterostructures can be used to encode long-lived valley polarizations.

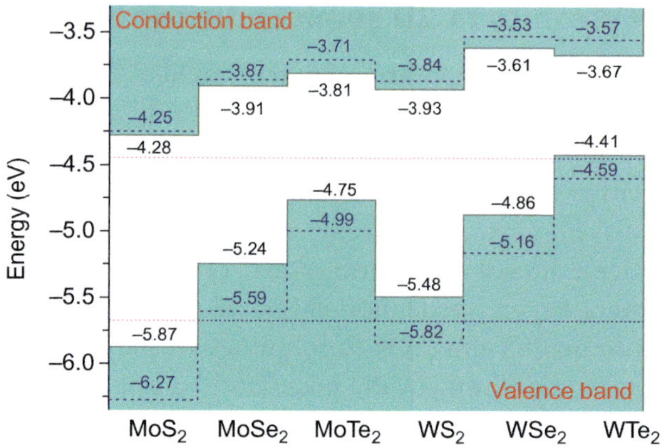

Fig. 10.10 Band offsets for TMDs relative to vacuum level. Solid (dashed) lines were calculated using the PBE (HBE06) method. *Adapted from Kang, J.; Tongay, S.; Zhou, J.; Li, J.; Wu, J. Band Offsets and Heterostructures of Two-Dimensional Semiconductors. Appl. Phys. Lett.* ***2013****, 102, 012111-012111-012114.*

One example of a type II heterostructure is $MoSe_2$-WSe_2 (Fig. 10.11A and B), which was first studied extensively by Rivera, Schaibley, Xu and colleagues.[50] Due to the type II band alignment, the lowest conduction band is in the $MoSe_2$ layer, and the highest valence band is in the WSe_2 layer. Upon injection of electrons and holes, electrons undergo charge transfer to the $MoSe_2$ layer, and holes charge transfer to the WSe_2 layer. If the attractive Coulomb interaction between the electrons and holes in opposite layers is sufficiently strong, an interlayer exciton can form. The energy of the interlayer exciton is determined by the band offsets between layers and the binding energy of the interlayer exciton. The structures were first studied by spatially dependent PL experiments (Fig. 10.11C). Here, the sample consists of three different regions—isolated $MoSe_2$, WSe_2 and the heterostructure (overlap). PL from the isolated $MoSe_2$ and WSe_2 regions show monolayer (intralayer) excitons well known from earlier studies. However, PL from the heterostructure region shows an additional low-energy peak, which is consistent with the prediction of a lower-energy interlayer exciton. Furthermore, since the electron and hole are spatially separated, these exhibit an out-of-plane dipole moment, which can be tuned by applying an out-of-plane electric field. The spatially indirect nature of the interlayer exciton also results in a significantly longer lifetime of 1–1000 ns for the interlayer exciton,[50, 51] compared to $\sim$1 ps for the intralayer exciton.[52, 53] This long lifetime provides an opportunity to realize spatial transport of interlayer excitons before these radiatively decay.

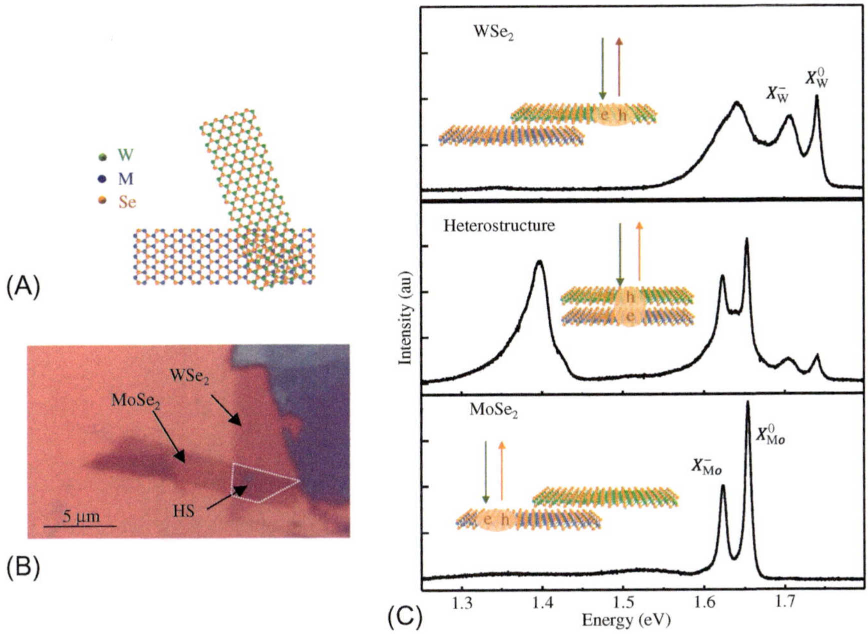

Fig. 10.11 (A) Depiction of an $MoSe_2$-WSe_2 heterostructure. (B) Optical microscopy image of a heterostructure. (C) Spatially resolved PL of three different sample regions. *Adapted from Rivera, P.; et al. Observation of Long-Lived Interlayer Excitons in Monolayer $MoSe_2$–WSe_2 Heterostructures. Nat. Commun.* ***2015****, 6, 6242.*

It is important to remember that the conduction and valence band valleys that give rise to the interlayer exciton are still located at the +K and −K valleys of the two different layers (Fig. 10.12A). This means that the electrons and holes that make up the interlayer exciton can still be used to encode valley polarization and can be read out either by monitoring polarization of the interlayer exciton PL[51] or by resonantly probing the intralayer exciton resonances using polarized light.[54, 55] In fact, it is possible to derive a polarization selection rule for TMD bilayer heterostructures.[51] In general, the selection rule is dependent on the twist angle between the two monolayers and has a periodic spatial dependence that results from the moiré pattern between the layers.[56–58] However, for nearly aligned monolayers (near zero-twist angle), $MoSe_2$-WSe_2 heterostructures show a PL polarization selection rule similar to that of single monolayers (Fig. 10.12B and C). Here, polarization helicity of the interlayer PL is shown to be dominantly co-polarized with the excitation laser's circular helicity. This provides a readout channel to measure valley polarization of the interlayer exciton.

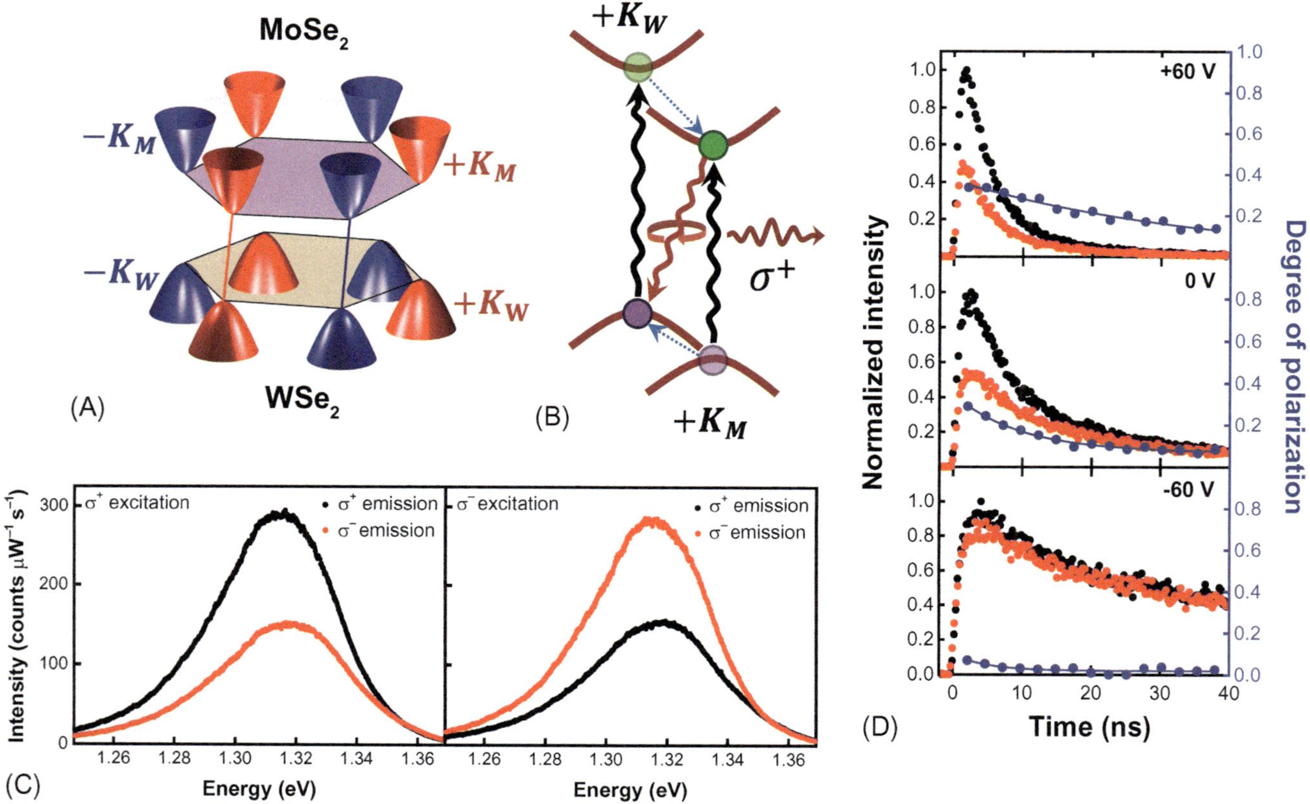

Fig. 10.12 (A) Band structure showing the alignment of $MoSe_2$ and WSe_2 bands. (B) For certain twist angles, the interlayer exciton can exhibit helicity-dependent PL similar to that of monolayers, shown in (C). (D) Temporally resolved PL for co-circularly (black) and cross-circularly (*red*) polarized interlayer exciton emission, as a function of sample gate voltage. The *blue* points show valley polarization η. *Adapted from Rivera, P.; et al. Valley-Polarized Exciton Dynamics in a 2D Semiconductor Heterostructure. Science* ***2016****, 351, 688-691.*

Another advantage of interlayer excitons over intralayer excitons for valleytronics is that since the electron and hole are spatially separated, the electron-hole exchange interaction (which causes depolarization in monolayer TMDs) is suppressed, which allows for enhanced valley polarization times (Fig. 10.12D).

10.5.1 Spatial transport of valley polarization

Since the interlayer exciton is a net-charge neutral quasiparticle, its spatial transport allows for the realization of pure valley current (without a charge current). Recall that due to the spin-valley locking effect,[2] a valley polarization in a monolayer TMD results in an accompanying spin polarization, so that a valley current can also accurately be called a spin-valley current. The first demonstration of valley-polarized interlayer exciton spatial transport used the $MoSe_2$-WSe_2 interlayer excitons described previously.[51] In the experiments, the spatial profile of the interlayer exciton's PL was imaged under circularly polarized excitation. The co- and cross-polarized PL spatial maps were measured as a function of excitation laser intensity (Fig. 10.13A and B). The signature of valley-polarized exciton transport was a dip in the center of the excitation spot at high intensity, which can be explained by a drift diffusion model for +K and −K polarized interlayer excitons.[51] This dip arises because the majority of polarized interlayer excitons (in one single valley) are subject to two different repulsive forces, one from the dipole-dipole interaction between interlayer excitons, and another from the electron-electron and hole-hole exchange interaction in individual layers. The exchange effect is suppressed for electrons in opposite valleys, which results in less drift and diffusion. Therefore, when subtracting the co- and cross-polarized PL images, a dip appears as a signature of the valley-polarized exciton transport (Fig. 10.13B).

A pure spin-valley polarization current was also demonstrated in WSe_2-WS_2 heterostructures using spatially and temporally dependent pump probe technique. The WSe_2-WS_2 heterostructure also exhibited a type II band alignment with a lower conduction band in the WS_2 layer. In this experiment, a polarized pump laser injected spin-valley polarized electrons and holes into the WSe_2 layer. The electrons then charge-transfer to the WS_2 layer where they recombine with holes in WSe_2, leaving an excess of valley-polarized holes in the WSe_2 layer that can mediate a valley polarization current (Fig. 10.13C and D). The spin-valley polarization current was mapped by focusing a polarized pump pulse on one side of the sample and

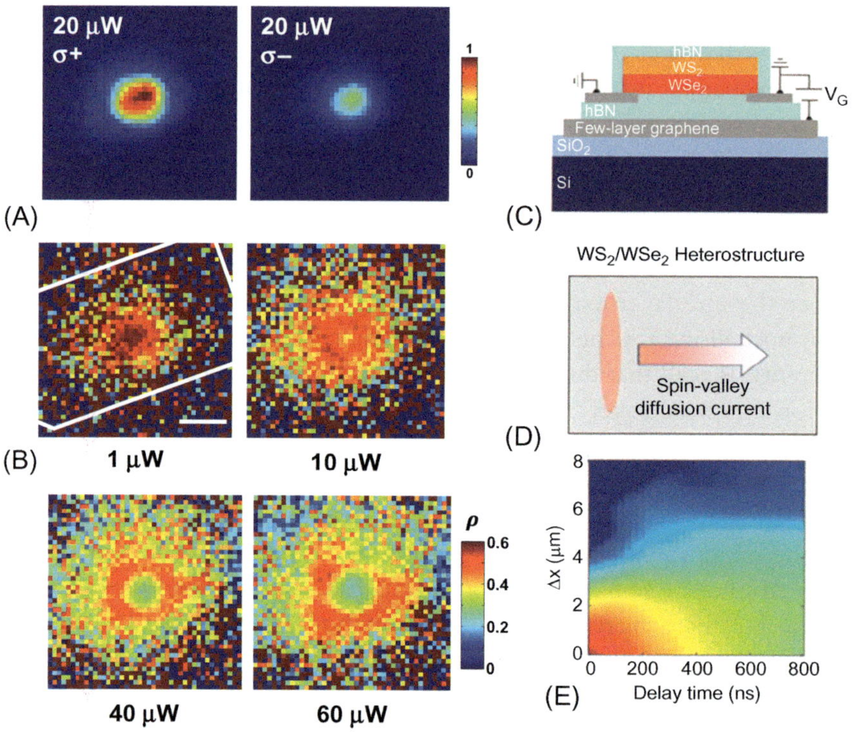

Fig. 10.13 Spin-valley transport. (A) Helicity-resolved interlayer PL from an $MoSe_2$-WSe_2 heterostructure. (B) Valley polarization as a function of increasing pump intensity showing the formation of a ring pattern from the valley-dependent drift diffusion of interlayer excitons. (C) Depiction of a WS_2-WSe_2 device used for imaging spin-valley currents (D). (E) Spatial/temporal map of spin-valley transport in WS_2-WSe_2 heterostructure showing several-microns-long diffusion lengths. *Panels (A, B) adapted from Rivera, P.; et al. Valley-Polarized Exciton Dynamics in a 2D Semiconductor Heterostructure. Science* ***2016****,* 351, *688-691 and (C–E) from Jin, C.; et al. Imaging of Pure Spin-Valley Diffusion Current in WS2-WSe2 Heterostructures. Science* ***2018****, 360, 893-896.*

then scanning a polarized probe pulse over the sample spatially for different time delays, resulting in a spatial-temporal map of spin-valley polarization (Fig. 10.13E), which shows a long-range (several microns) transport of polarization.

10.6 Conclusion and outlook

Although much progress has been made so far in controlling spin-valley polarizations, the next step is to develop more robust devices that can dynamically control the flow of spin–valley polarization, perhaps through electrostatic potentials, which have recently been used to demonstrate room temperature transport of interlayer excitons.[59] Such a control could be used to develop a valleytronic logic gate in which the polarization state of a valley polarization bit is controlled by another valley polarization control bit. The generation of valleytronic logic gates could enable the development of lower-power information processing devices that do not rely on the flow of charged particles. Another possibility is that valleytronic materials could enable new spin-valley-based information storage devices through integration with 2D ferromagnetic layers. The recent discovery of moiré excitons could also pave the way for quantum valleytronics based on localized excitons in 2D heterostructures.[56–63]

Acknowledgment

This work was supported by NSF grant # ECCS-1708562.

References

1. Xiao, D.; Yao, W.; Niu, Q. Valley-Contrasting Physics in Graphene: Magnetic Moment and Topological Transport. *Phys. Rev. Lett.* **2007**, *99*, 236809.
2. Xiao, D.; Liu, G.-B.; Feng, W.; Xu, X.; Yao, W. Coupled Spin and Valley Physics in Monolayers of MoS_2 and Other Group-VI Dichalcogenides. *Phys. Rev. Lett.* **2012**, *108*, 196802.
3. Consadori, F.; Frindt, R. Crystal Size Effects on the Exciton Absorption Spectrum of WSe_2. *Phys. Rev. B* **1970**, *2*, 4893.
4. Frindt, R. Optical Absorption of a Few Unit-Cell Layers of MoS_2. *Phys. Rev.* **1965**, *140*, A536.
5. Frindt, R.; Yoffe, A. Physical Properties of Layer Structures: Optical Properties and Photoconductivity of Thin Crystals of Molybdenum Disulphide. *Proc. R. Soc. Lond. Ser. A. Math. Phys. Sci.* **1963**, *273*, 69–83.
6. Joensen, P.; Frindt, R.; Morrison, S. R. Single-Layer MoS_2. *Mater. Res. Bull.* **1986**, *21*, 457–461.
7. Novoselov, K. S.; et al. Two-Dimensional Atomic Crystals. *Proc. Natl. Acad. Sci. U. S. A.* **2005**, *102*, 10451–10453.
8. Mak, K. F.; Lee, C.; Hone, J.; Shan, J.; Heinz, T. F. Atomically Thin MoS_2: A New Direct-Gap Semiconductor. *Phys. Rev. Lett.* **2010**, *105*, 136805.
9. Splendiani, A.; et al. Emerging Photoluminescence in Monolayer MoS_2. *Nano Lett.* **2010**, *10*, 1271–1275.
10. Mak, K. F.; Shan, J. Photonics and Optoelectronics of 2D Semiconductor Transition Metal Dichalcogenides. *Nat. Photonics* **2016**, *10*, 216–226.
11. Yao, W.; Xiao, D.; Niu, Q. Valley-Dependent Optoelectronics From Inversion Symmetry Breaking. *Phys. Rev. B* **2008**, *77*, 235406.
12. Mak, K. F.; He, K. L.; Shan, J.; Heinz, T. F. Control of Valley Polarization in Monolayer MoS_2 by Optical Helicity. *Nat. Nanotechnol.* **2012**, *7*, 494–498.

13. Cao, T.; et al. Valley-selective circular dichroism of monolayer molybdenum disulphide. *Nat. Commun.* **2012**, *3*, 887.
14. Zeng, H.; Dai, J.; Yao, W.; Xiao, D.; Cui, X. Valley Polarization in MoS_2 Monolayers by Optical Pumping. *Nat. Nanotechnol.* **2012**, 7, 490–493.
15. Jones, A. M.; et al. Optical Generation of Excitonic Valley Coherence in Monolayer WSe_2. *Nat. Nanotechnol.* **2013**, *8*, 634–638.
16. Chernikov, A.; et al. Exciton Binding Energy and Nonhydrogenic Rydberg Series in Monolayer WS_2. *Phys. Rev. Lett.* **2014**, *113,* 076802.
17. Novoselov, K. S.; et al. Electric Field Effect in Atomically Thin Carbon Films. *Science* **2004**, *306*, 666–669.
18. Mak, K. F.; et al. Tightly Bound Trions in Monolayer MoS_2. *Nat. Mater.* **2012**, *12*, 207–211.
19. Ross, J. S.; et al. Electrical Control of Neutral and Charged Excitons in a Monolayer Semiconductor. *Nat. Commun.* **2013**, *4*, 1474.
20. Dery, H. Theory of Intervalley Coulomb Interactions in Monolayer Transition-Metal Dichalcogenides. *Phys. Rev. B* **2016**, *94,* 075421.
21. Van Tuan, D.; Scharf, B.; Žutić, I.; Dery, H. Marrying Excitons and Plasmons in Monolayer Transition-Metal Dichalcogenides. *Phys. Rev. X* **2017**, 7, 041040.
22. Maialle, M. Z.; de Andrada e Silva, E. A.; Sham, L. J. Exciton Spin Dynamics in Quantum Wells. *Phys. Rev. B* **1993**, *47*, 15776–15788.
23. Kato, Y.; Myers, R.; Gossard, A.; Awschalom, D. Observation of the Spin Hall Effect in Semiconductors. *Science* **2004**, *306*, 1910–1913.
24. Xu, X.; Yao, W.; Xiao, D.; Heinz, T. F. Spin and Pseudospins in Layered Transition Metal Dichalcogenides. *Nat. Phys.* **2014**, *10*, 343–350.
25. Mak, K. F.; McGill, K. L.; Park, J.; McEuen, P. L. The Valley Hall Effect in MoS_2 Transistors. *Science* **2014**, *344*, 1489–1492.
26. Lee, J.; Mak, K. F.; Shan, J. Electrical Control of the Valley Hall Effect in Bilayer MoS_2 Transistors. *Nat. Nanotechnol.* **2016**, *11*, 421–425.
27. Wu, S.; et al. Electrical Tuning of Valley Magnetic Moment Through Symmetry Control in Bilayer MoS_2. *Nat. Phys.* **2013**.
28. Sie, E. J.; et al. Valley-Selective Optical Stark Effect in Monolayer WS_2. *Nat. Mater.* **2015**, *14*, 290–294. https://doi.org/10.1038/nmat4156.
29. Kim, J.; et al. Ultrafast Generation of Pseudo-Magnetic Field for Valley Excitons in WSe2 Monolayers. *Science* **2014**, *346*, 1205–1208.
30. Aivazian, G.; et al. Magnetic Control of Valley Pseudospin in Monolayer WSe_2. *Nat. Phys.* **2015**, *11*, 148–152.
31. Srivastava, A.; et al. Valley Zeeman Effect in Elementary Optical Excitations of Monolayer WSe_2. *Nat. Phys.* **2015**.
32. Rostami, H.; Asgari, R. Valley Zeeman Effect and Spin-Valley Polarized Conductance in Monolayer MoS_2 in a Perpendicular Magnetic Field. *Phys. Rev. B* **2015**, *91,* 075433.
33. Stier, A. V.; McCreary, K. M.; Jonker, B. T.; Kono, J.; Crooker, S. A. Exciton Diamagnetic Shifts and Valley Zeeman Effects in Monolayer WS_2 and MoS_2 to 65 Tesla. *Nat. Commun.* **2016**, 7, 10643.
34. MacNeill, D.; et al. Breaking of Valley Degeneracy by Magnetic Field in Monolayer $MoSe_2$. *Phys. Rev. Lett.* **2015**, *114,* 037401.
35. Mitioglu, A.; et al. Optical Investigation of Monolayer and Bulk Tungsten Diselenide (WSe_2) in High Magnetic Fields. *Nano Lett.* **2015**, *15*, 4387–4392.
36. Wang, G.; et al. Magneto-Optics in Transition Metal Diselenide Monolayers. *2D Mater.* **2015**, *2,* 034002.
37. Li, Y.; et al. Valley Splitting and Polarization by the Zeeman Effect in Monolayer $MoSe_2$. *Phys. Rev. Lett.* **2014**, *113*, 266804.

38. Huang, B.; et al. Layer-Dependent Ferromagnetism in a van der Waals Crystal Down to the Monolayer Limit. *Nature* **2017**, *546*, 270.
39. Seyler, K. L.; et al. Valley Manipulation by Optically Tuning the Magnetic Proximity Effect in WSe_2/CrI_3 Heterostructures. *Nano Lett.* **2018**, *18*, 3823–3828.
40. Blake, P.; et al. Making Graphene Visible. *Appl. Phys. Lett.* **2007**, *91*.
41. Zomer, P.; Guimarães, M.; Brant, J.; Tombros, N.; van Wees, B. Fast Pick Up Technique for High Quality Heterostructures of Bilayer Graphene and Hexagonal Boron Nitride. *Appl. Phys. Lett.* **2014**, *105,* 013101.
42. Dean, C.; et al. Boron Nitride Substrates for High-Quality Graphene Electronics. *Nat. Nanotechnol.* **2010**, *5*, 722–726.
43. Gorbachev, R.; et al. Detecting Topological Currents in Graphene Superlattices. *Science* **2014**, *346*, 448–451.
44. Cheng, R.; et al. Electroluminescence and Photocurrent Generation From Atomically Sharp WSe_2/MoS_2 Heterojunction p–n Diodes. *Nano Lett.* **2014**, *14*, 5590–5597.
45. Ross, J. S.; et al. Interlayer Exciton Optoelectronics in a 2D Heterostructure p–n Junction. *Nano Lett.* **2017**, *17*, 638–643. https://doi.org/10.1021/acs.nanolett.6b03398.
46. Cao, Y.; et al. Unconventional Superconductivity in Magic-Angle Graphene Superlattices. *Nature* **2018**, *556*, 43.
47. Kim, K.; et al. Tunable Moiré Bands and Strong Correlations in Small-Twist-Angle Bilayer Graphene. *Proc. Natl. Acad. Sci.* **2017**, *114*, 3364–3369.
48. Bistritzer, R.; MacDonald, A. H. Moiré Bands in Twisted Double-Layer Graphene. *Proc. Natl. Acad. Sci.* **2011**, *108*, 12233–12237.
49. Kang, J.; Tongay, S.; Zhou, J.; Li, J.; Wu, J. Band Offsets and Heterostructures of Two-Dimensional Semiconductors. *Appl. Phys. Lett.* **2013**, *102,* 012111-012111-012114.
50. Rivera, P.; et al. Observation of Long-Lived Interlayer Excitons in Monolayer $MoSe_2$–WSe_2 Heterostructures. *Nat. Commun.* **2015**, *6*, 6242.
51. Rivera, P.; et al. Valley-Polarized Exciton Dynamics in a 2D Semiconductor Heterostructure. *Science* **2016**, *351*, 688–691.
52. Moody, G.; et al. Intrinsic Homogeneous Linewidth and Broadening Mechanisms of Excitons in Monolayer Transition Metal Dichalcogenides. *Nat. Commun.* **2015**, *6*, 8315.
53. Moody, G.; Schaibley, J.; Xu, X. Exciton Dynamics in Monolayer Transition Metal Dichalcogenides. *J. Opt. Soc. Am. B* **2016**, *33*, C39–C49.
54. Schaibley, J. R.; et al. Directional Interlayer Spin-Valley Transfer in Two-Dimensional Heterostructures. *Nat. Commun.* **2016**, 7, 13747.
55. Jin, C.; et al. Imaging of Pure Spin-Valley Diffusion Current in WS2-WSe2 Heterostructures. *Science* **2018**, *360*, 893–896.
56. Kang, J.; Li, J.; Li, S. S.; Xia, J. B.; Wang, L. W. Electronic Structural Moire Pattern Effects on $MoS_2/MoSe_2$ 2D Heterostructures. *Nano Lett.* **2013**, *13*, 5485–5490.
57. Yu, H.; Liu, G.-B.; Tang, J.; Xu, X.; Yao, W. Moiré Excitons: From Programmable Quantum Emitter Arrays to Spin-Orbit-Coupled Artificial Lattices. *Sci. Adv.* **2017**, *3,* e1701696.
58. Seyler, K. L.; et al. Signatures of Moire-Trapped Valley Excitons in $MoSe_2/WSe_2$ Heterobilayers. *Nature* **2019**, *567*, 66–70. arXiv preprint arXiv:1809.04562.
59. Unuchek, D.; et al. Room-Temperature Electrical Control of Exciton Flux in a van der Waals Heterostructure. *Nature* **2018**, *560*, 340.
60. Jin, C.; et al. Observation of moiré excitons in WSe_2/WS_2 heterostructure superlattices. *Nature* **2019**, *567*(7746), 76.
61. Tran, K.; et al. Evidence for moiré excitons in van der Waals heterostructures. *Nature* **2019**, *567*(7746), 71.
62. Alexeev, E. M.; et al. Resonantly hybridized excitons in moiré superlattices in van der Waals heterostructures. *Nature* **2019**, *567*(7746), 81.

63. Seyler, K. L.; et al. Signatures of moiré-trapped valley excitons in $MoSe_2/WSe_2$ hetero-bilayers. *Nature* **2019**, *567*(7746), 66.

Further reading

64. Mak, K. F.; Xiao, D.; Shan, J. Light–Valley Interactions in 2D Semiconductors. *Nat. Photonics* **2018**, *12*, 451.
65. Zhong, D.; et al. Van der Waals Engineering of Ferromagnetic Semiconductor Heterostructures for Spin and Valleytronics. *Sci. Adv.* **2017**, *3*, e1603113.

CHAPTER 11

Perspective

Hui Ying Hoh[a,b], Han Zhang[a], Qiaoliang Bao[c]
[a]Key Laboratory of Optoelectronic Devices and Systems of Ministry of Education and Guangdong Province, College of Optoelectronic Engineering, Shenzhen University, Shenzhen, China
[b]Institute of Microscale Optoelectronics, Shenzhen University, Shenzhen, China
[c]Department of Materials Science and Engineering, ARC Centre of Excellence in Future Low-Energy Electronics Technologies (FLEET), Monash University, Clayton, VIC, Australia

11.1 Summary

We have explored several classes of 2D materials, with a focus on transition metal dichalcogenides (TMDs), black phosphorus (BP), and graphene. These materials possess attractive physical and optical properties and had demonstrated exciting possibilities. However, to truly harness these and realize commercial applications, more work remains to be done. For example, in the case of optical modulators, although the all-optical phase modulation scheme can be applied using 2D materials with high nonlinear refractive indices such as MoS_2,[1] WS_2,[2] and WSe_2,[3] graphene remains the most promising for realizing all-optical modulators. The same applies to flexible electronics, as most devices are mostly graphene-based. In the case of photodetectors, photovoltaic and light-emitting devices, a combination of materials are often utilized. Graphene is used in conjunction with other 2D materials such as MoS_2 and BP.

11.2 Challenges

In a 2012 review, Novoselov et al. asked, "Is graphene special and versatile enough to justify the inconvenience of switching to a new technology, usually a lengthy and expensive process?"[4] Often, the route to commercialization is not straightforward. New or rediscovered materials may remain in the laboratory and never see real use. Here, we consider the challenges.

11.2.1 Methods of production and sample quality

For niche applications such as photonics and optoelectronics discussed in this book, the quality of the material is critical. Disorders in a material can be intrinsic, such as point defects, grain boundaries, phase separation, or extrinsic disorder such as inhomogeneous strain and adsorbates or charge traps on a

2D Materials for Photonic and Optoelectronic Applications
https://doi.org/10.1016/B978-0-08-102637-3.00011-5

substrate. These are important considerations that can affect the performance of the material in a device. For example, a CVD-grown MoS_2 film consists of single crystals, which merge to form faceted tilt and mirror twin boundaries.[5] It has been observed that the mirror twin boundaries cause strong photoluminescence quenching, whereas tilt boundaries cause strong enhancement. Therefore, in a micron-sized sample, opposite effects due to the two types of boundaries are present. This may be seen as a complication, or an opportunity to fine-tune the optical properties of CVD-grown films, if growth conditions can be controlled to encourage one type of crystal boundary to form over the other.

When considering industrial applications, the market of applications is essentially driven or limited by progress in the methods of production. Always there is a trade-off between quality of product and cost and ease of production. Commercialization can only be realized when mass-produced materials exhibit similar exceptional properties as those of samples produced in research laboratories. More importantly, for industrial applications, batch-to-batch consistency is critical; thus the production method has to not only be scalable but also yield high-quality materials repeatedly.

Very recently, it has been reported that highly uniform, phase-pure semiconducting 2D nanosheets can be obtained via a scalable solution process.[6] Quaternary ammonium molecules are first electrochemically intercalated into a crystal, for example, MoS_2, resulting in a volume expansion of the crystal. The expanded MoS_2 is then sonicated in a suitable solvent to obtain exfoliated 2H-MoS_2 nanosheets, which are further processed into ink solutions for device fabrication. Thin-film transistors made with these nanosheets demonstrated excellent performance, with room temperature mobilities of about 10 $cm^2/V\,s$ and on/off ratios of 10,[6] surpassing those obtained for previous solution-processed MoS_2 thin-film transistors, while matching those obtained for CVD-grown MoS_2 films. More importantly, the authors showed that this approach can be applied to other 2D materials, including WSe_2, Bi_2Se_3, $NbSe_2$, In_2Se_3, Sb_2Te_3, and BP.

11.2.2 Heterostructures

Needless to say, a high-quality sample is wasted in a badly designed or poorly fabricated device. Extrinsic disorders such as trapped states and adsorbates such as water, atmospheric gases, and hydrocarbons come into play. A device may consist of heterostructures or an active layer with interconnects. In most cases, the heterostructures or various components are not assembled via layer-by-layer growth. This may be due to incommensurate lattice, or damage to material at harsh growth conditions resulting in defects. Therefore, the development of transfer techniques or deterministic placement methods is also imperative. Frisenda et al. discussed the fabrication of van der Waals (vdW) heterostructures, progress in transfer methods, and associated challenges in an extensive review.[7] vdW heterostructures that include hexagonal boron nitride (hBN), tungsten sulfide (WS_2), molybdenum sulfide (MoS_2), and graphene tend to be of high quality, comparable to molecular beam epitaxy (MBE) analogs made in ultrahigh vacuum. This is due to a "self-cleaning" mechanism, where adsorbates trapped between layers as bubbles can move and segregate into bigger bubbles, leaving large areas of "clean" surfaces. Annealing is also effective in removing or reducing the bubbles, in turn improving device performance. Unfortunately, only the few aforementioned materials undergo self-cleaning, and hence these are used for encapsulation purposes. Effective encapsulation with hBN and substrate passivation often leads to improved device performance,[8, 9] although in the case of TMDs, these usually contain more intrinsic disorders compared to graphene, and these factors need to be addressed first.

Recently, a multibit nonvolatile optoelectronic memory based on monolayer tungsten diselenide (WSe_2) sandwiched between two layers of few-layer hBN was reported.[10] The switching ratio of the memory can reach up to $1.1 \times 10,^6$ ensuring 130 distinct storage states. It also exhibits excellent performance for data retention, with a retention time of over 4.5×10^4 s. In addition, the WSe_2/hBN device is able to discriminate wavelengths in the full visible spectrum, enabling its potential as a filter-free color image sensor. The authors also fabricated a BP/hBN device in a similar fashion. It did not perform as well as an optoelectronic memory. However, an ultrahigh photoresponsivity of approximately 1.2×10^7 A W^{-1} is observed, and thus BP/hBN heterostructures may find a place in photodetector applications.

11.2.3 Stability

The stability of a material and a device fabricated from this material is another important consideration. Despite the recent progress on 2D materials beyond graphene, many devices demonstrated peak performance when made with samples with thicknesses ranging between 5 and 10 layers,[11] and performances suffer significantly for devices with thinner materials. This is because surface effects such as atmospheric adsorbates are more pronounced in thin crystals.

We will focus on BP since it is the most problematic 2D material; BP rapidly degrades under ambient conditions, particularly in the case of phosphorene—monolayer BP. To date, many techniques have been tested. These include surface modification such as covalent functionalization, surface coatings with hydrophobic polymer, and encapsulation using capping layers such as graphene, aluminum oxide (Al_2O_3), and boron nitride (BN). The recent progress on these methods is summarized in a review.[11] Since BP is unstable in air, the ideal method should allow capping layers to be grown *in situ*, so as to minimize exposure of the BP layer to air. For BP produced via liquid exfoliation, this can be done by selecting a solvent such as *N*-cyclohexyl-2-pyrrolidone (CHP), which forms a solvation shell around the exfoliated BP nanosheets.[12] The shell acts as a barrier and prevents oxidative species from interacting with the surface of BP sheets. It is also recently shown that platinum (Pt) nanoparticles grown on BP surfaces via *in situ* chemical reaction also enhances the stability of exfoliated BP sheets.[13] The Pt/BP heterostructures are stable for at least 15 days under ambient conditions. In addition, these exhibited excellent photocatalytic activity, outperforming three other common Pt catalysts. Thus, in situ reactions offer a promising route toward stabilizing exfoliated BP nanosheets. Moreover, the suspension of BP sheets can be stored, transported, or further processed following exfoliation and functionalization.

In the case of inorganic capping agents, BP heterostructures can be grown by incorporating several deposition sources, for example, a phosphide MBE source and an oxide atomic layer deposition (ALD) system. Similarly, in this way, BP will be protected from the ambient environment, and subsequent device fabrication may also be done in the same system. However, since BP has a low thermal stability of $<550°C$, much lower than the growth temperatures of many 2D materials, the variety of heterostructures that can be produced in this manner is limited. Moreover, ALD Al_2O_3 cannot be used for BP with less than five layers, as such thin samples are too unstable and are oxidized by ALD.[14] An oxygen plasma may be used for thinning bulk BP, forming a P_xO_y capping layer on which extra passivation layers such as ALD Al_2O_3 may be added.[15] Nevertheless, the P_xO_y layer itself is not sufficiently robust, and is found to degrade after 3 days. In another method, bulk BP is first exfoliated using a polydimethylsiloxane (PDMS) viscoelastic stamp onto SiO_2/Si substrates, inside a glovebox with controlled atmosphere.[16] The sample is next transferred using a special suitcase into an ultrahigh vacuum evaporator system, following which metallic Al is evaporated on the sample. Upon exposure to air, the deposited Al forms a layer of Al_2O_3 spontaneously. The passivated BP flakes are stable for more than a month, and this method has been found to be useful for thin flakes ($\leq$4 monolayers). One drawback, however, is that the use of a special suitcase is somewhat tedious, and perhaps impractical on a larger-scale production.

Another approach is to tap into the relatively more mature silicon-based photonics and optoelectronics, instead of replacing existing technology. A hybrid silicon-BP optical emitter, comprising few-layer BP flakes coupled to silicon (Si) photonic crystal resonators, was recently reported.[17] Silicon is transparent in the near-infrared (IR) region. BP with its layer-tunable direct bandgap in this IR range would complement silicon-based devices, especially for telecommunication applications. The solution-processed BP flakes were transferred to the Si nanocavity using a PDMS stamp, suggesting scalability of this method. More importantly, even without a passivating layer, the hybrid optical emitter is robust and found to be stable for a year under ambient conditions. Therefore, once a suitable capping layer and method that provides complete, effective, and long-term passivation is developed, meaningful progress on BP-based devices will be made.

11.2.4 Other issues

Once the issues relating to production are resolved, logistics must be taken into account. In a commentary on the commercialization of graphene, Zurutuza and Marinelli said, "The speed of industrial uptake will also be influenced by factors such as the optimization of supply-chain logistics (for example, material storage and transportation), the availability of intermediaries (for example, dispersions, solutions, master batches, pre-transferred films) to accelerate integration into products and systems, clear standards, and a robust framework for health and safety regulation."[18] For solution-processed materials, transporting and storing dilute solutions are wasteful and costly, since the bulk would be water. On the other hand, dry powders of nanomaterials may incur health concerns. Even for the most widely studied graphene, there are insufficient studies on health and safety risks related to it. Therefore, there is a need to study and regulate this aspect of 2D materials.

Resistance by existing technology is another consideration. In terms of electronic components based on silicon and III–V semiconductors, the manufacturing and processing technologies are mature, and replacing them will require large-scale modifications. For transparent conductors, the indium tin oxide market is well established, as discussed in the 2014 commentary.[18] Hence, even if all the challenges mentioned earlier are addressed, an overhaul is unlikely to occur in the near future.

11.3 Future directions

11.3.1 Novel device engineering

To fully exploit the remarkable properties of 2D materials, novel device designs or architectures may be more effective than incorporating these in traditional semiconductor devices. Not long ago, 3D arrays of MoS_2/graphene photodetectors were realized through geometric transformation.[19] Through a clever design, a 3D structure forms spontaneously via compressive buckling from a 2D assembly (Fig. 11.1A). The 3D device (Fig. 11.1B–C)

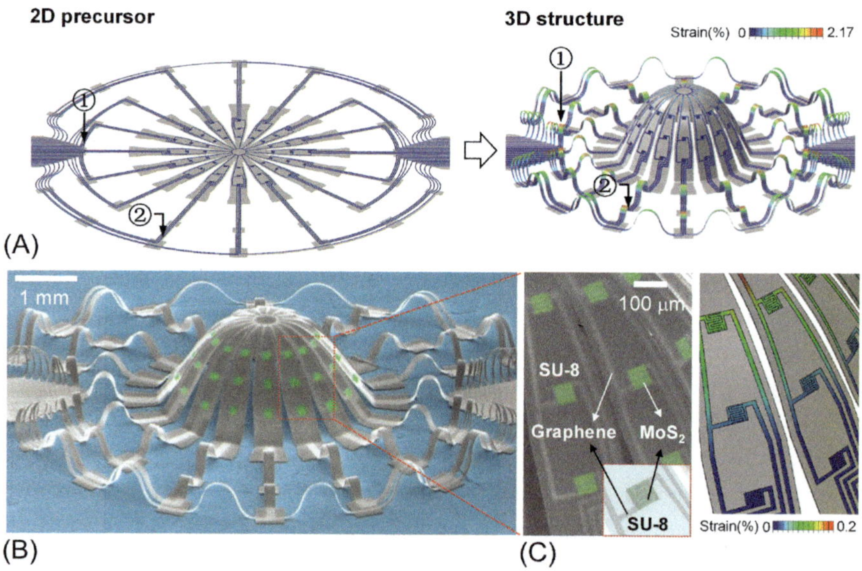

Fig. 11.1 From 2D to 3D: (A) Finite element analysis (FEA) of the MoS_2/graphene hemisphere structure and 3D interconnects before and after compressive buckling, showing the distributions of maximum principal strains. (B) Colorized SEM image of MoS_2 photodetectors consisting of MoS_2 (*green*), graphene (*light gray*), and an epoxy-based negative photoresist, SU-8 (*gray*), on the hemisphere with 3D interconnects. (C) Magnified view of the SEM image in (B) and corresponding FEA results for strain distribution. Inset: schematic of the unit device enclosed by the red box in the main image.[19] Reproduced under the Creative Commons Attribution 4.0 International License.

is sensitive to both the direction and intensity of illumination in real time, allowing additional capabilities in photodetection and imaging. This is a notable achievement as it is not accomplishable by typical 2D arrays.

In another recent work, patterned dielectric superlattices in conjunction with hBN-graphene-hBN heterostructures were exploited to engineer the bandgap of graphene.[20] Instead of transferring remote electrostatic modulations capacitively to the 2D electron gas as in conventional semiconductor heterostructures, the authors utilized a uniform, featureless gate electrode and modulated the dielectric interface in local vicinity. Through this approach, electrostatic boundaries can be as thin as 1 nm, limited by the encapsulating dielectric, hBN. Moreover, since the patterning of the superlattice is performed separately from the assembly of the hBN-graphene heterostructure, this technique can be extended to other 2D materials.

11.3.2 Other 2D materials

The 2D materials discussed in the previous chapters have gathered much attention in the past decade. On the other hand, others are just emerging, for example, molybdenum oxide (MoO_3). Recently, anisotropic polariton propagation along the surface of α-MoO_3 was reported.[21] α-MoO_3 is a natural vdW material, and prior to this work, there are few, if any, observations of such behaviors in natural crystals. The authors visualized and verified phonon polaritons with elliptic and hyperbolic in-plane dispersion in semiconducting α-MoO_3 flakes and disks. These polaritons have wavelengths comparable to those of graphene plasmon polaritons and BN phonon polaritons. More importantly, these exhibited ultra-low-loss character, with lifetimes of about 8 and 1.9 ps for elliptic and hyperbolic cases, respectively. The presence of ultra-low-loss polaritons, coupled with in-plane anisotropy, suggests a possible way to provide directional manipulation of light-matter interactions at the nanoscale.

MXenes are also emerging to be potential candidates for wide-ranging applications in nanophotonics and optoelectronics. In 2017, metallic Ti_3CNT_x MXene demonstrated a great potential as a saturable absorber for femtosecond mode-locked lasers.[22] Later, studies were extended to $Ti_3C_2T_x$ (T=O, OH, or F), and the results were promising,[23, 24] It was observed that saturation fluence and modulation depth of $Ti_3C_2T_x$ is dependent on the thickness of the film.[25] In addition, compared to other 2D materials, $Ti_3C_2T_x$ has a higher threshold for light-induced damage and up to 50% increase in nonlinear transmittance. Interested readers are encouraged to refer to a recent review, in which the nonlinear optical, optoelectronic, and plasmonic properties of MXenes are discussed.[26]

Topological insulators (TIs) may also find a place in photonic and optoelectronic applications, particularly in terahertz to infrared photodetection, due to the presence of Dirac fermions in their topologically protected surface states. Several groups have reported high-performance photodetectors based on bismuth selenide (Bi_2Se_3) TI operating at room temperature.[27–29] Using vdW epitaxy, scalable production of high-quality Bi_2Se_3 flakes with variable size and thickness was also demonstrated.[28] By controlling nucleation, flakes as large as 0.4 mm and thickness of up to 3 nm (three quintuple layers) were obtained. The Bi_2Se_3-based photodetector exhibits an excellent on/off ratio of 972.5, responsivity of 23.8 A/W, and external quantum efficiency of >2000% at 1456 nm (E-band of the telecommunication range). In another exciting work, a heterostructure device comprising WSe_2-graphene-Bi_2Se_3 displayed valley-locked spin polarization degree of freedom (DoF).[29] This device highlights an important approach: each 2D material used was chosen for a specific purpose. At the surface of WSe_2, an inversion-symmetry-broken 2D electron gas was induced. Valley-locked spin-polarized electrons are excited and the transferred photon angular momentum generates in-plane valley-locked spin DoF electrons. These photocarriers then diffuse in a submicrometer-long graphene layer. Graphene was used as it has a long spin diffusion length and no spin or valley preference for electrical transport. The spin DoF is measured via non-local electrical detection utilizing spin-momentum locking in Bi_2Se_3. Due to this unique coupling between current and spin polarization, the measured polarity of current is a true reflection of the valley-locked spin DoF electrons. The WSe_2-graphene-Bi_2Se_3 device showed a spin polarization degree >0.5, suggesting applications in spintronics and valleytronics. More importantly, by carefully selecting appropriate materials so as to exploit their outstanding properties, vdW heterostructures can find a place in many state-of-the-art technologies.

11.4 Outlook

Let us take a step back and look at the research activity of graphene, TMDs, and BP. We performed a search on Scopus, and the results are displayed in Table 11.1. From 2004 till November 2018, there have been more than 120,000 reports on graphene, while the activity in TMDs and BP is lagging

Table 11.1 Number of reports and patents on graphene, TMDs, and BP from 2004 to November 2018.

Searched terms	Number of reports (2004–November 2018)	Number of patents (2004–November 2018)
Graphene	124048	66802
Graphene, photonic	1673	2024
Graphene, optoelectronic	2719	2442
TMDs	4915	603
TMDs, photonic	185	50
TMDs, optoelectronic	819	102
BP	1802	474
BP, photonic	99	22
BP, optoelectronic	270	42

The numbers were obtained from a search (title-abstract-keyword) of the respective terms using the Scopus database. The terms "transition metal dichalcogenides" and "black phosphorus"—not their abbreviated forms, TMD and BP—were used in original searches.

significantly. Not surprisingly, graphene is still extremely popular. However, since interest in TMDs and BP started at least 5 years later, we expect to see a growing trend in the next decade, possibly matching that of graphene eventually. If we focus on photonic and optoelectronic applications of these 2D materials, a similar trend can be observed, as seen in Fig. 11.2, which depicts the cumulative number of reports on these materials.

To consider the industrial interest, we conducted a search on the number of patents filed using the same search method. The total number of patents from 2004 to 2018 with the keyword "graphene" is more than 66,000, among which <7% include the terms "photonic" or "optoelectronic." This result suggests that while industrial interest in graphene is strong, many groups are simply jumping onto the bandwagon but have yet to find a niche application for it. Nevertheless, we believe that this sheer amount of collective effort is bound to push some frontiers. In the case of TMDs, although the total number of patents is merely 603, 25% of these are related to photonic or optoelectronic applications. For BP, the proportion is 14%. Therefore, TMDs and BP hold a great potential for specific applications, if we can address the challenges discussed in the previous sections (Fig. 11.3).

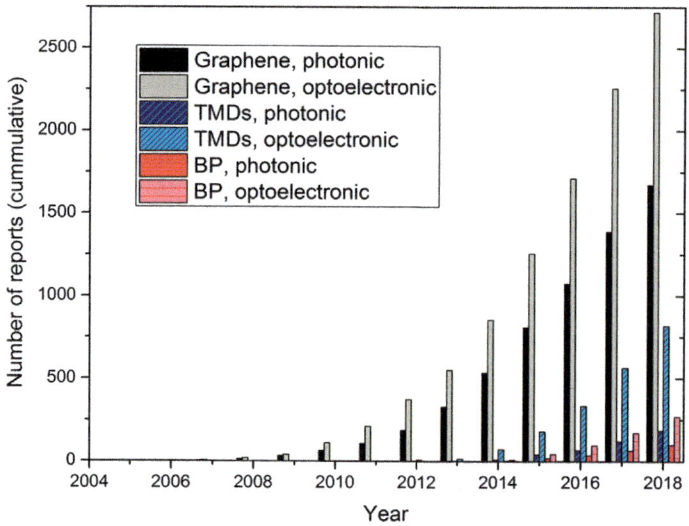

Fig. 11.2 Research activity in graphene, TMDs, and BP.

Considering the route to commercialization, we look to graphene again for an indication. Fig. 11.2 outlines the steps or "technology readiness levels" required to bring a material, in our case, graphene, from the laboratory to the real world. The first step is to understand the basic principles of the material, while the final and ninth step is the successful operation of a commercial product. Photonic and optoelectronic applications of graphene are currently in levels 2–6. We note that graphene-based devices, for example, graphene-LED light bulbs, are now available commercially.[30, 31] However, graphene is used for heat dissipation in these bulbs, rather than in novel ways that exploit its exceptional properties. This is why efforts on graphene-based LEDs are considered to be at the applied research and development stage. Nevertheless, graphene-LED bulbs are reported to be more energy-efficient and cheaper to manufacture; hence efforts to utilize graphene in LEDs will likely continue from an economic perspective. For the other 2D materials, more collaboration between academics and industrial scientists is required to bring a proof-of-concept to the next stage—from a prototype to an actual commercial product—as in the case of graphene.[32]

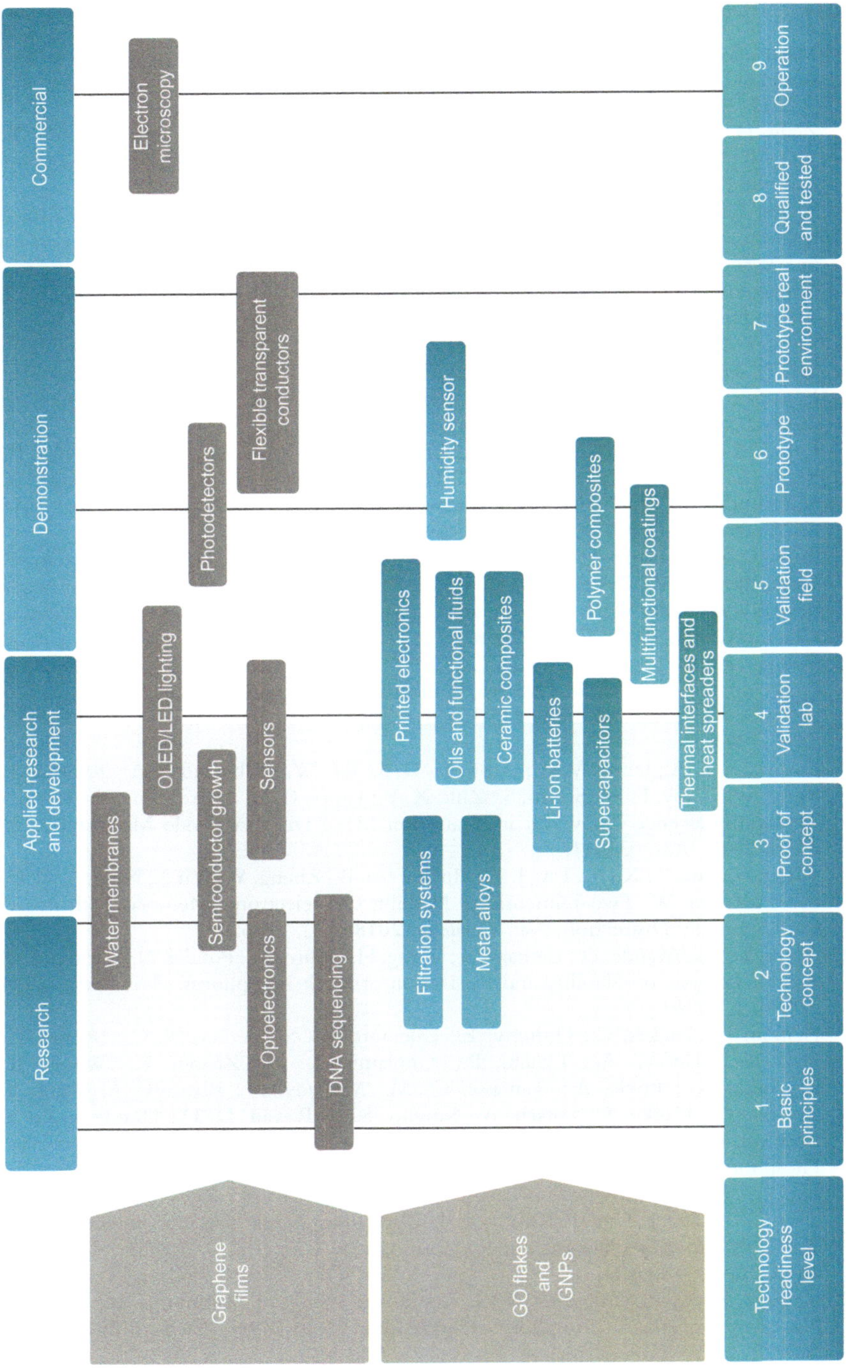

Fig. 11.3 Technology readiness and graphene applications.

References

1. Malard, L. M.; Alencar, T. V.; Barboza, A. P. M.; Mak, K. F.; de Paula, A. M. Observation of Intense Second Harmonic Generation From MoS_2 Atomic Crystals. *Phys. Rev. B* **2013**, *87*(20), 201401.
2. Zeng, H.; Liu, G.-B.; Dai, J.; Yan, Y.; Zhu, B.; He, R.; Xie, L.; Xu, S.; Chen, X.; Yao, W.; Cui, X. Optical Signature of Symmetry Variations and Spin-Valley Coupling in Atomically Thin Tungsten Dichalcogenides. *Sci. Rep.* **2013**, *3*, 1608.
3. Seyler, K. L.; Schaibley, J. R.; Gong, P.; Rivera, P.; Jones, A. M.; Wu, S.; Yan, J.; Mandrus, D. G.; Yao, W.; Xu, X. Electrical Control of Second-Harmonic Generation in a WSe_2 Monolayer Transistor. *Nat. Nanotechnol.* **2015**, *10*, 407.
4. Novoselov, K. S.; Fal'ko, V. I.; Colombo, L.; Gellert, P. R.; Schwab, M. G.; Kim, K. A Roadmap for Graphene. *Nature* **2012**, *490*, 192.
5. van der Zande, A. M.; Huang, P. Y.; Chenet, D. A.; Berkelbach, T. C.; You, Y.; Lee, G.-H.; Heinz, T. F.; Reichman, D. R.; Muller, D. A.; Hone, J. C. Grains and Grain Boundaries in Highly Crystalline Monolayer Molybdenum Disulphide. *Nat. Mater.* **2013**, *12*, 554.
6. Lin, Z.; Liu, Y.; Halim, U.; Ding, M.; Liu, Y.; Wang, Y.; Jia, C.; Chen, P.; Duan, X.; Wang, C.; Song, F.; Li, M.; Wan, C.; Huang, Y.; Duan, X. Solution-Processable 2D Semiconductors for High-Performance Large-Area Electronics. *Nature* **2018**, *562*(7726), 254–258.
7. Frisenda, R.; Navarro-Moratalla, E.; Gant, P.; Pérez De Lara, D.; Jarillo-Herrero, P.; Gorbachev, R. V.; Castellanos-Gomez, A. Recent Progress in the Assembly of Nanodevices and van der Waals Heterostructures by Deterministic Placement of 2D Materials. *Chem. Soc. Rev.* **2018**, *47*(1), 53–68.
8. Wang, J. I. J.; Yang, Y.; Chen, Y.-A.; Watanabe, K.; Taniguchi, T.; Churchill, H. O. H.; Jarillo-Herrero, P. Electronic Transport of Encapsulated Graphene and WSe_2 Devices Fabricated by Pick-Up of Prepatterned hBN. *Nano Lett.* **2015**, *15*(3), 1898–1903.
9. Obafunso, A. A.; Jenny, V. A.; Gabriella, D. S.; Jue, W.; Abhinandan, A.; Takeshi, T.; Kenji, W.; Tony, F. H.; Stefan, S.; Zhu, X. Y.; James, C. H. Approaching the Intrinsic Photoluminescence Linewidth in Transition Metal Dichalcogenide Monolayers. *2D Mater.* **2017**, *4*(3), 031011.
10. Xiang, D.; Liu, T.; Xu, J.; Tan, J. Y.; Hu, Z.; Lei, B.; Zheng, Y.; Wu, J.; Neto, A. H. C.; Liu, L.; Chen, W. Two-Dimensional Multibit Optoelectronic Memory With Broadband Spectrum Distinction. *Nat. Commun.* **2018**, *9*(1), 2966.
11. Abate, Y.; Akinwande, D.; Gamage, S.; Wang, H.; Snure, M.; Poudel, N.; Cronin, S. B. Recent Progress on Stability and Passivation of Black Phosphorus. *Adv. Mater.* **2018**, *30*(29), 1704749.
12. Hanlon, D.; Backes, C.; Doherty, E.; Cucinotta, C. S.; Berner, N. C.; Boland, C.; Lee, K.; Harvey, A.; Lynch, P.; Gholamvand, Z.; Zhang, S.; Wang, K.; Moynihan, G.; Pokle, A.; Ramasse, Q. M.; McEvoy, N.; Blau, W. J.; Wang, J.; Abellan, G.; Hauke, F.; Hirsch, A.; Sanvito, S.; O'Regan, D. D.; Duesberg, G. S.; Nicolosi, V.; Coleman, J. N. Liquid Exfoliation of Solvent-Stabilized Few-Layer Black Phosphorus for Applications Beyond Electronics. *Nat. Commun.* **2015**, *6*, 8563.
13. Bai, L.; Wang, X.; Tang, S.; Kang, Y.; Wang, J.; Yu, Y.; Zhou, Z.-K.; Ma, C.; Zhang, X.; Jiang, J.; Chu, P. K.; Yu, X.-F. Black Phosphorus/Platinum Heterostructure: A Highly Efficient Photocatalyst for Solar-Driven Chemical Reactions. *Adv. Mater.* **2018**, *30*(40), 1803641.
14. Gamage, S.; Li, Z.; Yakovlev, V. S.; Lewis, C.; Wang, H.; Cronin, S. B.; Abate, Y. Nanoscopy of Black Phosphorus Degradation. *Adv. Mater. Interfaces* **2016**, *3*(12), 1600121.

15. Pei, J.; Gai, X.; Yang, J.; Wang, X.; Yu, Z.; Choi, D.-Y.; Luther-Davies, B.; Lu, Y. Producing Air-Stable Monolayers of Phosphorene and Their Defect Engineering. *Nat. Commun.* **2016**, 7, 10450.
16. Galceran, R.; Gaufres, E.; Loiseau, A.; Piquemal-Banci, M.; Godel, F.; Vecchiola, A.; Bezencenet, O.; Martin, M. B.; Servet, B.; Petroff, F.; Dlubak, B.; Seneor, P. Stabilizing Ultra-Thin Black Phosphorus With In-Situ-Grown 1 nm-Al_2O_3 Barrier. *Appl. Phys. Lett.* **2017**, *111*(24), 243101.
17. Husko, C.; Kang, J.; Moille, G.; Wood, J. D.; Han, Z.; Gosztola, D.; Ma, X.; Combrié, S.; De Rossi, A.; Hersam, M. C.; Checoury, X.; Guest, J. R. Silicon-Phosphorene Nanocavity-Enhanced Optical Emission at Telecommunications Wavelengths. *Nano Lett.* **2018**, *18*(10), 6515–6520.
18. Zurutuza, A.; Marinelli, C. Challenges and Opportunities in Graphene Commercialization. *Nat. Nanotechnol.* **2014**, *9*, 730.
19. Lee, W.; Liu, Y.; Lee, Y.; Sharma, B. K.; Shinde, S. M.; Kim, S. D.; Nan, K.; Yan, Z.; Han, M.; Huang, Y.; Zhang, Y.; Ahn, J.-H.; Rogers, J. A. Two-Dimensional Materials in Functional Three-Dimensional Architectures With Applications in Photodetection and Imaging. *Nat. Commun.* **2018**, *9*(1), 1417.
20. Forsythe, C.; Zhou, X.; Watanabe, K.; Taniguchi, T.; Pasupathy, A.; Moon, P.; Koshino, M.; Kim, P.; Dean, C. R. Band Structure Engineering of 2D Materials Using Patterned Dielectric Superlattices. *Nat. Nanotechnol.* **2018**, *13*(7), 566–571.
21. Ma, W.; Alonso-González, P.; Li, S.; Nikitin, A. Y.; Yuan, J.; Martín-Sánchez, J.; Taboada-Gutiérrez, J.; Amenabar, I.; Li, P.; Vélez, S.; Tollan, C.; Dai, Z.; Zhang, Y.; Sriram, S.; Kalantar-Zadeh, K.; Lee, S.-T.; Hillenbrand, R.; Bao, Q. In-Plane Anisotropic and Ultra-Low-Loss Polaritons in a Natural van der Waals Crystal. *Nature* **2018**, *562*(7728), 557–562.
22. Jhon, Y. I.; Koo, J.; Anasori, B.; Seo, M.; Lee, J. H.; Gogotsi, Y.; Jhon, Y. M. Metallic MXene Saturable Absorber for Femtosecond Mode-Locked Lasers. *Adv. Mater.* **2017**, *29*(40).
23. Sun, X.; Zhang, B.; Yan, B.; Li, G.; Nie, H.; Yang, K.; Zhang, C.; He, J. Few-Layer $Ti_3C_2T_x$ (T = O, OH, or F) Saturable Absorber for a Femtosecond Bulk Laser. *Opt. Lett.* **2018**, *43*(16), 3862–3865.
24. Jiang, X.; Liu, S.; Liang, W.; Luo, S.; He, Z.; Ge, Y.; Wang, H.; Cao, R.; Zhang, F.; Wen, Q.; Li, J.; Bao, Q.; Fan, D.; Zhang, H. Broadband Nonlinear Photonics in Few-Layer MXene $Ti_3C_2T_x$ (T = F, O, or OH). *Laser Photonics Rev.* **2018**, *12*(2).
25. Dong, Y.; Chertopalov, S.; Maleski, K.; Anasori, B.; Hu, L.; Bhattacharya, S.; Rao, A. M.; Gogotsi, Y.; Mochalin, V. N.; Podila, R. Saturable Absorption in 2D Ti_3C_2 MXene Thin Films for Passive Photonic Diodes. *Adv. Mater.* **2018**, *30*(10).
26. Hantanasirisakul, K.; Gogotsi, Y. Electronic and Optical Properties of 2D Transition Metal Carbides and Nitrides (MXenes). *Adv. Mater.* **2018**, *30*(52), e1804779.
27. Tang, W.; Politano, A.; Guo, C.; Guo, W.; Liu, C.; Wang, L.; Chen, X.; Lu, W. Ultrasensitive Room-Temperature Terahertz Direct Detection Based on a Bismuth Selenide Topological Insulator. *Adv. Funct. Mater.* **2018**, *28*(31), 1801786.
28. Wang, F.; Li, L.; Huang, W.; Li, L.; Jin, B.; Li, H.; Zhai, T. Submillimeter 2D Bi_2Se_3 Flakes Toward High-Performance Infrared Photodetection at Optical Communication Wavelength. *Adv. Funct. Mater.* **2018**, *28*(33), 1802707.
29. Cha, S.; Noh, M.; Kim, J.; Son, J.; Bae, H.; Lee, D.; Kim, H.; Lee, J.; Shin, H.-S.; Sim, S.; Yang, S.; Lee, S.; Shim, W.; Lee, C.-H.; Jo, M.-H.; Kim, J. S.; Kim, D.; Choi, H. Generation, Transport and Detection of Valley-Locked Spin Photocurrent in WSe_2–Graphene–Bi_2Se_3 Heterostructures. *Nat. Nanotechnol.* **2018**, *13*(10), 910–914.
30. SERA Technologies. www.seratechnologies.com/what-is-a-graphene-led-bulb/.
31. Graphene Lighting PLC. http://graphenelighting.co/.
32. Park, S. The Puzzle of Graphene Commercialization. *Nat. Rev. Mater.* **2016**, 1, 16085.

Index

Note: Page numbers followed by *f* indicate figures and *t* indicate tables.

M

N

T

9780081026373